MODERN
LATIN AMERICA

MODERN LATIN AMERICA

FOURTH EDITION

Thomas E. Skidmore
Peter H. Smith

New York   Oxford
OXFORD UNIVERSITY PRESS
1997

Oxford University Press

Oxford   New York
Athens   Auckland   Bangkok
Bogota   Bombay   Buenos Aires   Calcutta
Cape Town   Dar es Salaam   Delhi
Florence   Hong Kong   Istanbul   Karachi
Kuala Lumpur   Madras   Madrid   Melbourne
Mexico City   Nairobi   Paris   Singapore
Taipei   Tokyo   Toronto

and associated companies in
Berlin   Ibadan

Library of Congress Cataloging-in-Publication Data
Skidmore, Thomas E.
Modern Latin America / Thomas E. Skidmore, Peter H. Smith.
—4th ed.   p.   cm.
Includes bibliographical references (p.   ) and index.
ISBN 0-19-510016-6 (cloth). — ISBN 0-19-510017-4 (pbk. : alk. paper)
1. Latin America—History.   I. Smith, Peter H.   II. Title.
F1413.S55   1997
980.03'3—dc20   95-47246

3 5 7 9 8 6 4 2
Printed in the United States of America
on acid-free paper

*For
David, James, Robert
and
Jonathan and Peter*

CONTENTS

ACKNOWLEDGMENTS

For this fourth edition we have updated our interpretation of modern Latin American history in the context of a rapidly changing international environment. The end of the Cold War has removed one of the defining characteristics of post-1945 Latin America and brought new issues to the forefront of the inter-American agenda, among them free trade, drug trafficking, and illegal migration. Concern with economic recovery after the devastation of the 1980s has focused attention on age-old questions of poverty, inequality, and social justice.

The continued transition from military rule to civilian rule has given reason for hope that Latin America may yet succeed in overcoming the fateful historical legacy outlined in this book. At issue is not only the consolidation of democracy but also the extent and quality of democratic governance.

Among those who provided useful suggestions and comments for this edition were Lynn Guitar and Eric Van Young. Able research assistance was provided by Julie Grey and Frances Mejia. Ms. Grey also assisted with editing and compiled the index.

Once again we welcome comments and suggestions from readers.

Providence, Rhode Island   T. E. S.
La Jolla, California   P. H. S.
March 1996

MODERN
LATIN AMERICA

WHY LATIN AMERICA?

"The U.S. will do anything for Latin America, except read about it," according to James Reston, for many years the dean of U.S. political commentators. Is there any reason why we should try to prove him wrong? There are several. First, our nation's economic interests are deeply involved in the region. Latin America is one of our major trading partners. It is the site of much U.S. investment and a source for oil and other critical raw materials. An acceleration of growth in key countries—such as Mexico and Brazil—may soon produce significant new powers on the world scene.

We also have political links. Revolutionary upheavals and repressive responses in Latin America directly challenge U.S. foreign policy. They raise difficult questions about how to protect and promote U.S. national interests (defined as not simply economic or strategic interests). Ronald Reagan dramatized this fact shortly after his 1980 election by meeting with Mexican President José López Portillo on the U.S.-Mexican border, in what was the U.S. president's first such conference with any other head of state. Later in the 1980s the United States divided bitterly over support for an exile army (the Contras) fighting to overthrow the Sandinista government in Nicaragua. President George Bush emphasized his own concerns by seeking a special relationship with Mexico and proposing a free-trade agreement that would tighten economic bonds between all of Latin America and the United States. His successor and political rival, President Bill Clinton, has followed up the free-trade initiative by hosting a hemispheric "Summit of the Americas" at Miami in December 1994.

There is another important consideration closer to home. Large sections of our country have become Hispanized by the influence of migrants from Mexico, Puerto Rico, Central America, and the Caribbean. This is in addition to the Hispanic descendants of the original Spanish-speaking population of the formerly Mexican Southwest. Migration, both historical and recent, then brought peoples and customs from Latin America to the American Southwest (from Texas to Califor-

nia), Florida, and New York. Many major U.S. cities now have more children from Spanish-speaking families than from any other group. Bilingualism has become a political issue forcing us to rethink the meaning of Spanish-speaking America, both within our borders and beyond.

Most U.S. citizens (or "North Americans," as we are commonly called in Latin America) know little about our neighboring societies to the south. Many believe that the United States can impose its will on the region through "big-stick" diplomacy or military might. Others do not even care. Looking for an "easy" foreign language, high school or college students choose Spanish and then assume everything associated with speaking Spanish must be "easy." Such ignorance can be dangerous, and one purpose of this book is to help reduce misinformation. In fact, this lack of knowledge is felt just as keenly in West Europe. British journalists used to tell how some of their number once had a contest to think up the most boring newspaper headline imaginable (it had to be

---

U.S. STEREOTYPES OF LATIN AMERICA

On December 10, 1940, the Office of Public Opinion Research conducted a nation-wide poll in which respondents were given a card with nineteen words on it and were asked to indicate those words that seemed to describe best the people of Central and South America. The results were as follows:

Dark-skinned	80%	Imaginative	23%
Quick-tempered	49%	Shrewd	16%
Emotional	47%	Intelligent	15%
Religious	45%	Honest	13%
Backward	44%	Brave	12%
Lazy	41%	Generous	12%
Ignorant	34%	Progressive	11%
Suspicious	32%	Efficient	5%
Friendly	30%	No answer	4%
Dirty	28%	No opinion	0%
Proud	26%		

Since respondents were asked to pick as many descriptive terms as they liked, percentages add to considerably more than 100.

Sources: John J. Johnson, *Latin America in Caricature* (Austin: University of Texas Press, 1980), p. 18; Hadley Cantril ed., *Public Opinion, 1935–1946* (Princeton: Princeton University Press, 1951), p. 502.

real and actually printed) and they came up with "Small Earthquake in Chile: Not Many Dead." Chile's complex history over the last thirty years puts that provincialism into painful relief.

By training and outlook, most North Americans and Europeans search for intellectual formulae that will yield clear-cut answers to our inquiries: the "Latin lover," the "Frito Bandito," the soulful Ché Guevara, the Brazilian mulatta carnival queens—these are the images that often first come to mind. But when we move beyond these caricatures (which have their own truth to tell), we find Latin America to be a complex region.

Latin America is not an easy place to understand, despite the fact that the same language, Spanish, is spoken everywhere—except Brazil (Portuguese), the Andes (Quechua and other Indian languages), the Caribbean (French, English, and Dutch), Mexico (scattered pockets of Indian languages), and Guatemala (over twenty Indian languages). The term Latin America covers a vast variety of people and places. Geographically, Latin America includes the land mass extending from the Rio Grande border between Texas and Mexico to the southern tip of South America, plus some Caribbean islands: a total area two and one-half times the size of the United States. Brazil itself is larger than the continental United States.

Physical features present sharp differences: from the Andean mountain range, stretching the full length of western South America, to the tropical forest of the Amazon basin; from the arid plains of northern Mexico to the fertile grasslands of the Argentine pampa. The people of Latin America contain elements and mixtures of three racial groups—native Indians, white Europeans, and black Africans. By 1992 the total population came to 453 million, compared with 255 million in the United States.

Latin American society displays startling contrasts—between rich and poor, between city and country, between learned and illiterate, between the powerful lord of the hacienda and the deferential peasant, between wealthy entrepreneurs and desperate street urchins. Politically, Latin America includes twenty-six nations, large and small, whose recent experience ranges from military dictatorship to electoral democracy to Fidel Castro's socialist regime in Cuba (see Map 1). Economically, Latin America belongs to the "developing" world, beset by historical and contemporary obstacles to rapid economic growth, but here too there is diversity—from the one-crop dependency of tiny Honduras to the industrial promise of dynamic Brazil.

Throughout their modern history Latin Americans have sought, with greater or lesser zeal, to achieve political and economic independence from colonial, imperial, and neo-imperial powers. Thus it is bitter irony that the phrase "Latin America" was coined by mid-nineteenth-century French, who thought that since their culture, like that of Spanish and

MAP 1   Contemporary Latin America (pop. as of 1992, in millions)

Portuguese America, was "Latin" (i.e., Romance language-speaking), France was destined to assume leadership throughout the continent.

As these observations suggest, Latin America resists facile categorization. It is a region rich in paradox. This insight yields a number of instructive clues.

First, Latin America is both young and old. Beginning in 1492, its conquest by the Spanish and Portuguese created a totally new social order based on domination, hierarchy, and the intermingling of European, African, and indigenous elements. The European intrusion profoundly and ineradicably altered the Indian communities. Compared

with the ancient civilizations of Africa and Asia, these Latin American societies are relatively young. On the other hand, most nations of Latin America obtained political independence—from Spain and Portugal—in the early nineteenth century, more than 100 years before successful anticolonial movements in other Third World countries. Thus by the standard of nationhood, Latin America is relatively old.

Second, Latin America has throughout its history been both tumultuous and stable. The Conquest began a tradition of political violence that has erupted in coups, assassinations, armed movements, military interventions, and (more rarely) social revolutions. Ideological encounters between liberalism, positivism, corporatism, anarchism, socialism, communism, fascism, and religious teachings of every doctrinal hue have sharpened the intensity of struggle. Despite the differing forms of political conflict, old social and economic structures have persisted. Even where modern revolutions have struck, as in Mexico (1910) and Bolivia (1952), many aspects of traditional society survive. The Cuban Revolution (1959) seems at first an exception, yet even in Cuba the pull of history has continued to be strong, as we shall see.

Third, Latin America has been both independent and dependent, autonomous and subordinate. The achievement of nationhood by 1830 in all but parts of the Caribbean basin represented an assertion of sovereignty rooted in Enlightenment thought. Yet a new form of penetration by external powers—first Britain and France, then the United States—jeopardized this nationhood. Economic and political weakness vis-à-vis Europe and North America has frequently limited the choices available to Latin American policymakers. Within Latin America, power is ironically ambiguous: it is the supreme commodity, but it has only a limited effect.

Fourth, Latin America is both prosperous and poor. Ever since the Conquest, the region has been described as a fabulous treasure house of natural resources. First came the European lust for silver and gold. Today the urge may be for petroleum, gas, copper, iron ore, coffee, sugar, soybeans, or for expanded trade in general, but the image of endless wealth remains. In startling contrast, there is also the picture of poverty: peasants without tools, workers without jobs, children without food, mothers without hope. An aphorism oft repeated in Latin America summarizes this scene: "Latin America is a beggar atop a mountain of gold."

One can easily think of additional contrasts, but these should illustrate the difficulty—and fascination—in trying to come to grips with Latin America. To understand Latin American history and society requires a flexible, broad-gauge approach, and this is what we try to offer in this book. We draw on the work of many scholars, presenting our own interpretation, but also acquainting the reader with alternative views.

Interpretations of Latin America

Most analysts of modern Latin America have stressed the area's political instability, marked frequently by dictatorship. North American and European observers have been especially fascinated with two questions: Why dictatorships? Why not democracy? This preoccupation is not recent. In 1930, for example, a U.S. economic geographer specializing in the region observed, "the years roll on and there arise the anxieties and disappointments of an ill-equipped people attempting to establish true republican forms of government." A year earlier an English scholar had noted that "the political history of the republics has been a record of alternating periods of liberty and despotism." Implicitly assuming or explicitly asserting that their style of democracy is superior to all other models of political organization, North American and European writers frequently asked what was "wrong" with Latin America. Or with Latin Americans themselves.

What passed for answers was for many years a jumble of racist epithets, psychological simplifications, geographical platitudes, and cultural distortions. According to such views, Latin America could not achieve democracy because dark-skinned peoples (Indians and blacks) were unsuited for it; because passionate Latin tempers would not stand it; because tropical climates somehow prevented it; or because Roman Catholic doctrines inhibited it.

Each charge has its refutation: dictatorial rule has flourished in predominantly white countries, such as Argentina, as well as among mixed-blood societies, such as Mexico; it has appeared in temperate climes, such as Chile, not only in the tropics, such as Cuba; it has gained support from non-Catholics and nonpracticing Catholics, while many fervent worshippers have fought for liberty; and, as shown by authoritarian regimes outside Latin America, such as Hitler's Germany or Stalinist Russia, dictatorship is not restricted to any single temperament. Such explanations not only failed to explain; when carried to extremes they helped justify rapidly increasing U.S. and European penetration—financial, cultural, military—of the "backward" republics to the south.

The scholarly scene improved in the late 1950s and early 1960s, when North American social scientists formulated "modernization theory." As applied to Latin America, this theory held that economic growth would generate the social change that would in turn make possible more "developed" politics. The transition from a rural to an urban society would bring a change in values. People would begin to relate to and participate in the voluntary organizations that authentic democracy requires. Most important, a middle class would emerge—to play both a progressive and moderating role. Latin America and its citizenries were not so inherently "different" from Europe and North

America. Instead they were simply "behind." Modernization adepts thought the historical record showed this process was well under way in Latin America.

Thus analysts went to work describing Latin American history in the light of modernization theory. One optimistic and widely read U.S. scholar found in 1958 that the "middle sectors" had "become stabilizers and harmonizers and in the process have learned the dangers of dealing in absolute postulates." The author of a late-1970s textbook on Latin American history saw "Latin American history since independence . . . as modernization growing slowly against the resistance of old institutions and attitudes."

Reality, however, proved harsher. Instead of spreading general prosperity, economic growth in the 1960s and 1970s (and it reached sustained high rates in Mexico and Brazil) generally made income distribution more unequal. The gap in living standards between city and countryside grew. Domestic capital's ability to compete with the huge transnational firms declined. Meanwhile, politics was hardly following the model predicted by many experts on modernization. The middle strata, relatively privileged, forged a sense of "class consciousness" which, in critical moments of decision, such as in Argentina in 1955 or 1976, Brazil in 1964, and Chile in 1973, led them to join the ruling classes in opposition to the popular masses. Politics took an authoritarian turn, producing military governments. And in stark contradiction of modernization theory, these patterns emerged in the most developed—and most rapidly developing—countries of the continent. What had gone wrong?

Two sets of answers came forth. One group of scholars focused on the cultural traditions of Latin America and their Spanish and Portuguese origins. These analysts argued, in effect, that antidemocratic politics was (and remains) a product of a Roman Catholic and Mediterranean world view that stressed the need for harmony, order, and the elimination of conflict. By failing to grasp these continuities in the Iberian experience, scholars had confused form with substance, rhetoric with reality. Latin America's constitutions were never as democratic as they appeared, party politics was not as representative as it might have looked. The North American and European academic community, afflicted by its own myopia and biases, had simply misread the social facts.

A second group of scholars accepted modernization theory's linking of socioeconomic causes with political outcomes but turned the answer upside down: Latin America's economic development was qualitatively different from that of North America and West Europe, and therefore it produced different political results. Specifically, these scholars argued, Latin America's experience was determined by the pervasive fact of its economic dependence. "By dependency," as one exponent of this viewpoint has explained,

we mean a situation in which the economy of certain countries is conditioned by the development and expansion of another economy to which the former is subjected. The relation of interdependence between two or more economies, and between these and world trade, assumes the form of dependence when some countries (the dominant ones) can expand and be self-sustaining, while other countries (the dependent ones) can do this only as a reflection of that expansion, which can have either a positive or a negative effect on their immediate development.

By its intrinsic character, "dependent development" generated inequities, allocating benefits to sectors participating in the world market and denying them to other groups. A typical case might involve a country whose economic growth relied on a single export crop, such as coffee or sugar. A national landowning elite, the planters, would collaborate with export-import merchants, often foreign, to sell the goods on an overseas market. Most profits would be restricted to those groups. The planters would use much of their money to import high-cost consumer goods from Europe or the United States, and the merchants (if foreign) would remit profits to their home countries. The export earnings would therefore provide precious little capital for diversifying the local economy, thus creating a situation that some observers have labeled "growth without development." Because of a labor surplus, field workers would continue to receive low wages; groups outside the export sector would get little benefit. Consequently, regional imbalances would intensify and income distribution would become more unequal than before. What growth occurred, moreover, would be subject to substantial risk. If the overseas market for coffee or sugar contracted—for whatever reason, as it did in the 1930s—then the entire economy would suffer. It would in this sense be "dependent" for its continued growth on decisions taken elsewhere and it would be, as our just-cited author says, "conditioned by the development and expansion of another economy."

The proponents of "*dependencia* theory," as it quickly came to be known, maintained that economic dependency leads to political authoritarianism. According to this view, the "dependent" location of Latin America's economics placed inherent limitations on the region's capacity for growth, especially in industry. The surest sign of this economic trouble is a crisis in the foreign accounts—the country's ability to pay for needed imports, as occurred dramatically in Mexico, Argentina, Chile, and Brazil in late 1982 and early 1983. Exports lag behind imports, and the difference can only be made up by capital inflow. But the foreign creditors—firms, banks, international agencies such as the World Bank—deny the necessary extra financing because they believe the government cannot impose the necessary "sacrifices." Backed against the wall, the country must take immediate steps to keep imports flowing in. Political strategy falls hostage to the need to convince the foreign creditors.

The most frequent solution in the 1960s and 1970s was a military coup. The resulting authoritarian government could then take its "hard" decisions, usually highly unpopular anti-inflation measures, such as increased public utility prices, cuts in real wages, and cuts in credit. Hardest hit are the lower classes. To carry out such policies therefore requires a heavy hand over the popular sectors. Thus, the coups and repressive authoritarian regimes that emerged in Brazil, Argentina, and Chile came about not in spite of Latin America's economic development, but because of it.

The 1980s replaced these authoritarian regimes with civilian leaders and elected governments. Explanations for this trend took many forms. Once thought to be dominant and monolithic, authoritarian regimes came to display a good deal of incoherence and fragility. Everyday citizens rose up in protest movements, formed civic organizations, and demanded popular elections. Confronted by severe economic crisis, people from Argentina and Chile to Central America sought to express their political rights. By the mid-1990s almost all countries of the region, with the conspicuous exception of Cuba, had elected governments. Whether or not these new regimes were fully "democratic," a point that led to much debate, they represented considerable improvement over the blatantly dictatorial patterns of the 1970s. Many observers expressed the optimistic hope that, at long last, Latin America was moving toward a democratic future.

Economic prospects brightened as well. Under pressure from international creditors throughout the 1980s, Latin American leaders imposed far-reaching measures designed to "liberalize" their national economies—reducing tariffs and other barriers to trade, selling state-supported companies to private investors, and curtailing deficit spending. Inflation declined and foreign investment increased. As a result, average growth in Latin America rose from a scant 1.5 percent per year in 1985–89 to 3.5 percent in the early 1990s. The unexpected onset of economic crisis in Mexico in late 1994 led to disenchantment and confusion—extending a so-called tequila effect to other countries of the region—but many analysts remained hopeful that, in the long run, the economic outlook remained positive.

Scholars approached these political and economic developments with intellectual caution. Instead of launching grand theories, such as modernization or dependency, political analysts stressed the role of beliefs, ideas, and human conviction. Some interpreted the turn toward democracy in Latin America and elsewhere as a global triumph of U.S. values, especially in light of the collapse of the Soviet Union. Others emphasized the importance of leadership and tactical maneuvers at the elite level. As for economics, some experts regarded the growth spurt of the early 1990s as vindication for pro-capitalist, free-market policy reforms. Others noted that the surge tended to reflect the ebb and flow of international investments, and that capital promptly vanished in the

face of crisis—leaving Latin America just as "dependent" as before. Of continuing concern, for many, was the problematic relationship between economic and political transformation. Does economic liberalization lead to political democracy? Or might it be the other way around? Recent developments in Latin America thus raised new questions and posed new challenges for the scholarly community.

Analytical Themes in This Book

This book is a survey of modern Latin American history, not a formulation of social theory; but we cannot escape the need for concepts in approaching our material. From modernization theory we take the causal premise that economic transformations induce social changes which, in turn, have political consequences. From the dependency school we borrow the ideas that:

1. a country's place in the international division of labor defines the shape of available paths to economic growth;
2. functional location on the "periphery" of the world system, as distinct from the commercial-industrial "center," and development at a stage when the North Atlantic system was already far advanced, meant that economic transformations in Latin America would be different from patterns traversed earlier in Europe and North America;
3. these differences in economic processes would produce different forms of social change—with respect, for example, to the nature of the "middle classes," the urban and rural working classes, and the relationship among these classes;
4. this combination of social and economic forces would define the options available to political leaders and help explain the alteration of democratic and authoritarian regimes;
5. within these constraints, some Latin American countries did much better than others in exploiting their own resources (especially agricultural) for economic development.

In other words, we intend to examine the relationship between economics and politics within an international context. We believe that this approach can be applied not only to the 1960s, 1970s, 1980s, and 1990s but also to the entire modern era. We shall be looking for these key features throughout the book.

But we see limits to the utility of this approach. The farther we go back in Latin American history, the harder it is to find the data on economic relations and social class behavior. We do not accept the all-embracing theoretical claims put forward by some analysts, and we do not share the view that only revolutions can break the cycles of dependency. More fundamentally, we believe that historical transformations

are complex processes and to understand them we need to adopt a multicausal approach. Ideas and ideology, for example, are not merely adornments or superstructures; they have important effects on the perceptions, attitudes, and actions of the people who make history. Anyone who has ever tried to compare the political traditions of Argentina and Brazil can vouch for this truth. Demographic factors, such as rapid population growth, also have far-ranging social and political effects. In our portrait of Latin American society, we hope to integrate an "international political economy" approach with consideration of cultural and other noneconomic forces.

Our narrative begins by describing first the Conquest and the colonial period, 1492–1825, when Latin America entered the periphery of the capitalist world-system through subordination to Spain and Portugal. We then describe how the disruption of this connection led to independence, followed by a phase of economic and political consolidation between 1830 and 1880.

In the late nineteenth century Latin America increased its links to the global system, this time providing raw materials (especially foodstuffs and minerals) to Great Britain, continental Europe, and the United States, and it is the formation of these export-oriented economies and their successors that occupies the major share of our attention. Why have we chosen this approach to modern Latin America? It is because by the best-known indicators—export growth, urbanization, life expectancy, literacy, political participation—Latin America around 1880 entered a period of more rapid change than anything it had experienced since the Iberian conquest in the sixteenth century.

Chapter Two is an overview of this process. It traces some common processes and patterns that accompanied the emphasis on exports and, later on, the drive toward "delayed" industrialization. We then turn to case studies: Argentina, with its traditional emphasis on beef and wheat, wracked by internal strife and military intervention before its recent turn toward democracy (Chapter Three); Chile, a leading source of nitrates and copper and, eventually, the site of an abortive socialist experiment (Chapter Four); Brazil, so well-known for coffee and, more recently, its rapid industrial growth amid political transition (Chapter Five); Peru, with its strong Indian tradition and its uncertain lurch toward nationhood (Chapter Six); Mexico, with its proximity to the United States, the scene of a popular upheaval in 1910 (Chapter Seven); Cuba, so dependent on sugar and so close to the United States, the one Latin American society that has undergone a full-fledged socialist revolution (Chapter Eight); the Caribbean, where sharp ideological struggles have gripped small island nations (Chapter Nine); Central America, a region of highly stratified societies long ignored until the political explosions of the late 1970s and early 1980s (Chapter Ten). We attempt to give full consideration to social and political themes in

these country studies, and each chapter can be read independently. Taken together they represent 84 percent of Latin America's population and roughly the same fraction of its territory.

In Chapter Eleven we summarize recent and current relationships between Latin America and the international community, particularly the United States, and in the Epilogue we speculate about likely developments toward and beyond the year 2000.

This book offers a picture of Latin American society, not a definitive catalog of facts. Our goal is to trace patterns and trends that help us to understand the complexities and variations in Latin America's paths to the present. We hope our presentation will stimulate discussion and debate, and we expect that students and colleagues will disagree with many of our interpretations. Above all, we want to introduce our readers to the excitement and fascination of the history of an area that is intriguing in its own right and has its own role to play on the world stage.

ONE

THE COLONIAL
FOUNDATIONS, 1492–1880s

When Europeans reached present-day Latin America they found three important civilizations: Mayan, Aztec, and Incan. That we should still call the native peoples of this hemisphere "Indians" perpetuates the error of sixteenth-century Spaniards who wanted to believe they had reached the spice-rich Indies.

The Mayan people, who occupied the Yucatán peninsula, southern Mexico, and most of present-day Guatemala, began to build their civilization around 500 B.C. The most famous achievements of this group were cultural—not only the building of exquisite temples but also pioneering accomplishments in architecture, sculpture, painting, hieroglyphic writing, mathematics, astronomy, and chronology (including the invention of calendars). Normally organized into a series of independent city-states, some with populations of 200,000 or more, the Mayans developed a complex social order. For reasons unknown, classic Mayan society collapsed, falling victim to domination (972–1200) and then absorption (1200–1540) by Toltec invaders from the central Mexican highlands. Yet, the direct descendants of the Mayans have survived in southern Mexico and Guatemala down to our own day.

Mexico's spacious central valley eventually became the seat of the Aztec empire. One of the Chichimec tribes that came from the north to subdue the Toltecs in the twelfth and thirteenth centuries, the Aztecs engaged in constant war with their neighbors, finally constructing the city of Tenochtitlán around 1325 (on the site of contemporary Mexico City). After gaining control of the entire valley of Mexico, they created a major empire—one that was just reaching its peak as Columbus touched shore in the Caribbean.

Aztecs were noted for their military organization and prowess at ceremonial city-building. Their art, except for their haunting poetry, was inferior in subtlety and craftsmanship to that of many other ancient Mexican civilizations.

In its final form, Aztec society was rigidly stratified. At the bottom

were slaves and at the top was a hereditary nobility. Education, marriage, and labor were meticulously programmed, and the economy was owned communally. Hereditary rulers, such as Moctezuma II, exercised immense political power. Despite centralization of authority, however, conquered states in neighboring areas were not incorporated into the empire. They were treated as tribute-paying vassals, and some— such as nearby Tlaxcala—were allowed to maintain a perpetual state of war with Tenochtitlán. One reason for this warfare was that the Aztec religion required human sacrifice, and prisoners of war could be served up for bloody rituals.

Incas adopted a very different pattern of organization. Their empire stretched for 3000 miles along the Andes, from northern Ecuador through Peru to southern Chile, and into the interior as well. After consolidating their hold in the Cuzco Valley in Peru, the Incas began expanding their empire in the early 1400s and continued until the Spanish Conquest in 1532. Once defeated, groups became integral parts of the empire. To strengthen support for the emperor, or Inca, local nobles from conquered areas were brought to Cuzco and treated as royal guests, while resistant elements in recently conquered zones were transferred to areas controlled by loyal followers. Political power belonged to a tightly organized, highly disciplined bureaucracy, with teams of local officials on the bottom and a single supreme ruler at the top. Incas were thereby able to command effective authority over most of the Andes.

Incas were master engineers, building a vast road system (for human and animal transit, since they did not use the wheel), an intricate irrigation system, and impressive terraced agriculture on mountain sides. The Incas also excelled in textile design and in treating head injuries, the latter made possible by extraordinary skills at trepanning the human skull.

Aside from the Mayans, Aztecs, and Incas, there were many other Indian cultures. In the area of modern-day Mexico alone there were over 200 different linguistic groups. Estimates of the size of Latin America's indigenous population have varied widely. One scholar has set the figure at 90 to 112 million, with 30 million each in central Mexico and Peru. Though this calculation may be too high, it is clear that by European standards of the late fifteenth century the Indian societies had grown very large. Then the Spaniards arrived.

The European Context

Europe's "discovery" of America (the Indians presumably knew where they were) was part of the remarkable European expansion in the fifteenth century. Europe was coming to know the rest of the world, as its navigators and explorers pushed back the frontiers of then-current knowledge of the globe. By the early 1600s they had woven a network

of communications all the way around the earth, and had established the economic dominance that would shape the modern world.

This burst of European expansion was made possible by a combination of factors. One was technical skill. Pilotage and navigation were notable examples, as was the ability to adapt coastal ships to the challenges of the open ocean. And another example was weaponry, which was to fortify the Europeans against the occasionally well-armed native peoples, as in Mexico.

A second factor was the economic base, which furnished capital for the maritime and military enterprise. Technology alone was not enough. Vikings had shown the technical ability to reach America but lacked the resources to carry out settlement and colonization, which required men and money. In short, the New World was not to be had by speculators of small resources or limited purpose.

Third, there had to be a European power interested in more than technical expertise and profit. It had to be ready to pursue the unknown with exceptional determination. Spain and Portugal fit this description. These Catholic monarchies, with their crusading ideal of converting heathen masses to the true religion, had a unique motivation. Spain in particular had come late to the consolidation of its territory against the infidel Muslim occupier. Portugal, although earlier rid of the Muslim intruder, was equally committed to the militant spread of the Christian faith. Their boldness set a precedent for European intruders into Latin America over the next four centuries. However much Latin America struggled, it was to remain an extension, at times a contradiction, of the Europe that had sailed west in the fifteenth century.

Spanish America: From Conquest to Colony, 1492–1600

It was no coincidence that Columbus reached America in the same year that the Spaniards liquidated the last Moorish stronghold in Spain. The reconquest down the Iberian peninsula saw the warring Christian nobles acquiring land and the crown strengthening its political control. The result by 1492 was a nobility and would-be nobility anxious for more conquests, and a crown ready to direct these subjects overseas.

Spaniards therefore reached the New World in a conquest spirit already well developed at home. Spain had presented moderate opportunity for upward social mobility, and there is considerable evidence to suggest that the New World conquerors—Hernán Cortés, Francisco Pizarro, and their followers—came to America in order to win social status as well as wealth. Spanish motivation was no doubt complex. Ferdinand and Isabella and successive monarchs thought the wealth of the New World could strengthen their hand in Europe. Many dedicated missionaries hoped to save the souls of heathen Indians. The conquerors had multiple purposes in mind: as one conquistador said, "We

MAP 2 Colonial Latin America: Political Organization

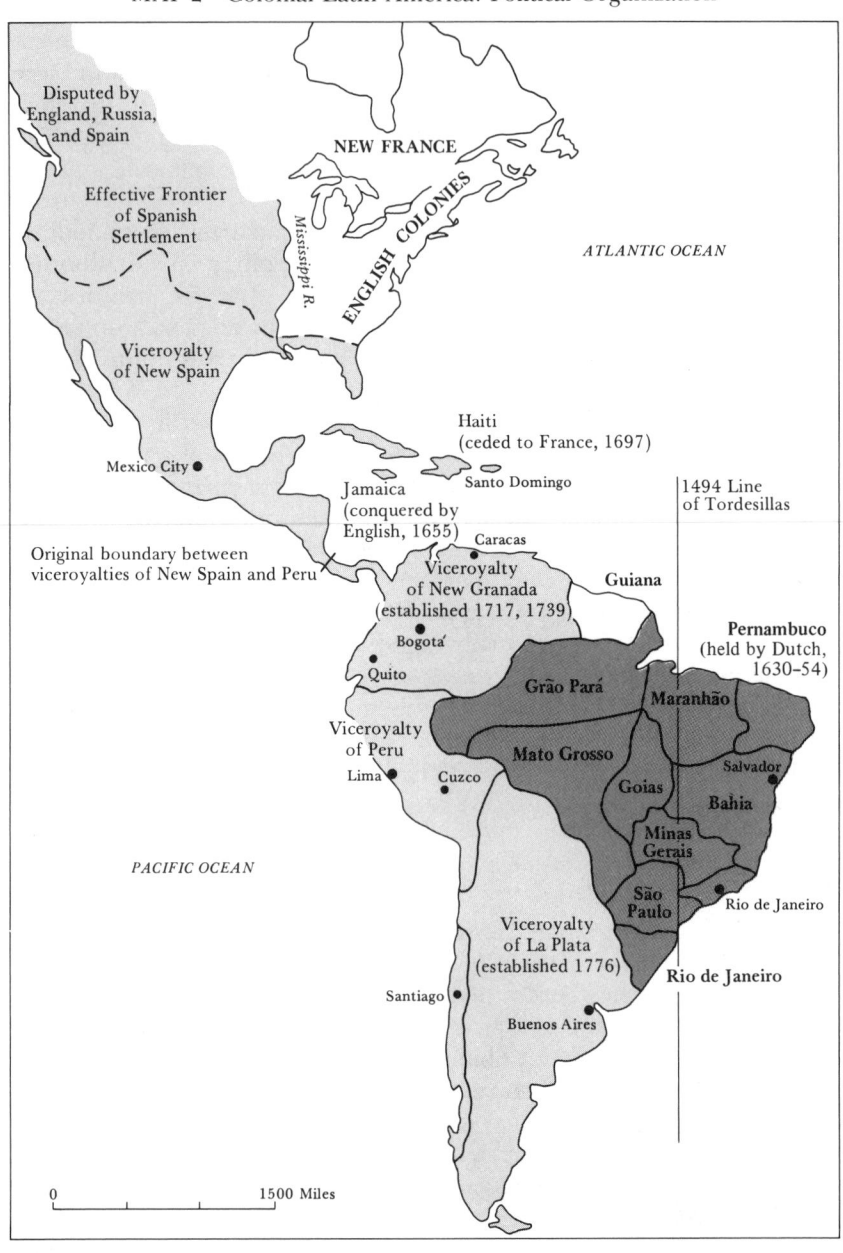

Disputed by England, Russia, and Spain

Effective Frontier of Spanish Settlement

NEW FRANCE

Mississippi R.

ENGLISH COLONIES

ATLANTIC OCEAN

Viceroyalty of New Spain

Mexico City ●

Haiti (ceded to France, 1697)

Santo Domingo

Jamaica (conquered by English, 1655)

1494 Line of Tordesillas

Original boundary between viceroyalties of New Spain and Peru

Caracas

Viceroyalty of New Granada (established 1717, 1739)

Guiana

Pernambuco (held by Dutch, 1630–54)

● Bogotá

● Quito

Grão Pará

Maranhão

Viceroyalty of Peru

Mato Grosso

Lima ●   Cuzco ●

Goias

Salvador

Bahia

Minas Gerais

São Paulo

● Rio de Janeiro

PACIFIC OCEAN

Viceroyalty of La Plata (established 1776)

Rio de Janeiro

Santiago ●

Buenos Aires ●

0 |———|———| 1500 Miles

☐ Spanish America: Viceroyalties

■ Portuguese America (Viceroyalty of Brazil): Captaincies–general ca. 1780

The viceroyalties of La Plata and New Granada originally belonged to the viceroyalty of Peru.

16

came here to serve God and the King, and also to get rich." But their central motive appears to have been the achievement of noble rank and wealth. (About one-third of the conquerors of Peru came from the lesser or "common" nobility; two-thirds were of plebeian origin. These were people with status to gain.) Thus driven, they set out for they knew not what. In a few short years they had toppled the mighty empires of the Aztecs and the Incas.

How did they do it? How could a few hundred Spaniards overthrow empires with millions of Indians? When Cortés set out from Cuba toward Mexico in 1519, he had only 550 men and 16 horses. Within two and a half years he and his battered Spanish contingent (bolstered by several hundred reinforcements) had reduced to rubble the magnificent Aztec capital of Tenochtitlán and accepted the surrender of their disheartened and bewildered god-king, Moctezuma. One explanation for their feat was the superiority of Spanish equipment and tactics—gunpowder (for muskets and cannons), horses, organization, and the confidence to stay constantly on the attack. Another factor was the initial propensity for the Aztecs to identify Cortés and his followers with the god Quetzalcoatl, whose return to the valley was predicted by myth. Important also was the role of non-Aztec peoples, such as the Tlaxcalans, who resisted and resented Aztec domination and who supplied the Spaniards with troops and advice on appropriate military tactics. Finally, and perhaps most important, an outbreak of smallpox, previously unknown in the Americas, ravaged an Indian population lacking natural immunity. By 1521, two years after the start of the Cortés campaign and less than thirty years after Columbus's first voyage, the Aztec empire had fallen under Spanish control. Cortés lost no time in asserting his authority. He extracted pledges of allegiance from neighboring chieftains and directed a vigorous reconstruction effort.

Some factors that favored the Spaniards in Mexico operated also in Peru, but Pizarro's task was simplified by the civil war then wracking the Incan empire: the Inca Atahualpa, preoccupied by the local conflict, never took Pizarro as seriously as warranted. The small Spanish band accomplished the takeover by 1533. They carted off as their booty a hoard of gold and silver large enough to fill a 12' × 17' room to the height of a man's extended arm. The dream of El Dorado had come true in the Andes.

The Spanish Conquest was centered in the Caribbean and the strongholds of the two great Indian empires—Aztec and Incan. Spaniards also launched explorations from the major population centers, but they lacked the men or resources to establish direct control in many such areas. Primary Spanish attention focused on their new realms in Mexico and Peru.

It did not take long for Spaniards to recreate many aspects of their own society in the Americas. They laid out typically Spanish designs for cities and created richly complex societies. Coopers, bakers,

scribes—people from all walks of life in Spain—came, under tight immigration control, to make their way in the New World.

Men dominated this diaspora. According to a study on Peru, for instance, white males outnumbered white females by at least seven to one. This not only created intense competition for the hands of Spanish women; it also led Spaniards to take Indian women as their consorts. Their mixed-blood children, often illegitimate, came to be known as *mestizos*. In time, the *mestizo* race would become the dominant ethnic component of much of Spanish America, including Mexico, Central America, and the Andean countries.

The Spanish crown soon realized it had a conflict of interest with the independent-minded conquerors and promptly created an elaborate bureaucracy, designed to keep the New World economy and society under firm control. In Spain the key institution for New World affairs was the Council of the Indies. Overseas the main unit of organization was the viceroyalty, headed by a viceroy ("vice-king") appointed by the king. The first viceroyalty was established in Mexico (then known as New Spain) in 1535, the second in Peru in 1544; two others were set up in the eighteenth century (see Map 2). The church had parallel structures, led by the archbishop and by the officials of the Inquisition.

In practice, this bureaucracy led to intense conflict over matters of jurisdiction, but the genius of the system was that stalemates, once they developed, could always be transmitted to a higher authority, such as the viceroy or the Council of the Indies. This meant that the various institutions would serve as watchdogs over each other (aside from periodic reviews and investigations of performance in office). Another feature of the system was, surprisingly, its flexibility. Virtually all groups had some measure of access to the bureaucracy. And though the crown retained ultimate authority, local officials possessed considerable autonomy, as shown by their occasional responses to royal decrees—*obedezco pero no cumplo* (roughly, "I accept your authority but will not execute this law"). Despite its seeming idiosyncrasies, the Spanish bureaucracy operated rather well in the New World, keeping the colonies under royal rule for nearly 300 years.

Underpinning this political structure was a set of values and assumptions that legitimized monarchical, elitist rule. They stemmed from the fundamental Roman Catholic premise, most clearly articulated by Thomas Aquinas, that there were three kinds of law: divine law, that is, God's own heavenly will; natural law, a perfect reflection or embodiment of divine law in the world of nature; and human law, man's thoroughly imperfect attempt to approximate God's will within society. Born in original sin, humanity was fallible by definition and it was only by the grace of God that some people were less fallible than others. The goal of political organization, therefore, was to elevate the less fallible to power so they could interpret and execute God's will in a supe-

rior way. And the ruler, once in power, was responsible to his or her own conscience and to God—not to the will of the people.

This rationale provided convincing justification for the supremacy of the Spanish monarch. Its theological origin revealed and fortified close links between church and state. Resuscitated in the postcolonial era, as it has often been, the code also furnished, as we shall see, a devastating critique of democratic theory. In time, political rulers would thus legitimize their power through residual aspects of traditional Roman Catholic doctrine.

The empire's economic structure reflected the prevailing mercantilist theory that economic activity should enhance the power and prestige of the state, measured on the basis of gold or silver bullion. The good mercantilist was supposed to run a favorable balance of trade, thus acquiring species or bullion in payment. Following this logic, Spain attempted to monopolize the access to wealth discovered in the New World. The first target was mining, first of gold and then mainly of silver. Another goal was to maintain complete control over commerce. Agriculture, by contrast, received little initial attention from crown officials (except for export products), and manufacturing, when later considered, was actively discouraged.

The central foundation for this economy was Indian labor, obtained from the natives by one form of coercion or another. They paid tribute to the crown and its appointed emissaries. Since cheap labor was so critical, the Spanish crown, colonists, and clerics fought bitterly for control of the Indians. In 1542, seeking to curtail the colonists, the king decreed the "New Laws," aimed at protecting the Indians by removing them from direct tutelage of the conquistadores and bringing them under the direct jurisdiction of the crown. By 1600 the crown had largely succeeded in this task, at least in legal terms. In reality, however, these changes altered only the legal form of oppression; the fact of oppression persisted.

For the Indians, the Conquest meant above all a drastic fall in population. Scholars have argued long and hard about the size of the indigenous population when the Spaniards arrived. The most reliable studies of central Mexico place the pre-Conquest populations, as of 1519, at around 25 million; for 1523 the figure is 16.8 million, for 1580 it is 1.9 million, and for 1605 it is 1 million—a total decline of 95 percent! Data on Peru are less complete, but they also show continuing decline, from 1.3 million in 1570 (forty years after the Conquest) to less than 600,000 in 1620, a drop of more than 50 percent. However uncertain the exact magnitudes, the Conquest clearly resulted in demographic calamity, largely attributable to diseases such as smallpox, measles, and influenza.

The Indian survivors saw their social order undermined and distorted. Forced to give their labor to the Spaniards, the Indians struggled to maintain their traditional social networks. The most fertile land

was seized by the conquerors—who, in many cases, converted the land to raising livestock. Indians saw the symbols of their old religion destroyed and they clung to such syncretistic practices as they could devise. Diseases took a heavier toll on men than on women, and the resulting gender imbalance further disrupted marriage patterns and family structure.

To offset the decline in Indian population, especially in tropical lowland regions, Spaniards began importing black slaves from Africa—a practice already familiar in Spain and Portugal and their Atlantic islands. Between 1518 and 1870, Spanish America imported more than 1.5 million slaves—over 16 percent of the entire Atlantic slave trade—mostly through Cuba and the northern tip of South America, destined for labor in the lowland coastal areas. Brazil, with its extensive sugar plantations, brought in about 3.7 million.

As we shall see later, Latin America produced largely multiracial societies, in contrast to the highly polarized biracial society that developed in North America.

The three ethnic components of colonial Spanish American population—Indians, Europeans, Africans—fit together in a social structure that divided itself along lines of race and function. The white sector, which included less than 2 percent of the sixteenth-century population, was the most powerful and prestigious. In the same period the mixed bloods included free blacks, *mestizos,* and mulattoes—all told, less than 3 percent of the total. The Indians, over 95 percent of the population, were placed in a unique position carefully limited and protected by a battery of royal laws.

There were additional important social relationships. One was the rivalry between whites born in Spain *(peninsulares)* and whites born in the New World *(criollos* or creoles). Another was the structure of occupations, such as the church, the army, merchants, and sheep farmers. These overlapping social categories produced in colonial Spanish America a complex stratification system in which social status was the major prize. Conflict between the *peninsulares* and creoles would eventually shape the struggles that led to independence from European rule.

Interaction between the racial groupings was less tension-filled, but still tenuous. Though interracial concubinage was widespread, interracial marriage was probably rare, and even then it followed gradations—whites might marry *mestizos,* and *mestizos* might marry Indians, but whites seldom married Indians. As civil and religious consecration was extended to interracial liaisons, especially those involving whites, they tended to blur social boundaries, legitimize aspirations for mobility, and foment uncertainty about the system of stratification. Movement definitely existed, both socially and geographically, and individuals could experience considerable change during their lifetimes.

Marriage and family customs generally assumed male domination of females. The cult of masculine superiority *(machismo)* appeared early in

Latin America, within a broad range of social and ethnic strata, and many women led restricted lives. But contrary to the stereotypical image, the standard family was not always headed by a male patriarch presiding over a large brood of children. More often than not families consisted of married couples reasonably close in age with two to four children.

But not all women married, and those who did often did not remain married for life. Data on the sixteenth century are sparse, but by 1811, according to census results, only 44 percent of the adult females in Mexico City were married. Many women were widows, and approximately one-third of the households in Mexico City were headed by single women. This was due in part to the lower life expectancy for men. For whatever reason, many Mexican women spent much of their lives as single women.

Spanish America: The Transformation of Colonial Society, 1600–1750

Spanish American colonies underwent profound changes soon after 1600. The first impetus came from Europe, where Spain began to lose the power it once had in the late fifteenth and sixteenth centuries. After the defeat of the armada by the English in 1588, the royal treasury repeatedly went bankrupt, the nobles challenged the crown, Catalonia erupted in revolt, and in 1640 Portugal—since 1580 governed by the Spanish monarch—successfully reasserted its independence. At the same time, Spain and Portugal began losing their monopolies on the New World. The English, Dutch, and French established settlements in North America and also gained footholds in the Caribbean.

With Spain's decline, the rest of seventeenth-century Europe sought to counterbalance France, now the leading power. The New World became a vital element in the European power equation. This became clear in the War of the Spanish Succession (1700–1713), which installed the Bourbons on the Spanish throne and gave the British the contract (*asiento*) for the slave trade to the Spanish colonies.

Far-reaching changes were also taking place within the colonies. The ethnic composition of society underwent profound transition. Continued immigration and natural increase turned the whites, mainly creoles, into a sizable segment of the population, perhaps 20 percent by 1825. Much more dramatic was the relative growth of the *mestizo* and mixed-blood category, from less than 3 percent around 1570 to approximately 28 percent by 1825. The shift in the Indian population was even greater, despite a slight recovery in absolute terms–down from over 95 percent to barely 42 percent. By the same year (1825) blacks accounted for about 12 percent of the Spanish American population.

As time passed the creoles began to assume active roles in key sectors of the economy, such as mining and commerce. Especially striking was their increasing ownership of land (something the early Spanish mon-

archs had discouraged) and, in some areas, the appearance of great landed estates, or haciendas. Typified by vast territorial holdings and debt peonage, the haciendas often became virtually autonomous rural communities governed by the owners or their foreman. Land titles were hereditary, and most were held by creoles. The hacienda in Spanish America brought a reversion to some of the values (although not the structure) of a classic feudal society. By the mid-eighteenth century, the crown was confronting a proud New World nobility.

The political role of the creoles was less obvious. In the late seventeenth and early eighteenth centuries they held many important political posts, mainly on the local or regional level, such as in town councils or *audiencias*. Upper-level positions were still reserved for *peninsulares*. With the decline of Spain as an imperial power, however, political institutions ceased to function as before.

Portuguese America: A Different World?

The history of Portuguese America contrasts with the story of colonial Spanish America. Under the royal House of Aviz, Portugal had established a far-flung empire with outposts in India, China, Africa, and some Atlantic islands. In fact, Portugal had become the European leader in exploration by shrewd use of its superior technical skills in cartography and navigation. In 1494 the Treaty of Tordesillas between Spain and Portugal granted Portugal the eastern half of South America (the dividing line could hardly be precise in the unknown territory), and in 1500 Pedro Alvares Cabral, the Portuguese sea captain credited with "discovering" Brazil, claimed that vast territory for his monarch (see Map 2).

This New World incursion differed from Spain's in two fundamental respects. First, there was no Indian civilization in Brazil comparable to the Aztecs or the Incas. The Tupí-Guaraní, the largest language group, lived along the coast from what is now Venezuela into southern Brazil and Paraguay, and Tapuias inhabited the interior. Some Indians were cannibalistic and most were seminomadic—which meant that Brazil would have to be settled gradually, rather than taken at a single blow. More important, it meant that the Portuguese, unlike the Spanish, did not face a highly organized, settled indigenous civilization. These Indians had built no imposing cities and they had no mythic explanations for this sudden alien intrusion.

Furthermore, there was no trace of silver or gold, and consequently no easy path to fabulous wealth. The first important economic activity was the export of brazilwood (hence the country's current name), prized in Europe for its qualities as a source of dye. And in time, contrasting sharply with most of colonial Spanish America, agriculture, especially cane sugar cultivation, predominated in the Brazilian colonial economy.

The fortress-like construction of this sixteenth century Dominican monastery in south-central Mexico aptly illustrates the alliance of church and crown in the conquest of New Spain. (Courtesy of the Library of Congress.)

The scarcity (compared to Spain) of human and mineral resources forced the Portuguese crown to resort to unusual means in trying to persuade or entice its subjects to occupy the New World holdings. In the 1530s the kings started making massive grants of effective power over (almost totally unexplored) land, usually to military men with prior experience in India or Africa or to handpicked personal favorites, and in either instance to men "of gentle blood." The land donations were huge, averaging about 130 miles along the coastline and running all the way west (as much as 500 miles or more) to the imaginary Line of Demarcation that divided Portuguese from Spanish America.

Not until 1549 did the crown begin to establish an effective imperial bureaucracy—but the purpose was to protect the area from French and British intrusions and not, as in the case of Spanish America, to reconquer the possessions from the conquerors. On the contrary, it was the lack of a Portuguese presence that forced Lisbon to act.

Partly because in its first century Brazil received a lower priority than Portugal's other overseas dominions (which were more profitable), monarchical control started out much looser than in Spanish America.

Even when the Portuguese crown tightened up after 1549, the royal institutions were largely limited to the Atlantic coast, where the taxes on exports could be easily collected. Power on the local level rested with the landowners and the town councils. Even the church was weak in sixteenth-century Brazil, compared to Mexico and Peru.

During the late sixteenth and early seventeenth centuries, landowners developed a lucrative sugar industry in the Brazilian Northeast. Having earlier made technological breakthroughs in sugar processing in their Atlantic islands, such as the Madeiras, the Portuguese had come to rely on the Dutch to retail the product in Europe. To grow sugar in America, however, required abundant labor. The Portuguese landowners first turned to the Brazilian Indians. As in Mexico and Peru, however, the Indians soon fell victim to devastating European diseases. The survivors often fled into the interior. Although the Portuguese continued to exploit the Indians until well into the eighteenth century, they had to look elsewhere for a satisfactory labor supply.

The obvious source was Africa. By the early 1500s the Spanish and Portuguese already had a half-century of experience with African slave labor, both at home and in their Atlantic islands, such as the Canaries (Spanish) and the Madeiras (Portuguese). It was not until the 1580s that the Portuguese saw enough potential profit to warrant importing African slaves. By 1650, however, northeastern Brazil had become the world's greatest source of cane sugar, produced predominantly with African slave labor. Sugar exports were estimated at £2.5 million a year, which made Brazil's coastal Northeast probably the richest single region in the entire Americas.

Other European powers wanted in on the sugar boom. The British and Dutch brought new technology, which eventually made the Caribbean into the world's preeminent sugarcane producer. The Dutch invaded Brazil itself in 1624 and controlled the sugar-rich Northeast until an alliance of Portuguese planters, merchants, and mixed-blood troops pushed them back into the ocean in 1654. Yet the Portuguese were never again to duplicate the near monopoly on New World sugar production they had enjoyed earlier in the century.

In the central and southern regions of Brazil the economy first centered on cattle-raising and, more importantly, on slave raids against the Indians (who were often shipped to the Northeast). As one Jesuit missionary lamented, "the true purpose" of these expeditions "was to capture Indians: to draw from their veins the red gold which has always been the mine of that province!" Carried out by the *bandeirantes,* whose legendary status in national history resembles a mixture of the California gold prospectors and the American cowboy, these forays extended Portuguese control over the Brazilian interior. Furthermore, they led to the discovery of mineral wealth, which had so long eluded the Portuguese. In the 1690s gold was found in Minas Gerais ("General Mines") and people flocked to the area. Diamonds were located in

1729. Mining reached its peak in 1750, with a yearly output of £3.6 million, although the low level of technological expertise contributed to a decline in mining near the late 1700s. The era also brought a brief export boom in cotton, but Brazil would have to await the nineteenth-century coffee boom before regaining much prosperity.

Brazil's colonial economy had been created for export. It was "the King's plantation." The resulting social structure reflected the investment the Portuguese crown had made. The most important single human consequence was the pervasive presence of African slaves. Over 2.5 million Africans had been brought to Brazil by 1810, nearly one-third of the Atlantic slave trade in that era. Blacks were a major component of Portuguese American society, in contrast to most areas of Spanish America.

As shown in Table 1-1, blacks amounted to nearly one-half of Brazil's total population around 1825, compared with 12 percent in Spanish America, and the mixed-blood group, mainly mulattoes, added another 18 percent. All in all, perhaps as much as two-thirds of the entire Brazilian population in the early nineteenth century were of partial or total black ancestry.

The multiracial colonial Brazilian society was highly stratified, according to the relatively little research done to date. Racial intermarriage was rare, accounting for no more than 10 percent of all marriages, and, as in Spanish America, it followed lines of gradation— whites might marry mulattoes, but they almost never married blacks. Concubinage and common-law relationships were more frequent among blacks than whites. As in Mexico City, about one-third of the family units in a sample of colonial Brazilian communities were headed by single women.

A second major facet of the social structure was internal division

TABLE 1-1   Racial Composition of Early Latin American Population

	Spanish America		Portuguese America	
	1570 (%)	*1825 (%)*	*1570 (%)*	*1825 (%)*
Whites (legally defined or by social convention)	1.3	18.2	2.4	23.4
Mixed-bloods (*mestizo* or mulatto)	2.5	28.3	3.5	17.8
Blacks	(included with mixed-bloods)	11.9	(included with mixed-bloods)	49.8
Indians	96.3	41.7	94.1	9.1
Total	100.1	100.1	100.0	100.1

Note: Some columns may not add up to 100 because of rounding.

Source: Adapted from Richard M. Morse, "The Heritage of Latin America," in Louis Hartz, ed., *The Founding of New Societies* (New York: Harcourt, Brace & World, 1964), p. 138.

within the white ruling stratum, particularly between Brazilian-born landowners and Portuguese-born merchants. This difference resembled the creole-*peninsular* conflict in Spanish America, and it had the potential for leading to an independence movement. As it turned out, European politics cut the process short. In any case, the looser crown control of Brazil had generated less resentment among the colonists than in most of Spanish America.

Portuguese America's integration into the Western economy as a peripheral area resembled that of Spanish America, but with notable differences: first, for two centuries Brazil lacked the gold and silver that obsessed the Spaniards in Mexico and Peru; second, Brazil's main contribution until the eighteenth century was agriculture, not mining; third, and perhaps most important, Portugal had developed a simpler system for ensuring revenues from its prize colony ("the milch cow," as Brazil was known in Lisbon). Unlike Spain, Portugal did not develop a vast bureaucratic network aimed at taxing and controlling the domestic market. Instead, it concentrated almost entirely on taxing Brazil's exports. As a result, Brazil offered less potential than Spanish America for breeding a powerful alliance of colonial interests which might rebel against the political authority of the mother country.

The Roots of Independence

The independence movements that led to the creation of most of present-day Latin America's nation-states owed their origins to events in Europe. Most were not radical and none brought cataclysmic changes in the social order. Much of the impetus proved to be conservative, thereby shaping the direction of the young republics in the early nineteenth century. Our story begins back in Europe.

The Bourbon monarchs of Spain, whose family had succeeded to the crown in 1713, had sought to reverse Spain's decline, both in Europe and America. Hoping to shore up New World defenses against rival European powers while also increasing revenues for the crown, the Bourbon kings imposed far-reaching administrative and political reforms. One was to create new viceroyalties—one at New Granada (first in 1717, then again in 1739) and another at Buenos Aires (1776).

In addition, Charles III (1759–88) replaced the complex administrative arrangement of the Hapsburgs with the intendancy system. In effect, this led to the replacement of the hated *corregidores* in Spanish America with "intendants"—local governors who were directly responsible to the crown, not to the viceroy. Almost all of the intendants were Spanish-born *peninsulares*, rather than American creoles, presumably thereby assuring loyalty to the monarch. The intendants greatly strengthened crown control over government, but also collided with prosperous creoles, many of whom had taken advantage of relaxed administration.

This trend could be seen in the administration of local courts. Desperately needing funds, the late seventeenth-century Hapsburg monarchs put court appointments up for sale, as had Philip IV earlier in the century. It was creoles who bought, and by 1750, out of ninety-three judges fifty-one were American-born. The Bourbon monarchs reversed this trend, and by 1807 only twelve out of ninety-nine judges were creoles. Ultimately, creoles would decide to look elsewhere for positions of authority and prestige.

One place they looked was to the town councils, or *cabildos,* which were barely functioning by the early eighteenth century. *Cabildo* offices did not always find eager buyers. With the arrival of the intendants, however, more efficient taxation gave the *cabildos* increased revenues—and they reasserted their role as local councils. The *cabildos* thus became institutional bases of creole authority.

Charles III also sought to increase royal power by tightening crown control of the church. The most dramatic step was the expulsion of the Jesuit order from all of Spanish America in 1767. Charles saw the Jesuits as a state within a state, a rival source of power and wealth. The best properties of the Jesuits were auctioned off and the proceeds, of course, went to the crown.

The military was another power source. To ward off outside threats and to crush any potential rebellion, the king decreed the establishment of colonial militias, an excellent source of prestige for status-hungry creoles. But it also altered the military balance. By 1800, for instance, there were only 6000 members of the regular Spanish army in the viceroyalty of New Spain—compared to 23,000 American-born members of the colonial militia. This was the foundation of the patriot army that would later fight for independence.

The Bourbons wanted especially to promote colonial economic development, in order to strengthen their hand in Europe. In 1778 Charles III promulgated a Decree of Free Trade, which meant that the twenty-four ports of Spanish America could now trade directly with any port in Spain or with each other (but *not* with any port outside the Spanish realm). Commerce would no longer be restricted to four colonial ports (Veracruz, Cartagena, Lima/Callao, and Panama) or tied to the Cadiz monopoly in Spain. Buenos Aires immediately began to profit from the measure. In fact, contraband trade had long flourished on these formerly forbidden routes. But the crown increased its customs receipts, since it could now place taxes on the goods that were once smuggled.

Partly for this reason, the colonial economy flourished under the Bourbons. The port of Buenos Aires, a small and lackluster town in 1776, grew to a city of 50,000 by the year 1800. Mexico was minting three times as much silver and gold in the 1790s as it had been in the 1740s. Commerce was thriving by the turn of the century.

The Bourbon policies appeared to be a success. Administration became more efficient, defenses improved, commerce swelled, and gov-

ernmental revenues increased. But creoles were upset by many of these changes, which threatened (and often managed) to reduce their status and influence. It was the challenge to creole status, more than the influence of Enlightenment thought or the example of British colonies in North America, that ultimately prompted the Spanish American dominions to opt for independence.

There had been colonial resistance, to be sure. In 1780 Túpac Amaru II, claiming to be a lineal descendant of the Incas, led an Indian revolt with an army of nearly 80,000 men. It took nearly two years of brutal fighting to stamp out the insurrections that swept over southern Peru and Bolivia. In 1781 the citizens of Socorro, in New Granada, violently protested against a tax increase and the disruption spread through much of the viceroyalty. Although patriotic Latin American historians have often described these events as "precursors" to the creole-led independence movements of the nineteenth century, this was not the case. As for Túpac Amaru II, some strands in that insurrection pointed toward independence, but on terms of Indian leadership that would never gain solid creole support. In the 1781 rebellion in New Granada, the protestors did not seek independence from the Spanish crown. They were protesting within the system, not against it.

Then how did independence come? Once again Latin America's fate was determined by dynastic politics in the Old World. After having tried and failed to help the French Bourbons save their crown, Spain allied with the revolutionary French regime in 1796, a pact that led directly to the annihilation of the Spanish navy at the Battle of Trafalgar (1805). Meanwhile, Napoleon Bonaparte, now France's dictator, in 1807 occupied Portugal, England's long-time ally. Napoleon reached the hills above the capital city of Lisbon just as the English royal navy was whisking the Portuguese royal house of Braganza and its court off to Brazil. Napoleon then turned to Spain. In 1808 he occupied Madrid, planting his brother, Joseph, on the Spanish throne. This was the act that prompted the colonies to revolt.

There was Spanish resistance to Joseph, as supporters of Ferdinand VII rallied to the cause. A junta was set up in Seville to rule in Ferdinand's name. In 1810 there followed a parliament, or *cortes*, dominated by Spanish liberals who had moved into the vacuum created by the monarch's absence. In 1812 it proclaimed a new constitution that asserted the authority of parliament, abolished the Inquisition, and restricted the role of the king.

The Colonial Response

When Napoleon put his brother on the Spanish throne, the creoles rejected him as an imposter, just as most Spaniards had done. Since Spain no longer had a government, the colonists argued, sovereignty

reverted to the people. Could this logic be extended to an argument for independence?

Yet there was nothing inevitable about the train of events that overtook Spanish America. Neither the European Enlightenment nor the example of the American Revolution alone would have fomented rebellions in Spanish America. Without Napoleon's intervention the Spanish American colonies might all have remained Spanish until well into the nineteenth century, as did Cuba.

One of the focal points of resistance to Napoleon was Buenos Aires, the seat of the newest viceroyalty, whose *cabildo* had already acquired remarkable authority. In 1806 an English squadron occupied the city of Buenos Aires, sending the viceroy fleeing to the interior city of Córdoba. A citizen's army drove the British out, and in 1807 it made short shrift of a second British attack. So it was the creoles, not the viceregal authorities, who successfully defended Buenos Aires from invasion. This demonstrated both the weakness of the crown and the capacity of the local citizenry.

Another lingering issue in the Río de La Plata region was free trade. The proclamation of 1778 had partially opened up trade for Buenos Aires, which could now ship goods directly to Spain—rather than along the long tortuous route overland to Panama and finally across the Atlantic. But it was England, not Spain, that offered the most promising market for hides and salted beef. A contraband trade therefore flourished and Argentine desire for open commerce with other European countries intensified.

In 1809, after Napoleon had ousted Spanish King Ferdinand VII, a young lawyer named Mariano Moreno called for a two-year experiment with totally free trade. Moreno argued that such a step would strengthen loyalties to the Spanish crown and provide increased revenues—since duties could be charged on legal trade, but not on contraband traffic. Late in the year the viceroy granted Buenos Aires limited freedom of trade with nations allied to Spain or neutral in the Napoleonic wars. Once again, the elite of Buenos Aires tasted political success.

When Napoleon's forces seized the centers of Bourbon resistance in Spain in 1810, leading citizens met and decided to create a "provisional junta of the Provinces of the Río de la Plata, governing for Ferdinand VII." Although it was not until 1816 that a congress formally declared independence, the pattern had been set.

In 1810 a similar movement emerged in Caracas, where the municipal *cabildo* deposed the Spanish captain-general and organized a junta to govern in the name of Ferdinand VII. As in Buenos Aires, the insurgent group consisted mainly of wealthy creoles. Their leaders had more decisive views. The most famous, Simón Bolívar, from the beginning wanted independence for America.

Born into a wealthy creole family in Caracas in 1783, Bolívar was

orphaned at the age of nine. He was then sent to Spain to complete his education, and after three years he returned to Caracas with a young Spanish bride, who within months died of yellow fever. Bolívar was devastated and never remarried. (He did not deprive himself of female companionship, however.) With his magnetic, charming, persuasive personality, he inspired loyalty and confidence among his followers. Familiar with the ideas of the Enlightenment, he vowed in 1805 to free his homeland from Spanish rule. In July 1811 the congress that convened to govern Venezuela responded to his vision by declaring independence.

But the pro-Ferdinand regency in Seville proved more resilient than had been expected, sending troops to crush this upstart rebellion. Together first with blacks, then with *llaneros* (cowboys) of the Venezuelan interior plains, Spanish forces defeated colonial troops under Francisco de Miranda. Bolívar himself managed to escape to New Granada. In 1813 he returned to Venezuela and won a series of startling military victories, triumphs that earned him the title of "the Liberator."

Events in Europe again intruded. In 1814 Ferdinand VII returned to the Spanish throne, annulled the liberal constitution of 1812 and restored himself in an absolute monarchy. Many creoles concluded that since the king was back there was no reason to continue their mobilization.

Bolívar now saw his men and munitions dwindle. After a series of defeats he was obliged in 1814 to flee again to New Granada and then to the English island of Jamaica. He hoped that Spanish America might become a single nation, but knew the odds were low. Here he was much influenced by the recent failures to establish republican government in Venezuela. Republican democracy "is overperfect, and it demands political virtues and talents far superior to our own. For the same reason, I reject a monarchy that is part aristocracy and part democracy although with such a government England has achieved much fortune and splendor." So Bolívar concluded: "Do not adopt the best system of government, but the one which is most likely to succeed."

In New Spain events took a different course. In a preemptive strike against creole patriots, *peninsulares* ousted the viceroy José de Iturrigaray in 1808 and promptly recognized the regency in Seville. Mexico City was firmly in royalist hands until 1821.

The provinces of New Spain, particularly north of the capital, were another story. By 1810 a group of prominent creoles, including a priest named Miguel Hidalgo y Costilla, were planning to seize authority in Ferdinand's name. When the plot was discovered, Hidalgo decided to act. On September 16, 1810, in the little town of Dolores, he gave an impassioned call to arms. And, curiously, it was not the local notables who rallied, but rather the long-suffering mixed-bloods and Indians. They flocked to the banner of the Virgin of Guadalupe, whom they

had long ago appropriated as their own. This "colored plebe" now formed a massive, angry, undisciplined army—"a horde," in the eyes of the startled creole elite.

Hidalgo's men stormed into the city of Guanajuato, where they massacred 500 Spanish soldiers and civilians, including the intendant, in an all-out assault on the municipal granary. After looting freely, they headed toward Mexico City. Hidalgo struggled to maintain control.

By November 1810 Hidalgo was on the outskirts of Mexico City with about 50,000 men in arms. In a decision that has prompted debate and speculation ever since, he then pulled back from the city. Surely he could have taken the capital. Why did he withdraw? Was he afraid of his own following? Instead Hidalgo moved north. After a defeat near Guadalajara in early 1811 he went on to Coahuila where he was captured and subsequently executed by a firing squad at Chihuahua.

Leadership of the ramshackle insurgency now passed to José María Morelos, another priest. Like Hidalgo, Morelos supported the abolition of Indian tribute and slavery, and even proposed agrarian reform. The latter was an explosive issue among the colonial elite. He insisted, too, that citizens had the right to choose their own form of government. Ultimately, Morelos envisioned "a new government, by which all inhabitants, except *peninsulares,* would no longer be designated as Indians, mulattoes, or *mestizos,* but all would be known as Americans." Thus Morelos combined nationalism with a commitment to social and racial equality.

In 1813 the Congress of Chilpancingo declared Mexico's independence from Spain (although it is September 16, the anniversary of Hidalgo's speech, that is celebrated as the country's independence day). The congress also decreed that slavery should be abolished and that Roman Catholicism should be the state religion. The constitution, adopted the following year, affirmed the ideal of popular sovereignty, created a system of indirect elections, and designed a powerful legislature alongside a weak three-person executive.

Meanwhile the Spaniards were winning military victories. One of the Spanish commanders was the young Agustín de Iturbide, later to play a central role in Mexican independence. In 1815 Morelos was captured, tried (by the Inquisition as well as by secular authorities), and executed. Others continued to fight for the cause, but Spaniards now held the upper hand.

Thus ended the first phase of the Spanish American independence movements. New Spain's Morelos and Hidalgo were both dead. Bolívar languished in exile in Jamaica. The junta in the Río de la Plata struggled to maintain unity and had yet to call for independence. By 1815, with Ferdinand back on the throne, the Spanish crown appeared to have snuffed out its colonial rebellion.

Achieving Independence

The Spanish military advantage in South America did not last long. In 1816 Bolívar returned to Venezuela and began duplicating his earlier victories. But now he had allied with José Antonio Páez, the brilliant leader of the fearless *llaneros*, who had earlier fought for the royalists. Now Páez was fighting for independence from Spain. Bolívar's cause was further bolstered by the arrival of reinforcements from England, which by 1819 numbered over 4000. Thus strengthened, Bolívar established firm control of Venezuela by early 1819.

After defeating Spanish forces in New Granada, Bolívar attempted in 1821 to create a new state of Gran Colombia, uniting Venezuela, New Granada, and Ecuador. It gained little support, so Bolívar marched southward, hunting for more royalists and Spaniards to defeat.

Meanwhile, José de San Martín was conducting an extraordinary military campaign in the south. The son of a Spanish officer, born on the northern frontier of present-day Argentina, he began a military career at the age of eleven. In 1812 he offered his services to the junta in Buenos Aires, having decided in favor of independence for the colonies. A soldier by training and outlook, he did not have the political acumen of Bolívar or the social commitment of Morelos, but he was a skillful military strategist.

As commander of the rebel forces, San Martín was ready by early 1817 to attempt one of the most daring exploits of this era: leading an army of 5000 across the Andes for a surprise attack on royalist troops in Chile. He caught the Spaniards completely off guard, won a major victory in the battle of Chacabuco, and triumphantly entered the city of Santiago. San Martín now prepared for the next step in his campaign, the liberation of Peru.

By 1820 San Martín reached the Peruvian coast. Lima was even more monarchist than Mexico City. As the capital of a major viceroyalty, the city had received numerous favors and privileges from the crown. Although the elevation of Buenos Aires to a viceroyalty in the Bourbon era had hurt Lima economically, its monarchist sentiment was still strong. Creoles and *peninsulares* both tended to favor the continuation of Ferdinand VII's rule. San Martín withheld his attack, noting: "I do not seek military glory, nor am I ambitious for the title of conqueror of Peru: I only wish to free it from oppression. What good would Lima do me if its inhabitants were hostile politically?"

Here, too, radical change in Spain catalyzed events. When Ferdinand VII succumbed to political pressure and suddenly endorsed the liberal constitution of 1812, the turnabout stunned his Lima supporters. They were especially distressed over the abolition of the Inquisition and the challenge to the dignity of priests. Many could accept limitations on monarchical authority, but not on the role and power of the church.

This turn of events in Spain drastically altered the climate of opinion in Mexico City and Lima. Independence from Spain was no longer a radical or even a liberal cause. Now it was a *conservative* goal, a means of upholding traditional values and social codes. As if acknowledging this fact, the *cabildo* of Lima invited San Martín to enter the city in mid-1821. On July 28 he formally proclaimed the independence of Peru.

After further skirmishes with royalist troops, San Martín went to Ecuador for a historic meeting with Simón Bolívar. Exactly what happened there has never been established. Bolívar may have set the tone when he offered a toast "to the two greatest men in America, General San Martín and himself." Apparently Bolívar rejected San Martín's proposal for a monarchy in Peru, insisted on the union of Gran Colombia, and declined San Martín's offer to serve under his command. In any case, San Martín then resigned all his offices and soon went to Europe, where he died in 1850.

In late 1823 Bolívar moved to Peru, where the Spaniards still maintained an imposing force. In 1824 the royalists were decisively defeated by colonial troops in the battle of Ayacucho. In 1825 Bolívar entered Upper Peru (present-day Bolivia) in the hope that Peru and Upper Peru might form a single nation, but he was too late. The regional leaders of Upper Peru were set on creating their own republic. They promptly did so, naming it for Bolívar and making him president for life.

After returning to Lima, Bolívar went on to Gran Colombia, hoping to patch up the failing union. By now he had grown bitter and vindictive, upset that his dreams had failed to materialize. In 1830 both Venezuela and Ecuador withdrew from Gran Colombia. Suffering from tuberculosis, Bolívar looked back in despair. "America," he said, "America is ungovernable. Those who have served the revolution have plowed the sea." On December 17, at the age of only forty-seven, the Liberator passed away.

In Mexico the defeat of Morelos in 1815 stalled the independence movement—until Ferdinand VII declared submission to the constitution of 1812, thus pushing prosperous and prominent creoles to the side of independence. The cause was led by the same Agustín de Iturbide who had led the royalists against Morelos. Ironically, the independence movement acquired a conservative tinge.

The opportunistic Iturbide persuaded the viceroy to give him command of royalist forces in the south. He then marched against a rebel leader with whom he immediately struck an alliance—for the sake of independence. In 1821 he issued a call for three "guarantees": of religion (the Catholic faith to be the official creed), of independence (presumably under a monarchy), and union (fair treatment for creoles and *peninsulares* alike). Iturbide took Mexico City and established an empire—with himself, of course, as emperor. It lasted only two years.

In Central America the landed creole class became as worried about

liberal dominance in Spain as had been their counterparts in Mexico. In 1822 the Central American landowners decided to cast their lot with Iturbide's empire and announced their annexation to royalist Mexico. When Iturbide abdicated in 1823, the modern-day Central American states, from Guatemala to Costa Rica (excluding Panama) became the independent United Provinces of Central America.

The Brazilian Path to Independence

Independence came to Brazil in a manner very different from that of Spanish America. That was partly due to the fact that Brazil was by 1800 far more populous and prosperous than the tiny mother country. By contrast, no single Spanish colonial territory equaled metropolitan Spain in economic or political power. When the colonials proclaimed independence, Spain fought back doggedly and Spanish Americans grew to hate the crown. The Portuguese, on the other hand, did not even have the military power to stop the Brazilians' move toward political autonomy.

The context of Brazilian independence pointed up another important difference. When the Napoleonic army invaded Iberia in November 1807 the entire Portuguese court was able to flee to Brazil, thanks to the British royal navy.

Upon arriving in early 1808, the Portuguese court found a colony which had no printing press, no academic faculties, and no commerce, save with the mother country. The newly arrived prince regent and later monarch Dom João VI promptly decreed the end of Portugal's commercial monopoly by opening Brazil's ports. His logic was obvious. Since Napoleon now controlled Portugal, the exiled Portuguese monarch could only continue to benefit from Brazil's foreign trade if the formerly exclusive link with Lisbon was severed. The prime beneficiaries were the British, who had, after all, brought the Braganza family and its retinue to Brazil.

Britain gained privileged access to Brazil in 1810 by official, fifteen-year agreements which: (1) gave Britain the lowest tariff (even lower than Portugal!) on goods entering Brazil; (2) committed the Portuguese crown to the gradual abolition of the African slave trade; and (3) guaranteed British subjects in Brazil the right of trial by British-named judges. These treaties soon caused deep resentment among the Brazilian elite.

The exiled Portuguese monarch now set about creating new institutions, such as a national library, a national museum, and a botanical garden, all in Rio de Janeiro. A French artistic mission was requested to speed professionalization in architecture, painting, and sculpture.

The crown also sought to attract foreign immigrants to Brazil. It had very limited success, and large-scale European immigration was not to begin until the late 1880s. There was a push to promote textile manu-

facture, including repeal of the 1785 royal decree that banned all industry. But such measures could not get at the deeper causes of Brazil's economic backwardness: lack of capital, technology, skilled labor, and a significant domestic market.

In late 1808 French troops were driven from Portugal, and an assembly (*Cortes Gerais*) was called to write a new constitution. The newly victorious Portuguese Liberals, interested in exploiting Brazil's wealth, pressed for the return of the royal court to Portugal. Dom João soon did return to Lisbon, leaving his son Dom Pedro behind in Brazil as the prince regent of the Combined Kingdoms.

Attention now focused on the *Cortes Gerais,* which approved measures that would have restored Lisbon's royal trade monopoly in Brazil by setting lower tariffs for imports arriving in Portuguese vessels and/ or trans-shipped from Portugal. The *Cortes* also approved measures returning the individual Brazilian provinces to direct and separate rule from Lisbon, thereby undermining the central rule created in Rio de Janeiro after 1808. However "liberal" the Portuguese Liberals were in Portugal, they abhorred the move toward autonomy of their American "co-kingdom."

The landowners and urban professionals who constituted the Brazilian elite had been preparing to confront the Portuguese recolonizers. Their passionate rhetoric overflowed the fledgling Rio press. They wanted the prince regent, Dom Pedro, to refuse to return to Lisbon. By June 1822 Dom Pedro had decided to convoke a Constituent Assembly in Brazil. The *Cortes* in Lisbon then demanded the prince regent's immediate return and took new steps to reverse Brazil's growing autonomy. The Brazilian plantation owners' pressure on Dom Pedro now paid off: On January 9, 1822, he defied the summons of the *Cortes*. "I shall remain!" he cried, giving birth to the only durable independent monarchy in modern Latin American history.

To win their independence the Brazilians had to fight, but not on the scale of the Spanish Americans. The fiercest combat came in Bahia, on the northeastern coast, and Grão Pará, in the eastern Amazon valley. In Bahia a junta proclaimed loyalty to Portugal and fought off the local pro-independence rebels. In 1823 the rebels triumphed, aided by Admiral Cochrane, one of the English military officers hired to give the rebel governments experienced help in combat. Another mercenary, Admiral Grenfell, led the victory over a similar loyalist junta in Grão Pará. His forces then mopped up a local rebel wing that was demanding more radical social change. In Brazil, as in Mexico, the elite was alert to repress any fundamental challenge to the socioeconomic establishment.

Portugal's military weakness partly explains why Brazil's struggle for independence proved far less bloody than Spanish America's. Equally important, the Brazilian rebels did not split over the issue of republicanism because, with a few exceptions, the elite preferred a monarchy

to a republic. Thanks to the exile of the court, the Brazilians could opt for an independent monarchy. Brazil thus entered independence with a unique legacy. Not least important, Brazilians did not associate independence with military prowess: no Brazilian Simón Bolívar or San Martín arose to dominate the patriotic imagination.

The Aftermath of Independence, 1830–1850

The new Spanish American republics faced formidable problems as they embarked on independence in the 1820s. The physical violence of the wars wrought economic disaster. The destruction probably reached its highest point in Venezuela, where *guerra a muerte* ("war to the death") took a heavy toll of human life and reduced the livestock population by more than one-half between 1810 and 1830. The early phases of the Mexican wars, particularly during the campaigns of Hidalgo and Morelos, took a similar toll on people and property. Uruguay, where José Artigas led bands of gaucho rebels against well-entrenched Spanish troops, also suffered grave losses. During the second phase of the movement the theater of operations shifted to other areas, especially Peru, where the fighting appears to have been less intensive than before—but the burden of supporting large armies was nevertheless heavy. The civilian labor force was decimated and, throughout the continent, capital was scarce.

The economies of the new nations were overwhelmingly based on agriculture and mining. This was equally true of most of the world outside West Europe. Yet Latin America differed from most of Africa, the Middle East, and Asia in that over the past two and a half centuries it had been partially brought into the world trading economy dominated by Europe. It was the exportable surplus from Latin American agricultural and mining production that linked it to the North Atlantic economy. With the creation of separate countries this basic economic structure remained intact almost everywhere, slowly to be modified in succeeding decades.

Trade had come to an almost complete standstill between 1810 and 1826. Commerce with Spain had stopped, and trade among the former colonies was also greatly reduced. Northwest Argentina, for instance, suffered from the loss of trade with Peru. Montevideo, still under Spanish control, could no longer function as a commercial entrepôt. Guerrilla warfare in New Spain and other areas made transport difficult and dangerous. Communications systems within and between the former colonies, never much favored by the Spaniards, fell into near-total disuse.

There was also the factor of postindependence regional conflict within major areas of Spanish and Portuguese America. Mexico was wracked by battles which kept that country divided and without effective national direction before 1850. Brazil, at the same time, collapsed

into a series of regionalist revolts that left the monarchy effectively neutralized until the 1840s. And in the Río de la Plata region the fierce rivalry between the province of Buenos Aires and the rest of the country was temporarily resolved only by the dictatorship of Juan Manuel de Rosas (1829–52). Everywhere the move was to assert economic autonomy by locality or region. That meant fragmentation. In Spanish America it meant that Bolívar's dream would be buried under the advance of nationalism. One after another of the new republics claimed economic independence. They would soon find the world market a sobering test.

In many parts of Spanish America the new governments had to deal with public debts, even before they could attempt to rebuild their economies. To sustain the fighting, to equip the armies, the insurgent regimes frequently had to obtain or borrow funds. Tax collection, to put it mildly, was difficult. As a result, the national treasuries were empty and government authorities had to turn elsewhere for funds. A prime source was Britain, where bankers supported the regimes with loans—particularly in Argentina, Chile, Peru, and Mexico. Thus the new governments immediately ran up debts to foreign lenders. Managing the foreign debt has remained, down to our own day, a major problem for Latin American governments.

Another area in which foreign capital invested was the African slave trade, which continued on a large scale to Brazil (until 1850) and Cuba (until 1865). Both had an export-oriented agriculture that made slave labor profitable during an era when it was being abolished elsewhere in the Western Hemisphere.

The years between 1830 and 1850 saw Latin America's exports to the North Atlantic economy increase. Key primary products were wheat and nitrates from Chile, tobacco from Colombia, hides, salted beef, and wool from Argentina, guano from Peru, sugar from Cuba, coffee from Brazil, and cacao from Venezuela. These same countries were heavily importing textiles and consumer goods, thereby often throwing local artisan producers out of work. It was the industrial producers in West Europe (especially Britain) competing against the small-scale Latin American producers who had survived from the colonial era. The result was a foregone conclusion.

This was all part of free trade, the dogma that had arrived in Latin America with Enlightenment philosophy and the postindependence commitment to the principles of liberalism. Applying this dogma was the most significant economic policy decision in nineteenth-century Latin America. Along with a rapid inflow of foreign (primarily European) imports came a small cadre of foreign merchants, especially British. They became key figures throughout Latin America in the import of goods and services, the latter including shipping, insurance, and financing.

Should we be surprised that manufactured goods from Europe

steadily displaced domestic products? Wasn't it inevitable that Europe's greater technology and economies of scale would prevail? Transportation costs should have helped protect local producers, but the supposed (or genuine) superiority of foreign-made goods posed a serious dilemma soon after independence and has continued down to today. Latin American economies often failed to make their own industry truly competitive. Why? Lack of a sufficient market was certainly a factor. But equally important was the system of values and the social hierarchy which made it possible for the elite to perpetuate a society based on an agrarian-oriented economy.

The economic record of the 1830–50 period is therefore one of slow adaptation to the world economy. Latin America was on the fringes of the North Atlantic economy, which was to expand rapidly in the nineteenth century. Both research and data on the economic history of this era are distressingly scarce—but it appears, on the basis of evidence available, that Latin America's republics took a passive stance. The dynamism came from outside.

The creation and maintenance of large armies in most Latin American republics also crucially affected the social order, because they created a channel for careers based on talent. As the fighting intensified and the stakes increased, creole rebel leaders had to recruit soldiers and commanders on ability, rather than on skin color or social status. Thus José Antonio Páez, a rough-hewn *mestizo*, became a valued military leader in Venezuela. In Mexico José María Morelos was *mestizo*. Other examples abound. Military prowess became a means by which members of marginal groups could gain social recognition. None of the newly independent governments retained legal disabilities for *mestizos* or other mixed-bloods, a fact which helped to blur once-rigid social lines.

But if the wars opened a social avenue for ambitious *mestizos* and others, mobility was limited. Economic resources, particularly land, remained in the hands of traditional creole families. Commerce was modest in the years right after the fighting, and many merchant families retained their control of trade. Industry barely existed. As a result, there was only one way for men of modest origin to get ahead: through the military, and from there into politics.

This social dynamic helps explain much of the political turbulence in Spanish America between the 1820s and 1850. The new republics finished the wars with large military establishments, often led by *mestizos* who had no alternative careers. To get ahead they had to stay in the army—or move into government. In the meantime, creole landowners, in many parts of the continent, did not compete for political power. They withdrew to their haciendas, which could function as self-sufficient units, and tried to increase their landholdings. In effect they left government to the soldiers and to the bosses known as *caudillos,* partly because political power did not seem worth the trouble. Later in

the nineteenth century, when governmental authority became a valued commodity, *hacendados* and *estancieros* came off their lands and took over.

So governments were toppled and run by *caudillos,* often soldiers (or ex-soldiers) who took power by force. Once in the presidential office, they usually found that sparse treasuries offered little reward for their followers. Their bands then dispersed, and new *caudillos* would come forward with new bands of followers. The governments did not have strong finances and as a result were highly vulnerable to being overthrown. From the 1820s until midcentury, political authority in Spanish America was weak; the state, as a central institution, did not wield much autonomous strength.

During this era another current emerged, a move to consolidate and centralize power. It usually came out in attempted dictatorships, not popular consensus. The first two decades after independence thus saw the appearance of real or would-be "strong men," like Diego Portales in Chile and Juan Manuel de Rosas in Argentina, who sought to impose their will on their countries, thereby strengthening the role of the state. The struggle between locally based power and the centralizers— military or civilian—became a basic theme in the political life of the new nations.

If the Wars of Independence opened narrow channels for *mestizos* and middle-range groups in Spanish America, they did very little for the Indian masses. In general, Indians had been ambiguous about the struggle: though they sided with Hidalgo or stayed neutral in Mexico, they supported royalists in southern Chile, and in Peru and Colombia they fought on both sides. The leaders of the new republics therefore did not feel indebted to the Indians. More important, the Indians now lost the special protection of caste status they had enjoyed under Spanish colonial law. Whatever its drawbacks, that status had been an oft-used refuge for the Indians. They also lost their communal lands (which had been inalienable) and were theoretically forced into the competitive market so praised by nineteenth-century liberals. In fact, they became even more isolated and poverty-stricken.

Independence left a somewhat different social legacy for Brazil. Instead of *displacing* a ruling elite, as happened in Spanish America, Brazil *acquired* a ruling elite: the Portuguese crown and its attendants. Brazil also acquired a monarchy that would last until 1889. But these political trends had little effect on the black slaves who labored on sugar plantations and elsewhere in the economy. In fact, the institution of slavery was not abolished at independence or by the 1850s, as in Spanish America (except for Cuba and Puerto Rico) and it would later become a central issue in Brazilian politics. In Brazil, as in other new nations, independence did not change life much for the poorest segments of the population.

The Pull of the International Economy, 1850–1880s

After 1850 Latin America moved from the postindependence consolidation phase to begin laying the foundations for its greater integration into the world economy. In political terms this required governments ready to create the infrastructure needed to export key primary products, such as guano from Peru, coffee from Brazil, minerals from Mexico, and sugar from the Caribbean. As the era of the *caudillo* gave way to the era of the administrators, the prime task was national unification.

The independent republics moved to strengthen the use of two elements in their economies: land and labor. Most governments sought to put land into the hands of entrepreneurs who would invest and make it bear fruit. In Brazil and Mexico that meant government pressure to sell off government (previously crown) land. The losers in Mexico and the Andes were the Indians, but such action could also hit white or *mestizo* owners who had failed to develop their lands.

To provide labor, the Latin American elites in several countries hoped for immigration from Europe. These years saw repeated proposals to attract European immigrants, who would supposedly contribute to national development with little further investment. In fact, the elite—in countries like Argentina and Brazil—soon found that immigration was a sensitive issue, both at home and in the countries sending the migrants. Before 1880 immigration was nowhere a major factor in increasing the labor force. But the strong elite impulse to recruit migrants demonstrated their belief that their countries' economic and social salvation was to be found in Europe. As will become apparent, this reflected Latin American doubts about their countries' viability.

The mid-nineteenth century also saw an effort to improve Latin America's transportation network. What was needed were railroads, canals, docks, and roads. Since the sixteenth century, cargo (including people) had traveled by pack-mule or burro. In only a few areas did navigable rivers or lakes offer an alternative. By midcentury Latin America was the target for numerous proposals to build railroads. The impetus usually came from foreigners, especially British and North American. But few railways were actually built before the 1880s, so the transportation networks remained about as bad as they had been on the eve of independence.

The rhythm of economic activity quickened, nonetheless, throughout Latin America after 1850. The stimulus came primarily from the dynamic economies of North America and West Europe, led by Britain. As Europe plunged ever more deeply into industrialization, it needed increasing imports of food, such as sugar, beef, grain, as well as primary commodities such as guano and nitrate fertilizers, wool, and industrial metals. These were the decades when economic ties—trade, investment, financing, technology transfer, migration—deepened between Europe and Mexico, Argentina, Peru, Chile, Brazil, and Cuba

(even though still a Spanish colony). By 1880 the stage was set for even greater economic expansion.

The economic upturn after 1850 had several important limitations, however. First, it resulted in very little growth of domestic industry. The rising Latin American need for metal tools, small machines, instruments, construction equipment, weapons, and similar light industrialized goods, was primarily met from Europe, not from home-country shops or factories. The trend was hardly surprising. The British, French, or U.S. products were of better quality than anything produced at home, although that advantage could have been narrowed if the domestic producers had had enough time and a sufficient market to upgrade quality. But that would have required government protection either through high tariffs or outright import prohibitions. No Latin American government was prepared or able to take such a step in these decades.

The reasons were several. First, imported products were superior and were thus strongly preferred by local consumers; second, most governments lived off tariff revenues, which tough protectionism would have cut off; third, the economically powerful groups, such as the landowners and cattlemen, were strongly committed to free trade, which their European customers preached as the only true road to prosperity; finally, Latin American merchants, who were strategically located in the largest cities, had an obvious stake in fighting protection. That motive was even greater when the merchant was a foreigner (usually British or French), as happened frequently by midcentury. Not surprisingly, the Latin American advocates of protectionism or state-aided industry could make little headway.

A second limitation on economic expansion between 1850 and 1880 was its reinforcement of the highly stratified socioeconomic structure inherited from the independence era: a thin elite at the top, a slightly wider "middle" group, and the other 90 percent or so at the bottom. The continued focus on agro-ranching and mining meant that most laborers would continue under working conditions and rates of pay that could never move them toward becoming the consumers that a "developed" economy both produces and needs.

Latin America was being pulled further into the international economy in a way that would sharply limit its economic development. The nature of that economic link has continued to trouble Latin Americans for the last century, and will be a recurrent theme in the rest of this book.

TWO

THE TRANSFORMATION OF MODERN LATIN AMERICA, 1880s–1990s

Latin America has undergone a series of far-reaching economic, social, and political changes since the late nineteenth century. National economies have become integrated into the global system centered in Europe and the United States, social groupings and relationships have changed, cities have burgeoned, politics have witnessed reform and upheaval and sometimes stagnation. Variations on these themes have led to great diversity in national experience, and following this chapter we present eight case studies: Argentina, Chile, Brazil, Peru, Mexico, Cuba, the Caribbean, and Central America. As we shall see, these countries illustrate the complexity of Latin America's modern history.

Yet there have been important uniformities as well as differences, and the purpose of this chapter is to offer an outline of the patterns and processes of change. It does not depict the history of any single country; rather, it presents a composite portrait that can provide a basis for understanding the context in which individual countries developed. It will also enable us to compare countries and generalize about the historical forces at work throughout the continent.

If we are to understand modern Latin America it must be placed within the context of global economic expansion, beginning with the Conquest of the sixteenth century. Within this system, Latin America has occupied an essentially subordinate or "dependent" position, pursuing economic paths that have been largely shaped by the industrial powers of Europe and the United States. These economic developments have brought about transitions in the social order and class structure, and these changes in turn have crucially affected political change. We thus begin with a set of simplified causal relationships: economic changes produce social changes which furnish the context for political change.[1]

1. It is for this reason that each of the case studies in following chapters contains an overview section on "economic growth and social change"—with the exception of Mexico, where the Revolution of 1910 exerted such a strong political impact on that country's history and obliged us to employ a different format.

Phase 1: Initiation of Export-Import Growth, 1880–1900

It was the Industrial Revolution in Europe that precipitated change in the economies of nineteenth-century Latin America. As shown in Chapter One, Latin America had seen its links with the world economy reduced after achieving independence from Portugal and Spain. Latin America's native landowners converted their holdings into autonomous, self-sufficient entities, rather than producing goods for domestic or foreign markets. Mining had come to a standstill, partly as a result of the destruction of the Wars of Independence. Manufacturing was modest, mostly done by artisans in small establishments.

By the late nineteenth century, however, industrialization in Europe was producing a strong demand for foodstuffs and raw materials. English and European laborers, now living in cities and working in factories, needed to purchase food they could no longer cultivate. And captains of industry, eager to expand their output and operations, were seeking raw material, particularly minerals. Both incentives led governments and investors in Europe to begin looking abroad—to Africa, to Asia, and, of course, to Latin America.

As a result, the major Latin American countries underwent a startling transition in the late nineteenth century, especially after 1880. Argentina, with its vast and fertile pampas, became a major producer of agricultural and pastoral goods—particularly wool, wheat, and, most notably, beef. Chile resuscitated the production of copper, an industry that had fallen into decay after the independence years. Brazil became famous for the production of coffee. Cuba produced coffee as well as sugar and tobacco. Mexico came to export a variety of raw material goods, from henequen (a fiber used for making rope) to sugar to industrial minerals, particularly copper and zinc. Central America exported coffee and bananas, while from Peru it was sugar and silver.

The development of these exports was accompanied by the importation of manufactured goods, particularly from Europe. Latin Americans purchased textiles, machines, luxury items, and other finished products in relatively large quantity. There thus occurred an exchange, though the prices of Latin American exports were far more unstable than the prices of Europe's exports.

As development progressed, investment flowed into Latin America from the industrial nations, particularly England. Between 1870 and 1913 the value of Britain's investments in Latin America went from 85 million pounds sterling to 757 million pounds in 1913—an increase of almost ninefold in four decades. By 1913 British investors owned approximately two-thirds of the total foreign investment in Latin America. One of the most consistent British investments was in railroad construction—especially in Argentina, Mexico, Peru, and Brazil. British, French, and North American investors also put capital into mining ventures, particularly in Mexico, Chile, and Peru. This meant that the

Latins themselves would not have to invest there. It also meant that control of key sectors of the Latin economies was passing into foreign hands.

Thus was established, in the late nineteenth century, an "export-import" form of economic growth that stimulated development in the raw-material sectors of the Latin American economies. The impetus and capital came largely from abroad. With the adoption of this alternative, Latin America took a commercial road to "dependent" economic growth—dependent, that is, on decisions and prosperity in other parts of the world.

The rapid expansion of Latin America's export economies was accompanied, even preceded, by the victory of an intellectual rationale that justified Latin America's integration into the world economy. That rationale was liberalism, a faith in progress and a belief that it would come in economics only through the free play of market forces, and in politics only through a limited government that maximized individual liberty. Latin American liberalism, like most ideologies in Latin America, was an import. Its principal sources lay in France and England. Unlike those countries, however, Latin America had not undergone significant industrialization by the middle of the nineteenth century. Latin America therefore lacked the social structure that had nurtured liberalism in Europe, a fact that was bound to make Latin American liberalism different.

In the second half of the eighteenth century, Spanish America and Brazil underwent an abortive experiment in state capitalism. The upheaval caused by the French revolutionary wars had disrupted Spain's mercantilist monopoly in America. Havana had been captured by the English and its ports thrown open. The astounding increase in trade impressed all observers. The logic was inescapable: since smuggling had grown into a huge percentage of total trade throughout Portuguese and Spanish America, why not legalize free trade and gain taxes from the increase in government-monitored commerce?

The apologists of economic liberalism in Latin America quoted freely from European theorists who justified free trade and the international division of labor as "natural" and, indeed, optimal. Any deviation from its dictates would be folly—reducing trade and thereby reducing income. It is important to see that the great majority of those critics who attacked *political* institutions of monarchical governments (which they considered "illiberal") did not disagree with the ideology of *economic* liberalism. In Brazil, for example, Tavares Bastos indicted the imperial government on the charge that it stifled local political life but extolled the virtues of free trade and faithfully repeated European doctrines of laissez-faire.

One can say that throughout the later nineteenth century economic liberalism remained unshaken in Latin America. Attempts at protective tariffs were beaten back by politicians who argued that Latin America

was not well suited, either by its resources or by its relative bargaining position, to violate the principles of free trade.

The key debates about economic policy were largely restricted to the elites, who are defined here as that very small stratum (less than 5 percent of the population) that had the power and wealth to control economic and political decision making on the local, regional, and national levels.

The elites' commitment to liberalism was reinforced by their deep concerns about the supposed racial inferiority of their native populations. They implicitly acknowledged the correctness of racist theories by constantly urging heavy European immigration as the solution to their lack of skilled labor. They preferred immigrants from northern Europe (although the vast majority in fact came from Portugal, Spain, and Italy), hoping that habits of self-reliance and entrepreneurship—the hallmarks of the liberals' ideal—could be reinforced in Latin America.

Added to the racist doubts was a general sense of their own inferiority. Until the First World War, Latin American elites frequently described themselves as little more than imitators of European culture. Many doubted that their countries could ever achieve a distinctive civilization. For the tropical countries, their worries over racial determinism were reinforced by doubts about their climate, which European theorists constantly told them could never support a high civilization. This environmental determinism thus reinforced racial determinism, and the combination appeared to disqualify tropical lands as a stage on which the liberal dream could be realized.

Within Latin America the rapid growth of the export economies led to subtle but important transformations in the societies. First in sequence and importance was the modernization of the upper-class elite. Given these new economic incentives, landowners and property owners were no longer content to run subsistence operations on their haciendas; instead, they sought opportunities and maximized profits. This led to an entrepreneurial spirit that marked a significant change in the outlook and behavior of the continent's elite groups. Cattle raisers in Argentina, coffee growers in Brazil, sugar barons in Cuba and Mexico—all were seeking efficiency and commercial success. They were no longer a relatively enclosed, semifeudal elite; they became aggressive entrepreneurs.

New professional or "service" groups emerged to play additional economic roles. Particularly important was growth and change in the commercial sector. Merchants played an essential part in this transformation, as in the colonial era, but now many were foreigners—and this time they were tying the Latin American economies to the markets overseas, particularly in Europe. Another development concerned professionals, lawyers, and others who represented both foreign and domestic groups in their commercial transactions. Lawyers had always

been important, but during the export-import phase they assumed crucial new roles in helping determine the institutional framework for the new era.

These economic and social transitions also led to political change. With so much at stake, Latin American elites—particularly landowners—began to take a clear interest in national politics. No longer content to stay on their fief-like haciendas, they began to pursue political power. The era of the traditional *caudillo* was coming to an end.

This elite quest for political authority in late nineteenth-century Latin America took two basic forms. In one version, landowners and other economic elites took direct control of the government—as in Argentina and Chile. They sought to build strong, exclusive regimes, usually with military support, often proclaiming legitimacy through adherence to constitutions strongly resembling U.S. and European models. In both Argentina and Chile there was mild competition between political parties that tended, at least in this early phase, to represent competing factions of the aristocracy. But there was more agreement than disagreement about basic policy issues, and little serious opposition to the wisdom of pursuing export-oriented economic growth. Competition was restricted and voting was often a sham. One might think of such regimes as expressions of "oligarchic democracy."

A second pattern involved the imposition of dictatorial strongmen, often military officers, to assert law and order—again, for the ultimate benefit of the landed elites. Porfirio Díaz of Mexico, who took power in 1876, is the most conspicuous example—but the pattern also appeared in Venezuela, Peru, and other countries. In contrast to oligarchic democracy, where elites exercise direct political power, here it was the indirect application of elite rule through dictators who often did not themselves come from the upper ranks of the society.

In either case, the emphasis was on stability and social control. Dissident groups were suppressed and the struggle for power was contained within restricted circles. Indeed, one of the basic goals of these regimes was to centralize power, if necessary stripping it from regional *caudillos,* and to create powerful and dominant nation-states. These goals were not easily achieved, given the residual fragmentation of society and the social structure, but progress was made in the larger countries. In Argentina, for example, centralism triumphed with the establishment of the city of Buenos Aires as a federal district in 1880 (much as Washington, D.C., is under the direct jurisdiction of the federal government in the United States). In Mexico, the effective and often ruthless policies of Porfirio Díaz led to enhanced national power at the expense of local strongholds. And in Brazil, the imperial government of Dom Pedro II made significant headway toward the establishment of an effective nation-state (thereby also provoking a regionalist backlash that contributed to the empire's overthrow in 1889).

One intent of these centralizers was to promote further economic

development along the export-import lines of growth. Political stability was viewed as essential to attract foreign investment, which, in turn, could stimulate economic growth. And when the investment came, it helped strengthen the forces of law and order. Railroads are an example: foreign investors would be reluctant to put funds into a country threatened by political disorder; but once the railroads were built, as in Mexico, they became important instruments in consolidating central rule, since they could be (and were) used to dispatch federal troops to put down uprisings in almost any part of the nation.

Phase 2: Expansion of Export-Import Growth, 1900–1930

The success of these policies became apparent in the late nineteenth and early twentieth centuries, when export-oriented Latin American economies embarked on eras of remarkable prosperity. Argentina became so wealthy from its beef-and-wheat economy that the figure of the Argentine playboy became a watchword in fashionable European society—a free-spending young Latin in high-spirited pursuit of elegance. Plantations appeared and expanded in Mexico, producing henequen in the Yucatán and sugar in the central zones, especially in areas south of Mexico City; mining was also profitable, and the nascent industry of oil production was beginning to grow into a significant activity. Copper exports continued to grow from Chile, which produced some fruits and wheat for international markets as well. Technological improvements led to increased production of sugar in the Caribbean, especially in Cuba, as U.S. owners stepped up their investing in modern sugar mills. Brazil lived off coffee and natural rubber exports. The United Fruit Company expanded its huge banana plantations in Central America. In all these countries the money economy had become more sensitive to trends in the international economy, where exports earned the foreign exchange to buy badly needed imports. Any major shock to the world economy would be bound to produce rapid and dramatic effects in the commercialized sectors. Although industrialization had not gone far yet, factories existed in such sectors as textiles, leather goods, beverages, food processing, and construction materials. The most dynamic service sectors were transportation, government bureaucracy, commerce, and finance.

Consolidation of the export-import model of growth prompted two fundamental changes in the social structure. One was the appearance and growth of middle social strata. Occupationally, these consisted of professionals, merchants, shopkeepers, and small businessmen who profited from the export-import economy but who did not occupy upper-strata positions of ownership or leadership. Most often found in cities, middle-sector spokesmen were relatively well educated and were seeking a clearly recognized place in their society.

The second major change concerned the working class. In order to

sustain expansion of the export economies, elites tried to import labor from abroad (as the Argentine Juan Bautista Alberdi once put it, "to govern is to populate"). As a result, Argentina began, in the 1880s, an aggressive policy to encourage immigration from Europe: the tide of arrivals in Argentina over the next three decades was so great, even allowing for the returnees, that it has been called by one historian the country's "alluvial era." Brazil also recruited immigrants, primarily to work in the coffee fields of São Paulo. Peru and Chile received a number of immigrants, though far fewer in both relative and absolute terms than Argentina. Cuba remained a special case, since the importation of black slaves from Africa had long since determined the composition of the country's laboring class (this had also been true in parts of Brazil, particularly in the Northeast where sugar plantations thrived on slave labor). Mexico presents an interesting exception to this pattern. Alone among the major countries, Mexico never sought large-scale immigration from abroad. One reason is obvious: the country continued to have a large peasant Indian population, making it unnecessary to import new recruits for the labor force.

The appearance of incipient working classes led to new organizations with important implications for the future. Workers often established mutual aid societies and, in some countries, started labor unions. The nature of the Latin American economies set the context for worker activism. First, because exports were crucial, workers in the infrastructure that supported exports—especially railways and docks—held a vital position. Any work stoppage posed an immediate threat to the country's trade viability and therefore to its capacity to import. Second, the relatively primitive state of industrialization meant that most workers were employed in very small firms, usually with less than twenty-five employees. Only a few industries, such as textiles, fit the modern image of huge factories with mass production techniques. The unions in question were usually organized by craft, rather than by industry. The exceptions were rail workers, miners, and dockers, who, not coincidentally, were among the more militant workers.

The years between 1914 and 1927 saw a surge of labor mobilization. It was the high point of the anarchist, anarcho-syndicalist, and syndicalist influence, when the capital cities of every major Latin American nation were rocked by general strikes. Latin America suddenly seemed to be joining in the class confrontations shaking Germany and Russia, as well as the United States and much of the rest of Europe. It is at these critical moments—mass protests, general strikes, intensified ties between the unionized and the nonunionized—that we can see clearly the nature of the working class, its organizations, and the manner in which the dominant elites chose to respond.

What we shall need to compare, as the country studies unfold, are the similarities and differences in the patterns of interaction among employers, workers, and politicians, along with landowners, profession-

als, and the military. If there are similarities in urban labor's mobilization during the decade after the great protests that began at the end of the First World War, there were striking differences in the elites' responses. In particular, we shall find that the legal framework of labor relations received much more attention in Chile than in Argentina and Brazil.

Another major change during the 1900–1930 era was in the balance between rural and urban sectors of society. The importation of labor and migration from the countryside combined, in this era, to produce the growth of large-scale cities. By 1900 Buenos Aires had established itself as "the Paris of South America," a large and cosmopolitan city with about 750,000 inhabitants. All told, just about one-quarter of the Argentine population lived in cities with 20,000 or more inhabitants at the turn of the century; the same held true for Cuba. About 20 percent of the Chilean population resided in similar settlements, while the corresponding figures for Brazil and Mexico (the latter with a substantial indigenous population) were just under 10 percent. For Central America the figures were also under 10 percent and in Peru only 6 percent. The general fact is that expansion of the export-import economies brought with it the urbanization of Latin American society.

Because of national or ethnic origin, however, laboring classes did not gain much of a foothold on political power in the early twentieth century. Immigrants in Argentina and Brazil were not entitled to vote unless they went through naturalization, so politicians could afford to ignore them. In Mexico, workers of peasant background had little chance of influencing the Díaz dictatorship. And in Cuba, of course, the history of slavery left its own painful legacy.

What this meant, at least in the short run, was that Latin American elites, while promoting the export-oriented expansion, could count on a responsive labor force without an effective threat of political participation by the working class (although strike actions had proved worrisome). From then until the 1920s or 1930s it seemed, for many, like the best of both worlds.

And as a result, elites in several countries permitted enough political reform to allow effective pursuit of power by members and representatives of the middle sectors. The idea was to gain the allegiance of the middle sectors and therefore strengthen the structure of elite control and power. Accordingly, the early twentieth century was a period of political reform in some of the larger countries: a voting law in 1912 in Argentina opened suffrage to large sectors of the population and permitted a middle-class party, the so-called Radical Party, to win the presidency in 1916. Changes in Chile, actually beginning in the 1890s, saw the imposition of parliamentary rule on a previously presidentialist system. In Brazil, the overthrow of the monarchy in 1889 opened a period of limited electoral politics. Cuba, after gaining independence from Spain in 1898 (and, many would say, then ceding it to the United

States), remains a special case. And even in Mexico, where a large-scale revolution broke out in 1910, the generalization holds: the original goal of the revolutionary movement was not to transform Mexican society but merely to gain access to the political system for excluded fragments of the country's middle class.

The reformist movements often produced a "co-optative democracy"—in which effective participation spread from the upper class to the middle class, to the continued exclusion of the lower class. Such transitions usually reflected the attempts of ruling socioeconomic elites to co-opt the middle sectors into supporting the system, though they sometimes had unforeseen and unintended consequences—as in the case of Mexico, where events transpired to create a full-blown revolution. For the most part the goals were limited.

One significant side effect was the creation of a cadre, in various countries, of professional politicians. Party politics created careers for men (women did not even have the vote anywhere in Latin America before 1929) who could devote their entire adult lives to the pursuit of political power. As often as not they represented the interests of the reigning aristocracy, but they also formed an identifiably separate social group. As prominent actors in the civilian political scene, they would also become targets for the disdain and wrath of the military establishment.

For many countries of Latin America, or at least for the elites, the reformist formula worked fairly well. European demand for raw materials during and for several years after World War I led to continued and sustained prosperity. The export-import model of growth appeared to offer a functional and profitable means of integrating Latin America into the global system of capitalism. Political adaptations seemed to assure the long-run hegemony of national elites.

In fact, liberalism—both political and economic—was soon found wanting. Its failure illustrates the phenomenon so familiar throughout modern Latin America: unsuccessful cultural borrowing, or "alienation" as latter-day nationalists have described it. Copy as they would the legal institutions and philosophical phrases of classical liberalism, Latin Americans found their reality did not lend itself to the simple application of dogma. They failed to understand that originally European liberalism was the ideology of a rising social class whose growing economic power gave it the political leverage to put its ideology into practice.

Does this mean anything more than that Latin America lacked a large middle class? Only in part. More fundamentally, Latin America had remained an agrarian economy whose export sector was matched, in most countries, by a huge subsistence sector. Liberalism had a chance only because, after 1850, a small but growing sector of the society *thought* it saw its interests as different from those of the traditional sectors.

Specifically, the professionals—lawyers, doctors, military officers, civil servants, and merchants—all constituted an urban interest. They rapidly absorbed the European liberal ideas without gaining the relative economic power which their counterparts had gained in France and England. Thus, even if they had seen their economic interests as antagonistic to those of the traditional agrarian sector, they would have been in a weak position. But often they did not. Their livelihood was usually tied to the agrarian sector even though they lived in cities. Their clients, customers, and employers were men whose incomes depended largely upon commercial agriculture. And the prosperity of this agriculture, in turn, depended upon foreign trade.

At this point Latin American liberals were stymied by their economic liberalism. Believing its abstract principles and uncomfortably aware of their patent weakness vis-à-vis their principal creditors and trading partners—the United States and England—they could not think their way to "illiberal" economic solutions. Furthermore, the latter would have cost them *personally* in the short run. For example, protective tariffs for industry undoubtedly would have saddled urban consumers with higher-priced, lower-quality goods. Protection would also have imperiled the profits of the export-import merchants, who were a powerful pressure group. Liberals were therefore disinclined to advocate industrialization, which alone could have increased their number sufficiently to give them the political power that might have made possible the realization of liberal political ideals.

Economic and political liberalism cross-cut in yet another way. Nonliberal ideas in economics, such as protective tariffs and controls on foreign investments, were often associated in practice with antiliberal political ideas. Thus the connection was easily established: deviation from liberal principles in economics must mean authoritarian government, and it was therefore held in low regard.

A further argument used against advocates of economic unorthodoxy (i.e., nonliberal measures) was politically difficult to answer. Whenever proposals for government support of national industry were made, opponents charged, often to good effect, that a small group of selfish investors was seeking to profit at the expense of the public. In addition, local entrepreneurs were almost invariably underfinanced and inexperienced. Just as everywhere in the developing world, these businessmen faced formidable competition from goods imported from the industrialized economies. Without protection and subsidies they had little hope.

Latin American liberals were debilitated for yet another reason. It was their uncertainty about an underlying assumption of liberalism: faith in the rationality and enterprise of their individual countrymen. In Brazil, for example, politicians had spent years justifying slavery on the grounds that it was a necessary evil for Brazil's tropical, agrarian economy. Only enslaved Africans could do the work. Now the argu-

ment came back to haunt liberals. Slavery's legacy was a labor force that fell far short of the rational world envisaged by Bentham and Mill. The event that transformed this atmosphere was the spectacular collapse of the world capitalist economy in 1929–30.

Phase 3: Import-Substituting Industrialization, 1930–1960s

The Great Depression had initially catastrophic effects on the economies of Latin America. The precipitous economic decline in Europe and the United States suddenly shrunk the market for Latin American exports. The international demand for coffee, sugar, metals, and meat underwent sharp reduction—and Latin American leaders could find no alternative outlets for their products. Both the unit price and the quantity of Latin American exports dropped, with the result that their total value for the years 1930–34 was 48 percent below what it had been for 1925–29. Once again, events at the industrialized center of the world-system had decisive (and limiting) effects on Latin America and other Third World societies.

The ensuing world depression put great pressure on the political systems of Latin American countries, many of which suffered military coups (or attempted coups). Within a year or so after the October 1929 stock market crash in New York, army officers had sought or taken power in Argentina, Brazil, Chile, Peru, Guatemala, El Salvador, and Honduras. Mexico was enduring a special constitutional crisis of its own, and Cuba succumbed to a military takeover in 1933. It would be an exaggeration to say that the economic effects of the depression alone caused these political outcomes; but they cast into doubt the viability of the export-import model of growth, helped discredit ruling political elites, and made the populace more prepared to accept military regimes. From the early 1930s onward, the military reasserted its traditional role as a principal force in Latin American politics.

Latin American rulers had two options in responding to the global economic crisis. One was to forge even closer commercial linkages to the industrialized nations in order to secure a steady share of the market, whatever its size and dislocations. Argentina took such an approach, for example, struggling to preserve its access to the British market for beef. In 1933 Argentina signed the Roca-Runciman Pact, through which it would retain acceptable quotas in the English market in exchange for Argentine guarantees to purchase British goods and to insure profits on British businesses in Argentina. In this way, some countries tried to maintain the workability of the export-import model despite the reduced demand brought on by the depression.

An alternative tack, not necessarily inconsistent with the first, was to embark on industrialization. One of the goals of this policy, often supported by the military, would be to achieve greater economic independence. The idea was that by building its own industry, Latin America

An American cartoonist depicts the Latin American military as a continuing threat to democratic institutions. (Roy B. Justus, *Minneapolis Star,* 1963. Reprinted with permission of the Minneapolis Star and Tribune Company.)

would be less dependent on Europe and the United States for manufactured goods. For the military, this meant arms. By producing industrial as well as agricultural and mineral goods, the Latin American economies would become more integrated and more self-sufficient. And, as a result, they would be less vulnerable to the kinds of shocks brought on by the worldwide depression.

An additional purpose was to create jobs for the working classes, which had continued to grow in size and importance since the beginning of the twentieth century. Concentrated almost entirely in cities, the Latin American proletariat was still struggling to organize and sustain union movements. And in contrast to the previous generation, it was now trying to exert power as a social force. In some countries, such as Chile, union movements were relatively free of arbitrary government involvement. Elsewhere, as in Mexico and Brazil, politicians recognized labor as a potential political resource and took a direct hand in stimulating (and controlling) labor organizations. Whether perceived as ally or threat, the urban working class was seeking secure employment, and Latin American leaders saw industrialization as one way to respond.

But the most plausible *form* of industrial development was not simply to copy the paths traced by, for example, nineteenth-century England. Instead, Latin America's economies started producing manufactured goods that they had formerly imported from Europe and the United States. Hence the name for this type of development, "import-substituting industrialization" (ISI).

From the late 1930s to the 1960s, at least in major countries, ISI policies met with relative success. Argentina, Brazil, and Mexico developed significant industrial plants that helped generate economic growth. There were limitations and drawbacks in this form of development (discussed below), but the immediate result was to generate momentum for the national economies.

The social consequences of industrialization were complex. One result, of course, was the formation of an entrepreneurial capitalist class or, more specifically, an industrial bourgeoisie. In Chile, members of this group came principally from the families of the landed elite. In Mexico and Argentina, they comprised different social types and therefore presented a potential challenge to the hegemony of traditional ruling elites. But the basic point remains: industrialization, even of the ISI type, created a new power group in Latin American society. Its role was to be much debated as the century continued.

Of particular importance here was the role of the state in stimulating ISI growth. In contrast to the largely laissez-faire policies of nineteenth-century England and the United States, Latin American governments actively promoted industrial growth. They did so in various ways: erecting tariff barriers and raising the price of imported goods to the point where local industrial firms could successfully compete in the marketplace; creating demand by favoring local producers in government contracts (involving, for example, purchases for the military); and, most important, establishing government-run companies and investing directly in industrial firms. Through protection and participation, the state in Latin America furnished critical impetus for the region's industrial growth.

As industry progressed, the working classes also grew in strength and importance. Whether autonomous or state-directed, union movements increased rapidly and the support (or control) of labor became crucial for the continuation of industrial expansion. The workers were needed to provide the labor under conditions that would yield profits for their employers. Organized labor was emerging as a major actor on the Latin American scene.

The political expression of these socioeconomic changes took two forms. One was the continuation of co-optative democracy, through which industrialists and workers gained (usually limited) access to power through electoral or other competition. One example was Chile, where political parties were reorganized to represent the interests of new groups and strata in society. Pro-labor and pro-industrialist parties

entered the Chilean electoral process, leading eventually to the tragic confrontation of the 1970s. Under this system, they were co-opted into the governing structure, and as long as this arrangement lasted their participation lent valuable support to the regime.

The more common response involved the creation of multiclass "populist" alliances. The emergence of an industrialist elite and the vitalization of the labor movement made possible a new, pro-industrial alliance merging the interests of entrepreneurs and workers—in some cases, directly challenging the longstanding predominance of agricultural and landed interests. Each of these alliances was created by a national leader who used the power of the state for his purpose. Thus, as we shall see later, Juan Perón built a multiclass, urban-based populist coalition in Argentina during the 1940s; in Brazil, Getúlio Vargas began to do so in the late 1930s; and, under somewhat more complicated circumstances, Lázaro Cárdenas turned to populist solutions for Mexico during this same period.

Most populist regimes had two key characteristics. For one thing, they were at least semiauthoritarian: they usually represented coalitions against some other set of interests (such as landed interests) that were by definition prevented from participation, and this involved some degree of both exclusion and repression. Second, as time would tell, they represented interests of classes—workers and industrialists—that were bound to conflict among themselves. The maintenance of such regimes therefore depended in large part on the personal power and charisma of individual leaders (such as Perón in Argentina and Vargas in Brazil). It also meant that, with or without magnetic leadership, the regimes would be hard to sustain in times of economic adversity.

Phase 4: Stagnation in Import-Substituting Growth, 1960s–1980s

The 1960s presaged an era of crisis for Latin America. The economic strategy which grew out of the post-1929 industrialization policies had begun to run into serious troubles, both economic and political. On the economic front, the problems arose in part from the nature of ISI development itself.

First, industrialization through ISI was structurally incomplete. To produce manufactured goods, Latin American firms continually had to rely on imported capital goods (such as machine tools) from Europe, the United States, and now Japan. If such capital goods could not be imported, or were too expensive, local firms were placed in jeopardy. Latin Americans slowly realized that ISI growth did not end the dependency of Latin America on the industrialized nations. It merely altered the form of that dependency.

This inherent difficulty became acute because of the uneven terms of trade. Over time, the world market prices of Latin America's princi-

pal exports (coffee, wheat, copper) underwent a steady decline in purchasing power. That is, for the same quantity of exports, Latin American countries could purchase smaller and smaller quantities of capital goods. Industrial growth therefore faced a bottleneck. And the answer did not lie in increasing the quantity of their traditional exports, since that might simply depress the price.

Second, the domestic demand for manufactured products was inevitably limited. Industries ran up against a lack of buyers, at least at prevailing prices and credit terms. Brazilians could buy only so many refrigerators (especially in view of the highly unequal distribution of income, which kept the popular masses from even contemplating such purchases). This problem of limited markets might have been met through the formation of multinational or regional trade associations, or something like a Latin American common market; there were efforts in this direction, but they did not resolve the issue. The industries in the major Latin countries tended to be more competitive than complementary, and such rivalries posed serious political obstacles to the formation of such associations. As time went on, industrial firms in Latin America continued to face the problem of limited markets.

Third, and closely related, was the relatively high degree of technology involved in Latin American industry. This meant that it could create only a limited number of jobs for workers. In other words, Latin American industrial development in this era had picked up the capital-intensive technology typical of the advanced industrial economies: in comparison to nineteenth-century models of growth, it entailed more investments in machines and fewer in manual labor. Individual firms saw this as necessary to survive against the economic competition. One of the unintentional results, however, was to place a ceiling on the size of the domestic market for consumer goods, since relatively few wage earners could afford to purchase them. A second result was the failure to reverse the growing unemployment, which, by the 1960s, began to pose a serious threat to the prevailing social orders.

As pressure mounted, ruling elites in several countries imposed highly repressive regimes, often through military coups—as in Brazil (1964), Argentina (1966), and Chile (1973). In all cases the most important decisions were made (or were subject to veto) by the top ranks of military officers. In view of economic stagnation the military and elites believed they had to stimulate investment. And in order to accomplish this, they reasoned, they would have to dismantle, perhaps even crush, the collective power of the working class. The more organized the working class, the more difficult this task became.

Each of these military-dominated governments assumed the power of control over decisions concerning labor's most vital interests—wages, working conditions, fringe benefits, and the right to organize. Labor had to reconcile itself to the measures approved by the military-dominated government bureaucracies that set labor policies. Outright

strikes were virtually nonexistent in Chile between 1973 and 1979 or in Brazil between 1968 and 1978. Moves to organize strikes in those countries in those years invited harsh repression, although some relaxation occurred in Brazil beginning in early 1978. Argentina's stronger tradition of union initiative was harder to suppress, but labor leaders there were forced to show great prudence also. All three military regimes created a "command economy" when it came to labor relations.

Why this heavy hand toward labor? Viewed from the short term, all three cases could be explained by the need to undertake unpopular anti-inflation policies. These regimes came to power when inflation and balance of payments deficits had made their economies dangerously vulnerable. In all three cases international credit, both public and private from the capitalist world, had been virtually closed off. All three were required to launch stabilization programs. Since no capitalist country in recent years had succeeded in carrying out economic stabilization without producing a drop in real wages (usually very large), and since Argentina, Brazil, and Chile had all had extensive experience with organized labor's resistance to stabilization programs, it was no surprise these military governments sought tough controls over labor.

All three cases of antilabor policies, however, had more profound causes. These governments proclaimed themselves to be "antipolitical." All blamed their countries' distress on the alleged incompetence, dishonesty, or treachery of politicians. These military groups were most aggressive against the radically leftist politicians and labor leaders. Few channels of political opposition were left open. Just as Chile had once been the most democratic system, its military regime became the most draconian, abolishing all political parties and burning the electoral rolls. The generals repudiated the open pluralistic political competition for which that country had been famous. Chile was to enter an era "free" of politics.

Argentina's military government took stern measures in 1976, suspending Congress and all political parties, thereby signifying a hiatus in competitive politics. Brazil's military guardians, having come to power in a less radicalized political atmosphere than their counterparts in Argentina and Chile, also found themselves in their second year (1965) pushed to abolish the old political parties (replaced by two government-sanctioned new ones). A more repressive phase (although with fewer deaths than in Argentina or Chile) after 1968 was followed by a gradual "opening" after 1978.

Regimes pursuing this path became known as "bureaucratic-authoritarian" states, and they had several characteristics. One was the granting of public office to people with highly bureaucratized careers—to members of the military, the civil service, or large corporations. Second was the political and economic exclusion of the working class and the control of the popular sector. Third was the reduction or near-elimination of political activity, especially in the early phases of the re-

gime: problems were defined as technical, not political, and they were met with administrative solutions rather than negotiated political settlements.

Finally, bureaucratic-authoritarian governments sought to revive economic growth by consolidating ties with international economic forces—revising, once again, the terms of dependency on the global world-system. Specifically, leaders of these regimes often forged alliances with multinational corporations (vast international companies such as IBM, Philips, Volkswagen). To establish credit and gain time, they also needed to come to terms with their creditors, such as U.S. and European banks and international lending agencies (such as the World Bank and the Inter-American Development Bank). Tasks of this kind have commonly been delegated to the most internationalized members of the original coalition, frequently young economists trained at U.S. institutions—often identified by derisive nicknames, such as the "Chicago boys" in Chile.

Mexico, as we shall see in Chapter Seven, represents a different situation, since the state had acquired effective control over the popular sectors before the economic turndown of the 1960s, and the country was therefore able to make the transition from "populist" authoritarianism toward a modified version of "bureaucratic" authoritarianism without a brutal military coup. That control of the popular sectors was being tested anew during the protracted post-1982 economic crisis. Central America demonstrates the volatility of social conditions where economic development took place under traditional dictatorship, without giving rise to incremental reform. And Cuba, with its socialist revolution, offers still another pattern of transition and change.

Phase 5: Crisis, Debt, and Democracy, 1980s–1990s

Economic growth during the 1970s often relied on external borrowing. In 1973–74 and again in 1978–79, concerted action by petroleum-exporting countries led to abrupt increases in the world price of oil. Unable to spend all their windfall profits (technically known as "rents") in their own countries, Middle Eastern potentates made massive deposits in international banks. Logically enough, the banks sought to lend out this money to capital-starved but creditworthy clients—at profitable interest rates. Prominent bankers in Europe and the United States decided that Latin American countries looked like good potential clients, especially if their governments were committed to maintaining law and order.

Thus began a frenetic cycle of lending and borrowing. Between 1970 and 1980 Latin America increased its external debt from $27 billion to $231 billion, with annual debt-service payments (interest plus amortiza-

tion) of $18 billion. Complications quickly ensued. Commodity prices declined, real interest rates climbed, bankers became reluctant to make still more loans. Countries of the region experienced increasing difficulty in meeting obligations on the debt, and in August 1982 Mexico declared its inability to pay. The U.S. government frantically put together a rescue package for Mexico, but it provided only short-term respite. Merely to cover interest payments, major Latin American debtors—Argentina, Brazil, and Mexico—had to pay yearly the equivalent of 5 percent of their gross domestic product (GDP). Caught in a squeeze between declining export earnings and rising debt-service obligations, Latin America plunged into a decade-long economic crisis.

Over the course of the 1980s, international authorities—the U.S. government, the private bankers, and, especially, the International Monetary Fund (IMF)—imposed strict terms on Latin American borrowers. If the governments would undertake fundamental economic reforms, they could qualify for relief from debt burdens. These reforms almost always included opening up economies to foreign trade and investment, reducing the role of the government, promoting new exports, and taking steps against inflation. This set of "neo-liberal" ideas called for "structural adjustments" in economic policy and amounted virtually to a complete repudiation of time-honored ISI strategies.

Faced with little choice, most Latin American governments accepted IMF-sponsored conditions, at least formally. Smaller countries, such as Chile and Bolivia, first succeeded in carrying them out. Mexico made significant headway by the late 1980s, as did Argentina and Peru in the early 1990s. Brazil, the largest country of all, would resist IMF formulae until the mid-1990s.

By 1990, as more loans were made just to cover current interest payments, Latin America's total external debt climbed to $417.5 billion. From 1982 through 1989 Latin America transferred over $200 billion to industrialized nations, equivalent to several times the Marshall Plan. Real economic output (GDP) per capita declined in 1981, 1982, 1983, 1988, and 1989—and showed a cumulative decline of almost 10 percent over the decade.

In this context of economic crisis, Latin America turned away from authoritarianism—and, in many cases, toward democracy. The coalitions behind bureaucratic-authoritarian regimes turned out to be relatively fragile. Local industrialists felt threatened by multinational corporations, and the military's instinct to annihilate any militant opposition aroused protest from intellectuals, artists, and middle-sector representatives. Under the weight of the debt crisis, too, some military leaders chose to return to the barracks—and let civilians take over what seemed to be an "unsolvable problem."

Pressure also welled up from below. One conspicuous feature of

Latin American politics throughout the 1980s was a rise in civic participation, as ordinary citizens began to insist on their rights and demand
accountability from governments. In part this resulted from the uniting
of opposition forces produced by the brutality of military repression.
Second was an increasing commitment to the electoral process, as people clamored for free and fair elections. Third, as a consequence of all
these processes, there appeared a new cadre of civilian, middle-class,
well-educated presidents. This occurred most clearly in Brazil, Argentina, and Chile.

Most of these regimes were not complete democracies. In many
countries the military still wielded considerable power from behind the
scenes and could exercise a veto power over major policies. After years
of repression (including physical elimination) by military dictatorships,
the Marxist left by the 1990s was largely divided, demoralized, and
discredited by the collapse of communism in East Europe and the Soviet Union, and in some countries the left was still denied effective
participation in politics. Key topics, such as land reform, stood no
chance of serious consideration. Human rights suffered continuing violations. And many crucial decisions, especially on economic policy, were
made in high-handed and authoritarian fashion.

By the early 1990s, Latin America at last began to reap rewards for
accepting stringent reform policies. Excluding Brazil (which postponed
its reforms until 1994), average inflation throughout the region
dropped from 130 percent in 1989 to 14 percent in 1994. Partly in
response, international investors looked with favor on Latin America.
The inflow of private funds from abroad—mainly from Europe, Japan,
and the United States—climbed from only $13.4 billion in 1990 to the
impressive total of $57 billion in 1994. (In 1993 alone, U.S. investors
bought more foreign equities around the world—about $68 billion—
than during the whole decade of the 1980s.) And as a result, average
growth in Latin America increased from only 1.5 percent in 1985–89
to the respectable level of 3.5 percent in the early 1990s.

Problems lingered nonetheless. Most of this new private funding
came in the form of "portfolio" investments (that is, purchases of paper
stocks or bonds) rather than "direct" investments (such as plants or
factories). Portfolio investments tend to be highly mobile and notoriously volatile, and they can leave host countries almost instantly. So
when the U.S. Federal Reserve started increasing interest rates in early
1994, investors began to anticipate improved returns in the U.S. market. This expectation led to a 14 percent drop in capital flows to Latin
America in 1994. And when Mexico crashed in December 1994, foreign investors fled markets throughout the region in what came to be
known as the "tequila effect." The implication was painfully clear: despite impressive and often courageous efforts at economic reform,
Latin America still remained vulnerable to the vagaries of world financial markets.

There were structural problems as well. One was the persistence of poverty. According to international standards, nearly one-half the population of Latin America (46 percent) qualified as "poor" in the early 1990s. A second long-term problem was inequality. Ever since data on the subject first became available in the 1950s, Latin America has displayed the most uneven distribution of income in the world—more so than Africa, South Asia, and the Middle East—and this situation was getting progressively worse. By the early 1990s, the richest 10 percent of households in Latin America were receiving 40 percent of total income, while the bottom 20 percent was getting less than 4 percent. Social equity thus came to pose a major challenge for the region.

By the mid-1990s, the political spectrum in Latin America encompassed a broad span (even aside from socialist Cuba). At one end was what might be called "electoral authoritarianism," which took its harshest form in Guatemala. At the other end was "incomplete democracy," with many cases lying in between. After a long struggle with tyranny, Chile once again resumed its position, along with Costa Rica, as perhaps the most democratic country in the region—despite the continuing autonomy of the armed forces. Showing a remarkable degree of political openness, Argentina and Brazil transferred presidential power through free and fair elections. Partly as a result of the military dictatorships, however, political institutions (especially the judiciary, the legislature, and the civil service, as well as governmental ministries and institutes) were notoriously weak in these and other countries. Peru faced perhaps the starkest institutional vacuum of the entire region. A key question therefore emerged: Would Latin America's fragile democracies have the strength and ability to govern? Could they develop the institutional capacity needed to consolidate recent reforms and to attack the problems of poverty and inequality?

In summary, the evolution of leading societies in Latin America has followed a pattern in which economic, social, and political developments are linked. Adherence to a general pattern has varied from country to country, but it is nonetheless possible to discern the outline of a common historical experience since the late nineteenth century. (Table 2-1 presents a simple summary.) This set of patterns, one should remember, derives from the history of the largest and most economically developed nations of Latin America. Some of the economically less developed areas, such as Central America or Paraguay, have undergone only some of these transitions—and their trajectory is seriously affected by the timing of their start. Just as global factors have conditioned the historical experience of the larger countries, so will global factors condition the future development of less advanced countries. There is no guarantee, in other words, that the history of Argentina and Brazil foreshadow the future of Honduras and Paraguay—any more than knowing the history of the nineteenth-century United States enables us to predict the evolution of Chile or Mexico.

TABLE 2-1   Patterns of Change in Latin America

	Economic Development	*Social Change*	*Typical Political Outcome*
Phase 1 (1880–1900)	Initiation of export-import growth	Modernization of elite, appearance of commercial sector and new professionals	Oligarchic democracy or integrating dictatorship
Phase 2 (1900–1930)	Export-import expansion	Appearance of middle strata, beginnings of proletariat	Co-optative democracy
Phase 3 (1930–1960s)	Import-substituting industrialization	Formation of entrepreneurial elite, strengthening of working class	Populism or co-optative democracy
Phase 4 (1960s to early 1980s)	Stagnation in import-substituting growth; some export-oriented growth in the 1970s	Sharpening of conflict, often class conflict	Bureaucratic-authoritarian regime
Phase 5 (early 1980s to the present)	Scarcity of foreign exchange (worsened by foreign debt) leading to stagnation or recession	Increased mobilization of middle and lower class groups	Incomplete electoral democracy (with military veto)

Women and Society

If we are to judge by the conventional accounts, women have played only minor roles in the economic and political transformation of Latin America. A look at prominent public positions appears to confirm this impression. Why should this have been so? To answer, we need first to look into Latin American culture. A central cultural norm has consisted of the notion of *machismo,* a celebration of the sexual and social expressions of masculine power and virility. For centuries this idea has provided both a prescription and a justification for varied forms of aggression and assertiveness which have been linked, in turn, to the protection of honor. *Machismo* appears to have its origins in medieval conceptions of knighthood, including chivalry and *caballerosidad,* and it has undergone steady adaptation in the face of social change. In any event, it still lives.

The other side of this male-oriented stereotype has been, for women, the cult of *marianismo.* Named after the Virgin Mary, this myth exalts the virtues associated with womanhood—semidivinity, moral superiority, and spiritual strength. For it is women, according to the Latin

American conception, who are the custodians of virtue and propriety. They are pictured as having infinite capacities for humility and sacrifice, and, as mother figures, they demonstrate unfailing tolerance for the impulsive (often childlike) antics of the *macho* menfolk. The typical female image thus becomes one of saintliness and sadness, often identified with the rituals of mourning: a wistful figure, clad in black and draped with a mantilla, kneeling before the altar and praying for the redemption of the sinful males within her sheltered world.

Reality, of course, has not always complied with the mythologies of *machismo* and *marianismo*. But both cults have been integral parts of Latin American society, and they have consistently been used and exploited by members of both sexes.

The social role of females has typically been confined to the private sphere, particularly the family, and here women have often reigned supreme. Among the lower classes especially, women have from the colonial times onward often been the heads of household—because husbands have either moved elsewhere or died. And among the upper-class elite, extended families have frequently been dominated by forceful matrons, grandmotherly figures who wielded unchallenged authority over such family matters as marriage, place of residence, and inheritance.

Over time the boundaries of acceptable social behavior for women have broadened a great deal. In the nineteenth century women of culture often hosted literary discussions, or *tertulias,* where guests would engage in discourse about novels and *belles-lettres.* Some, like Clorinda Matto de Turner and Mercedes Cabello de Carbonero in Peru, became distinguished writers (a tradition first set by a seventeenth-century Mexican nun, Sor Juana Inés de la Cruz). But restrictions persisted. Mariquita Sánchez, hostess of a well-known salon in Buenos Aires, described the female plight in ironic verse:

> The only things we understand
> Are hearing a mass, reciting a prayer,
> Arranging our clothes,
> And patching and repair.

During the twentieth century the process of change accelerated. Within middle-class strata especially, young women ceased to be chaperoned on social occasions (partly because the family had less at stake in the event of an inconvenient marriage). Women have entered the job market and made their mark as teachers, professors, dentists, doctors, even lawyers. In large metropolitan cities, their lifestyle is often barely distinguishable from that of women living in Paris or New York.

Nonetheless, Latin American women have been slow in entering (or have been prohibited from entering) the public arena. As Table 2-2 reveals, women obtained the right to vote rather late in many countries, mostly in the 1930s or 1940s (and as late as 1961 in Paraguay). Survey

TABLE 2-2   Female Suffrage in the Americas

	Year in which National Woman Suffrage Was Recognized
United States	1920
Ecuador	1929
Brazil	1932
Uruguay	1932
Cuba	1934
El Salvador	1939
Dominican Republic	1942
Guatemala	1945
Panama	1945
Argentina	1947
Venezuela	1947
Chile	1949
Costa Rica	1949
Haiti	1950
Bolivia	1952
Mexico	1953
Honduras	1955
Nicaragua	1955
Peru	1955
Colombia	1957
Paraguay	1961

Source: Elsa M. Chaney, *Supermadre: Women in Politics in Latin America* (Austin: University of Texas Press, 1979), p. 169.

research indicates that many women interpret the franchise as a matter of civic duty, rather than partisan involvement. On many occasions, it appears, they have voted in deference to their husbands' preferences.

But not always. In 1958, for instance, women in Chile tipped the balance in favor of the conservative presidential candidate (men having given a plurality to the radical opponent). And in 1970 in the same country, lower-class women provided an important base of electoral support for the victorious left. More research on the subject is needed (it was easy in Chile, where by law women and men vote in separate booths), but there is every indication that increasingly women are asserting independent stands in key elections.

They have also shown their influence in other ways. In Argentina, women formed an impressive bloc in the Peronist movement of the 1940s and 1950s. They are active in the rituals of Mexican politics. They have participated in key demonstrations: one was a pot-clanging protest against Salvador Allende's government in Chile; another, starting in the late 1970s, was the weekly vigil of the "mothers of the Plaza de Mayo" seeking information on relatives and loved ones who had "disappeared" in Argentina. Women have taken part in revolutionary movements—in Mexico, Cuba, and Nicaragua—and they assumed posi-

Ill, gaunt, and nonetheless compelling, Eva Perón waves to the crowd during a motorcade for the inauguration of her husband into his second presidential term in June 1952; she died the following month. (United Press International.)

tions of prominence and leadership in many of the grass-roots organizations that emerged in the 1980s and 1990s.

Yet after decades of progress women have acquired relatively few important political posts, about 8–10 percent of legislative and ministerial positions by the mid-1990s. The first female president (Isabel Martínez de Perón, 1974–76) reached office through the death of her husband. And while in office Latin American women have often projected a distinctly female approach to their jobs. Listen, for instance, to Evita Perón, perhaps the most powerful woman in the history of the Western Hemisphere:

> In this great house of the Motherland, I am just like any other woman in any other of the innumerable houses of my people. Just like all of them I rise early thinking about my husband and about my children. . . . It's that I so truly feel myself the mother of my people.

Thus did Evita, willful and politically ambitious, harken to the themes of *marianismo*.

In the context of the constraints (and advantages) afforded by their culture, Latin American women have not developed a major feminist movement, although the beginnings have been made in Brazil and else-

where. For the most part, women have operated within prevailing socioeconomic and political categories. As Elsa M. Chaney predicted in 1979: "Latin American women probably will not replicate United States or Western European patterns of women's liberation. They have their own reality. . . . Whatever they do, Latin American women themselves will decide their own course of action in the context of their own culture and aspirations."

A Framework for Comparison

One of the aims of this book is to furnish a basis for the comparative analysis of contemporary Latin America. This involves three steps: first, identifying the patterns and processes shared in common by Latin American societies; second, identifying the differences between their individual historical experiences; third, and most difficult, ascertaining the reasons for those differences.

Thus far we have presented a general scheme for depicting socioeconomic and political transition in Latin America. To comprehend the similarities and differences between various countries we need to ask a common set of fundamental questions. We have, accordingly, approached the case studies that follow with several inquiries in mind:

1. How has the class structure evolved? What are the most important social classes? Are any classes missing? In some instances economic changes have conspired to create some groupings and to prevent the formation of others. Not every Latin society has had a peasantry or an industrial elite, for example, and the absence of a social grouping can have as important an effect on the social order as can the presence of others.
2. Which social classes have the most power? Who controls the economy and who dominates the political scene? How much effective competition takes place between existing groups?
3. Which groupings form alliances? Are there any social coalitions? On the basis of what interests? Are industrial workers more likely to join an alliance with entrepreneurial leaders, for example, than with peasants who share their lower-class status?
4. How autonomous is the state? Does the government represent the interests of only one social class (or coalition), or does it stand apart from such allegiances? If it is run by the army, for example, do military leaders strive to remain above and beyond the conflicts of civil society?

International factors have played key roles in the shaping of Latin American history, particularly in regard to economic issues. This dimension gives rise to additional questions:

5. At any given moment, what kind of activity is taking place at the center of the international economy? What are the shape and form of industrialization in Europe and the United States, and what kind of limitations and incentives do they pose for the countries of Latin America?

6. How does the relationship between the economies at the center and the periphery of the system affect the composition and arrangement of social classes within Latin American societies? Would a country exporting beef (Argentina), for example, be likely to have a different class structure from a country exporting copper (Chile)? What implications might this have?

7. How have Latin American countries managed to take advantage of their place in the global economy? Recent experience has suggested that the possession of oil, for example, can provide such countries as Mexico, Venezuela, and even Ecuador with economic opportunity and international influence (and with long-run difficulties too). Have there been similar cases in the past?

8. What are the predominant political factors in the international scene? The presence (or absence) of an East–West Cold War, to take one obvious example, could help determine the plausible range of choices available to Latin American policymakers. In specific moments, too, this concern can translate into another factor: geographical proximity to the United States. Because of presumed "national security" considerations of a geopolitical nature, the United States may well grant more leeway to a country like Argentina than, let us say, to nations in the Caribbean basin.

We will offer a comparative analysis of key countries in the Epilogue. Our immediate task is to turn to individual case studies.

THREE

ARGENTINA
Prosperity, Deadlock, and Change

Present-day Argentina was originally a backwater in the Spanish American colonial empire. Unlike Mexico or Peru, the region of the Río de la Plata lacked precious metals (although "Argentina" means silver in Spanish). Nor did the region have a settled native population. The Indians were few in number and nomadic, so the Spaniards had no ready large-scale source of labor. The area's greatest resource was its fertile land—some of the richest in the world, with loam over a foot deep in places. A further asset was the location of Buenos Aires, which was situated to become a great port if sufficient dredging were done. Yet these conditions produced no dynamic economy in the colonial era. Neither the technology nor the market for exploiting the fertile pampas was at hand. The port served largely to channel profitable contraband trade into Spanish South America.

Although Argentina's colonial economy was modest, its geography is important for understanding the region's later development. The most prosperous center was in the northwest of what is now Argentina—an area tied to Peru. Until 1776 Argentina was included within the viceroyalty of Peru, and its economic development was closely linked to the northward shipment of cotton, rice, wheat, and leather goods. The coastal region around Buenos Aires was less active. Its greatest industry was smuggling, since Lima's high customs rates made it attractive to avoid the authorities by using the route via the Río de la Plata. Only in 1776 did Buenos Aires assume importance, when it was made the seat of a new viceroyalty. Power then began to shift from the northwest to the southern coast, as Buenos Aires became the entry port for European imports, some of which directly competed with production in the northwest.

The Wars of Independence shook the viceroyalty of La Plata, but did not leave the damage to property that hit Mexico (and Uruguay). Anti-Spanish sentiment in La Plata united the local elite and produced what became an enduring myth of military prowess, as General José de San Martín defeated the troops loyal to the Spanish crown. With

independence achieved in the 1820s, the landowning aristocracy contemplated their realm with satisfaction. Small-scale industries and trading communities continued to exist in the central and northwestern interior. The coastal areas had no such artisan base, but the adjacent ranchers produced hides and salted meat for export, and the city of Buenos Aires was growing into an active seaside port. Most important, Buenos Aires and the interior to the north and west grew steadily farther apart.

The Struggle over Nationhood

The decades after independence saw a battle among Argentines about the direction of their new country's economic development. The competing groups came from different regions. One faction was made up of the "unitarians," mainly from the province (and city) of Buenos Aires. They wanted to nationalize the port city of Buenos Aires: strip it of its autonomy, then make it into a base from which to reduce provincial barriers to trade and thereby open the entire country to international commerce.

The second group was the "federalists," who were from the interior. They agreed on the need to nationalize the city of Buenos Aires because they wanted the city's customs receipts distributed to all the provinces. At the same time, they wanted to maintain provincial autonomy, especially the ability to levy interprovincial tariffs and thus protect local industries.

The third group was also called the "federalists," but they were of a very different kind: they were from the province of Buenos Aires and opposed nationalization of the port city of Buenos Aires, since that would mean loss of their province's existing monopoly over the city's customs revenues. They also wanted free trade. In effect, the third group was advocating the status quo.

The conflict among these three groups lasted through the 1830s and 1840s. In the end the issue was decided by one of Latin America's famous dictators: Juan Manuel de Rosas, a politically ambitious cattle rancher from the province of Buenos Aires. A physically imposing figure who could intimidate the tough gauchos, Rosas won the governorship of his province in 1829. His ambitions did not stop there. He wanted to rule all of Argentina, and pursued his goal with policies favoring the *estancieros* (ranch owners), thus furthering the consolidation of a landed aristocracy. Rosas was an ardent Buenos Aires federalist. More important, he had the military leadership and the following to achieve his goal of subduing rival *caudillos*, "so that not one of this race of monsters may remain among us," as he proclaimed. As he consolidated his personal authority, he extended the power of the province of Buenos Aires over the country, in effect building up a nation on the principle of federalism. At the same time, Rosas built a powerful gov-

ernment machine, complete with a kind of enforcement squad *(Mazorca)* that terrorized all who dared oppose the dictator, even if only because they failed to wear the official color of red.

Along with his military and political skills, Rosas was such an ardent nationalist that many foreigners saw him as a xenophobe. He sought to apply in his foreign policy the tactics that had worked so handsomely in domestic politics. "Nourished in the monopolistic maxims of Spanish colonial law," noted a visiting French diplomat in 1850, "he neither understands nor permits trade except when it is hemmed in by protective tariffs and rigorous regulations." Unfortunately for Rosas, he had succeeded in arousing a powerful opposition alliance that included Brazil and Uruguay, as well as General Justo José de Urquiza, who commanded the force that defeated Rosas in 1852. The Argentine strongman then immediately left for exile in England. Despite his ignominious fall, Rosas had succeeded in creating a united Argentina out of the disparate provinces. From the moment of his defeat Rosas became a legendary figure. Argentine nationalists adopted him as a prototypical Argentine patriot, who pursued national development against the alien forces seeking to subvert Argentina's rise to full nationhood. Rosas resembles Chile's Diego Portales and Mexico's Agustín de Iturbide, who also became strong-arm autocratic rulers in the decade after independence. Rosas ruled far longer than the other two.

During the Rosas era many Argentine intellectuals, such as Domingo Sarmiento and Esteban Echeverría, fled the repressive regime. They detested the gaucho leader's heavy-handed political success. Sarmiento described Rosas as the man "who applied the knife of the gaucho to the culture of Buenos Aires and destroyed the work of centuries—of civilization, law and liberty." These intellectuals dreamed of the day when they might capture the control of Argentina and steer it onto the course of liberal representative government. With the fall of Rosas in 1852 they got their chance.

Power was taken by Justo José de Urquiza, a federalist from the interior. He began by calling a constitutional convention, which promulgated a constitution in 1853, closely following the U.S. example. It was to be a federal system, with the president chosen by an electoral college whose members would be elected by popular vote. The federal congress had two houses, with the Chamber of Deputies elected by direct vote and the Senate elected by provincial legislatures. The provinces were to retain all powers not specifically granted to the national government, although a loophole provision authorized the central government to intervene in the provinces whenever it deemed necessary.

The controversy over the status of the city of Buenos Aires was far from settled, however. Protesting the nationalization of the city in the new constitution, the province of Buenos Aires refused to join the new confederation. Defeated in a brief civil war in 1859, the province was

forced to capitulate. Only two years later the province rose in revolt, led by Bartolomé Mitre, and captured control of the confederation.

In 1862 Mitre was inaugurated as president and he launched a new drive to unify Argentina. For the next two decades the liberals continued in power. Mitre was followed in the presidency by Domingo Sarmiento, author of *Facundo* (1845), the most famous literary attack on the *caudillo*-style gauchos. His real-life target was Rosas, whose rule Sarmiento had spent exiled in Chile. Sarmiento believed ardently in North-American-style public education and urged Argentines to follow the U.S. model. A diversion on the way was the drawn-out Paraguayan War (1865–70), in which tiny Paraguay held off Argentina, Brazil, and Uruguay for five years. At issue was the ill-advised attempt by the Paraguayan dictator, Francisco Solano López, to exercise a monopoly of control over access to the all-important Paraná River basin.

The third liberal president was Nicolás Avellaneda. In his term (1874–80) Argentina undertook its last major territorial conquest—the "Indian wars." The provinces to the south and west of Buenos Aires had long been plagued by Indian raids. Now an army force under the command of General Julio Roca subdued or exterminated the Indian bands. The year 1880 also brought a solution to the long troublesome question of the status of the city of Buenos Aires. It was detached from the province, rather like the District of Columbia in the United States, but its citizens were given full voting rights in national elections.

About 1880 Argentina entered a remarkable era of sustained economic growth. Since ousting Rosas, the liberals had enjoyed power long enough to lay the basis for their country's rapid integration into the world economy. Their leader now was General Roca, hero of the "Conquest of the Desert." The symbolism could hardly have been better: the Indian-fighter presiding over the Europeanization of a South American republic.

The political elite had few doubts about its mission. Like their counterparts in Brazil and Mexico, the Argentine politicians and intellectuals saw themselves as applying the true principles of European science and philosophy. They believed in both economic and political liberalism. They quoted the pseudoscience of Herbert Spencer, arguing that if an aristocracy ruled Argentina, it was the result of natural selection. With the Indian and the gaucho safely subdued, the elite confidently looked forward to enriching itself, and, by liberal logic, thereby enriching their country.

Overview: Economic Growth and Social Change

Argentina's economic success in the 1880–1914 era was based on its ability to supply agricultural goods needed by the North Atlantic industrial world. With the Industrial Revolution, West Europe, especially En-

gland, was becoming a net importer of foodstuffs. Argentina had a comparative advantage in producing two key goods: meat and grain. Important technological advances had made it practical to ship foodstuffs the many thousands of miles from Buenos Aires to London and Antwerp. One was the steam vessel, with its faster and far more certain pace than the sailing ship. Another was the process for chilling meat (the plants in Argentina were called *frigoríficos*), which made possible a tastier and more substantial product than meat preserved by drying (beef jerky).

Thanks to one natural resource Argentina was ideally suited to furnish the needed foodstuffs: its pampas were among the most fertile lands in the world. But it lacked two other essential factors, capital and labor. England, Argentina's principal customer, soon sent the capital. It came in the form of investment in the railroads, the docks, the packing houses, and the public utilities. It also came in the form of English firms to handle shipping, insurance, and banking. Virtually the entire infrastructure of the export sector was financed by the British. This inflow of capital was exactly what the Argentine political elite saw as essential for their country's development.

The other missing economic factor was labor. Here the solution also came from Europe, although not from Britain. The badly needed workers streamed into Argentina from southern Europe, especially Italy. Between 1857 and 1930 Argentina received a net immigration (immigrants minus emigrants) of 3.5 million. This meant that during that time about 60 percent of the total population increase could be attributed to immigration. Of these immigrants about 46 percent were Italian and 32 percent Spanish. The demographic effect of immigration on Argentina was greater than in any other major country of the Western Hemisphere. By 1914 approximately 30 percent of the population was foreign-born. (At the same time in the United States, another haven for European emigrants, only about 13 percent of the population was foreign-born.) The result was to give Argentina a distinctly European quality, with a resulting tension among Argentines as to their real national identity.

This immigrant labor force was a textbook example of the mobility of labor. The workers were hired and fired on primarily economic grounds, with a remarkable movement of laborers back and forth between Italy and the Argentine pampas (earning them the nickname of *golondrinas* or "swallows"). There was a fluid movement between city and country within Argentina, too, with Buenos Aires always attracting a large share of the foreigners.

This era also saw a modest industrialization, which, however, posed no threat to the basically agro-export orientation of the economy. Much of the industry processed products from the countryside, such as wool and beef, thereby leading the processors to identify their interests with agriculture.

The rapid economic growth of the 1880–1914 era had profound social implications. At the outset there was a landed elite on top, gauchos and wage labor on the bottom. With the epic tide of immigration, the national population swelled from 1.7 million in 1869 to 7.9 million in 1914. As the economy boomed, new economic niches appeared. The immigrants came first to the farms, but then often moved to the cities. The Italians and Spaniards arrived to become colonists, tenant farmers, and rural laborers. Other jobs opened in the urban sector: in transport (especially the railroads), processing, and the service industries (banking, government). Exploitation of Argentina's agricultural wealth produced an intricate network of economic interests and contributed to the creation of a complex rural-urban economy.

Argentina's economy thus entered its "golden age," an era of increasing prosperity based on the exportation of meat and grain and on the importation of manufactured goods from abroad. From the 1860s to 1914 Argentina's GDP grew at an annual average rate of a least 5 percent (exact data for pre-1900 are sketchy). This is one of the highest sustained growth rates ever recorded for any country. Trade grew steadily from 1880 onward, while statesmen optimistically propounded the virtues of progress and commercial growth. It seemed to many that Argentina, so blessed with natural resources, was on its way to unending expansion.

But the country paid a price for this success. Argentina's integration into the world economy meant that sharp fluctuations abroad would have severe repercussions at home. A decline in European demand for foodstuffs brought a decline in exports, which could provoke a downturn throughout the entire Argentine economy—a fate shared with Australia, another exporter of temperate agricultural products. And as shown in Figure 3-1, the peso value of Argentine exports varied considerably during the period from 1915 to 1939: up during World War I, down in the early 1920s, then up and down and up and down again when the Great Depression struck in the 1930s. This is one way in which Argentina, like other exporting countries of Latin America, came to be economically dependent on the industrialized center of the world-system. Through the reliance on trade, the condition of the economy was largely determined by trends and decisions outside of Argentina.

But the international market for meat and grain was relatively stable, at least in comparison to the demand for such goods as sugar and coffee (as we shall see below). The meat trade, in particular, held fairly firm throughout the 1930s. After recovering from drought and other setbacks, demand for wheat and other cereals also bounced back. So Argentina was hit hard by the depression, but not as quickly—or as completely—as some other countries of Latin America.

Another form of economic dependency appeared in the financial realm, since the Argentine banking system was periodically tied to the

FIGURE 3-1   Argentine Exports, 1915–39 (in thousands of pesos)

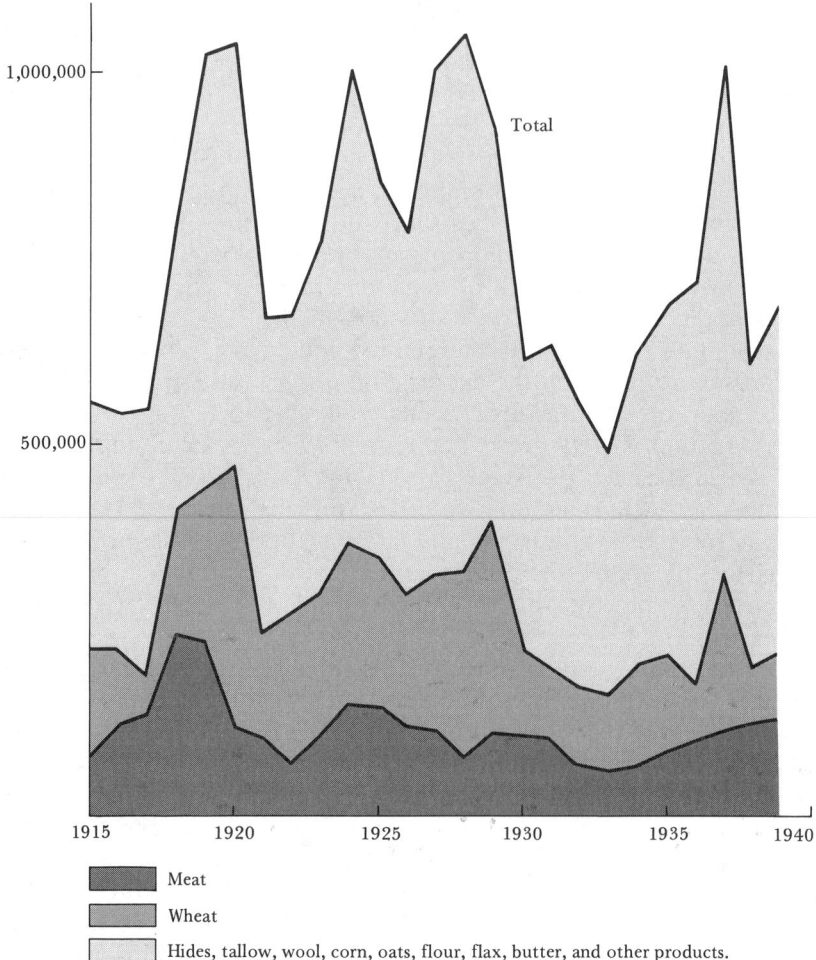

Meat

Wheat

Hides, tallow, wool, corn, oats, flour, flax, butter, and other products.

Source: Vicente Vázquez-Presedo, *Estadísticas históricas argentinas (comparadas)*, vol. II (Buenos Aires: Ediciones Macchi, 1976), pp. 190–93.

gold standard. Short-term fluctuations in trade caused sharp changes in Argentina's gold reserves, thus contracting or expanding the domestic money supply and making Argentina's economy a hostage to international currency movements.

Still another link to the world economy posed further problems: the large role played by foreign capital and foreign businessmen. Argentina's booming economy, along with government guarantees, attracted capital from abroad. From 1900 to 1929, 35 percent of the country's total fixed investment came from foreigners. Britain was the prime investor, followed by France and Germany. This high degree of foreign

economic involvement later became a prime target for the economic nationalists. Such dependence on foreign resources also contributed to ongoing Argentine self-doubt: a recurrent unwillingness to believe that Argentines could ever hope to achieve a more self-sufficient economy, especially in industry.

Argentina's export-import growth also created inequities at home, especially among geographical regions. While prosperity blessed the pampas and Buenos Aires, parts of the interior stagnated. Central and northwestern provinces such as Jujuy, La Rioja, Santiago del Estero, and Salta suffered economic decline and social decay. Only Mendoza, Tucumán, and Córdoba escaped this fate, thanks to their wine and sugar production. Throughout the nineteenth century the interior had fought to prevent its demise at the hands of Buenos Aires. It lost, and the price of its defeat was poverty.

Of course inequities also existed even within the prosperous regions. In the rural sector the wealthy *estancieros* built elegant chalets, while foreign-born tenant farmers and displaced native workers scratched out a meager existence. In the cities, especially Buenos Aires, elegantly attired aristocrats met at their European-style clubs while workers struggled to protect their families from the inflation that always seemed to outpace their wages. The Argentine boom, like so many others in capitalist countries at this time, did facilitate considerable upward mobility. But it also fostered huge income discrepancies, which were ultimately bound to create social and political tensions.

There were also tensions in the cultural realm. As Argentina experienced its impressive burst of economic growth, it became increasingly obvious that the Argentines did not yet have a clear-cut sense of nationality. Their nationhood was ill-defined because the flood of immigrants, mostly concentrated in Buenos Aires, had accentuated the longstanding contrast between the densely settled, Europeanized, cosmopolitan capital city and the rough-hewn, cattle-oriented society of the vast and lightly populated interior. This contrast had been dramatized by the bitter battle between the liberals and Rosas, whom they saw as the incarnation of the primitive gaucho, who could never build, only destroy, a civilized nation.

In the early twentieth century this liberal dogma was challenged by a new generation of nationalist writers such as Ricardo Rojas, who sought, in his words, "to awaken Argentina out of its coma." Rojas pointed to the Indian and the soil as the true sources of Argentine nationality. He and other nationalists looked to *Martín Fierro*, the classic poem about the gaucho (first appearing in 1872), as a highly authentic document in the creation of a national consciousness.

One of the more famous works in the tradition of Rojas's nationalism was Ezequiel Martínez Estrada's *X-Ray of the Pampa* (1933), a lengthy indictment of the Argentine elite for having tried to cloak their country's reality in the trappings of European civilization. But Martínez Es-

trada added another dualism: "Race track and soccer field are two political parties: the former stands for adventure, instability, ambition,
centralism, and monarchy; the latter represents work, passionate struggle, committee disputes, chaos, and democracy. The Jockey Club and
the sports associations support opposing trends and recruit antagonistic forces."

Meanwhile, the capital city was developing its own culture. The major foreign influence was Italian, and in the dock area there emerged
a unique dialect, *lunfardo,* a mixture of Spanish and Italian. It was a
strictly working-class phenomenon, as was also the tango, the famous
sensuous dance and accompanying music, which originated in the sub-
urbs (some said the brothels) and only became respectable in Argentina
after it took Paris by storm on the eve of World War I. By the 1920s
the tango had been adopted by the most fashionable Argentine circles
and the music was brought to a high art form by Carlos Gardel, Argentina's immortal tango composer and singer. His death in an air accident
in 1935 caused the greatest outpouring of grief Buenos Aires had ever
seen. As the 1930s continued, North American popular culture penetrated urban Argentina at a quick pace, much to the fury of Argentine
nationalist intellectuals.

One of the most crucial social effects of Argentina's expansion
turned out to be a nonevent: *the country never developed a peasantry,* at
least not in the pastoral areas of the pampa and the crucial coastal
provinces. The Conquest of the Desert in the 1870s virtually eliminated
the Indian population, and the land was promptly distributed in large
tracts appropriate for raising cattle and sowing grain. In contrast to the
policy in the Great Plains of the United States, Argentina did not give
its land to family farmers or individual homesteaders. Cattle ranching
did not require a large-scale work force, since barbed wire was sufficient to contain the herds; and though wheat was often grown by foreign colonists who rented land, they did not constitute an influential
social group. As a result, a classic peasantry—such as those of Mexico,
Chile, or northeastern Brazil—did not exist in Argentina.

This fact would have far-reaching implications. It meant, for instance, that land reform would never become the vital and symbolic
issue that it was in such countries as Mexico. It was not that land was
so evenly distributed in Argentina; it was that there were no long-time
rural dwellers to lay historic claim to the soil.

Further, the absence of a peasantry meant that there was no peasantry to form a power base, and perhaps even more crucial, there was
no peasantry available for coalitions with other social groups. Landowners, when challenged, could not resort to the time-tested (if paradoxical) alliance with the peasantry that frequently occurred in other
countries, and urban laborers could not enlist peasants in broad-based
warfare between the "popular classes" and the social system at large.

In the big cities, however, wage laborers were numerous and prone

This difference in living conditions reflects the social inequality that accompanied Argentina's economic expansion after the 1880s. *Above,* luxurious residences of aristocratic families in Buenos Aires; *below,* temporary shacks for workers in the city's port area around 1910. (Courtesy of the Archivo General de la Nación, Buenos Aires.)

to organization. According to one estimate, manual workers accounted for nearly 60 percent of the population in the city of Buenos Aires in the early twentieth century. And about three-fifths of the working class, in turn, consisted of immigrants who retained citizenship in their home countries—mainly Italy and Spain.

The first efforts at organizing Argentine labor were influenced by European precedents. In the 1870s and 1880s European anarchist and socialist exiles began vigorous organizing, and in 1895 the Socialist Party was founded. The Socialists were molded on the European model: a parliamentary party, clearly committed to an electoral and evolutionary strategy. As of 1900 one might have expected the Socialist Party to become a major political voice for the Argentine working class. Yet it failed to attract the immigrant workers. Socialist pleas for reform through the political system fell on deaf ears.

The urban working class proved receptive to another message; it came from the anarchists. Their *Federación Obrera Regional Argentina* (FORA) caught the workers' imagination with its call for direct action. The FORA-sponsored local and general strikes were effective enough to worry the government, which assumed that any labor problems must be the work of foreign agitators. Congress therefore passed the *Ley de Residencia* ("Residence Law") in 1902, which empowered the government to deport all foreigners whose behavior "compromised national security or disturbed public order"—by participating in strikes, for example. It was an instinctive reaction of the lawmakers, who refused to believe that class conflict could arise among truly patriotic Argentines.

A few of the reigning politicians decided to try a different approach. In 1904 Joaquín V. González proposed a general code for labor relations, which would have closely controlled labor action. Conservatives quickly attacked the bill because they opposed in principle labor's right to organize, while labor, led by the anarchists, opposed it because they feared that any government system would be manipulated against them. The combination killed the bill, thereby precluding the creation of any official labor relations system.

Political leaders continued their efforts to gain control of labor. In 1907 the Congress created a Department of Labor and invited the leading labor confederations, including the FORA, to participate in a tribunal created to adjudicate labor conflicts. But worker organizations refused to support the "corrupt bourgeois government" in this effort. Once again the Argentine labor movement escaped incorporation into a government-dominated system of labor relations.

Meanwhile the anarchists continued to organize. Their efforts reached a climax in 1910, the centennial of the declaration of Argentine independence, when a great public celebration was planned to glorify Argentina's progress. But the anarchist leaders had a different purpose in mind. As militant opponents of the liberal elite, they wanted to raise their protest against the farce of the European-oriented model

of progress. The oligarchy and the middle class were indignant over the threatened disruption of a patriotic ritual. The protestors filled the streets and plazas, but were crushed and scattered by masses of police. The anarchist-led challenge was more than met. Government repression completely carried the day. The reaction against the protestors spilled over into the Congress, which approved a new law (*Ley de Defensa Social*, or "Social Defense Law"), making even easier the arrest and prosecution of labor organizers. A hunt began for urban protest leaders.

This was the death knell for Argentine anarchism, but not of urban protest. Strike activity in Buenos Aires reached another peak in 1918–19, and it would oscillate from time to time after that (Figure 3-2). Even more significant is the persistence and continuity of worker agitation into and through the 1950s. Fairly early in this century, organized labor emerged as a key actor in Argentine society.

The Political System: Consensus and Reform

The liberal politicians later known as the "Generation of 1880" (so labeled for their emergence that year) drew their political power from several sources. First, they were themselves members of, or very close to, the landowning class that produced Argentina's riches. Second, they managed to monopolize the instruments of state power: they controlled the army and the elections, resorting to vote fraud when necessary. They also controlled the only authentic political party, the *Partido Autonomista Nacional* (PAN). The most important national decisions were made by *acuerdo*, or informal agreement among officials of the Executive. In this respect the Argentine liberals ignored one key aspect of the British/U.S. example—the central role of the legislature, which in Argentina had been rendered inconsequential in this period.

At first glance this political system seems to have admirably served the agro-export interests that profited from the post-1880 expansion. But the aristocrats in control did not go unchallenged. The spreading prosperity, which created new wealth holders in both country and city, helped feed political discontent among three groups: (1) newly prosperous landowners of the upper Littoral; (2) old aristocratic families, often from the far interior, who had failed to profit from the agro-export boom; and (3) members of the middle class doing well economically but excluded from political power.

These three groups joined forces to create the Radical Party, destined to play a major role in twentieth-century politics. In 1890, just as the country entered a short but severe economic crisis, they attempted an armed revolt. An *acuerdo* ended the rebellion, but some intransigent leaders founded the Radical Civic Union (*Unión Cívica Radical*, or UCR) two years later. Able to make no electoral progress against the fraud routinely practiced by the ruling politicians, they resorted twice more

FIGURE 3-2 Strikes and Strikers in Buenos Aires, 1907–72

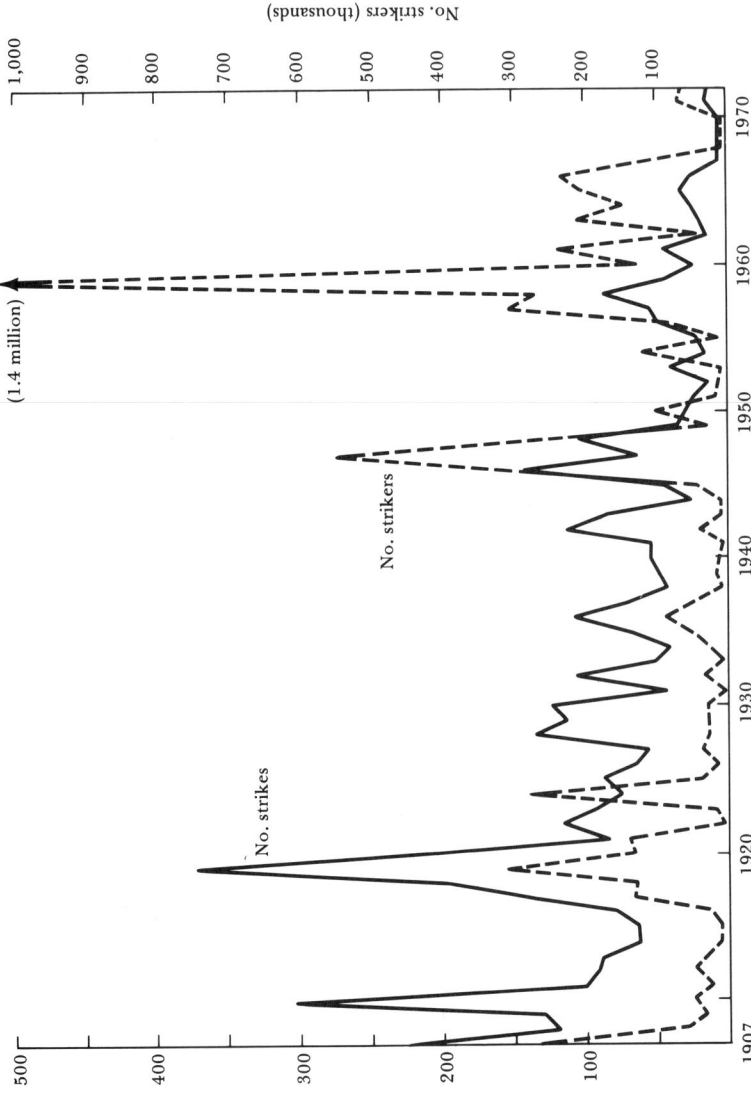

Sources: Departamento Nacional del Trabajo, *Estadística de las huelgas* (Buenos Aires, 1940), p. 20; Dirección Nacional de Investigaciones, Estadística y Censos, *Síntesis estadística mensual de la República Argentina*, 1, no. 1 (January 1947), 7; Vicente Vazquez-Presedo, *Estadísticas históricas argentinas (comparadas)*, vol II. (Buenos Aires: Ediciones Macchi, 1976), p. 47; Guido di Tella and Manuel Zymelman, *Las etapas del desarrollo económico argentino* (Buenos Aires: Paidós, 1967), pp. 537–38; International Labor Office, *Year Book of Labour Statistics*, 11 (1949–50), 379; 16 (1956), 456; 25 (1965), 692; 33 (1973), 752.

to armed revolt. Both attempts failed. Nonetheless the Radicals, led first by Leandro Alem and then by Hipólito Yrigoyen, maintained their stubborn pursuit of political power. As for economic goals, the Radicals were by and large successful participants in the agro-export economy and were strongly committed to it. They simply wanted a share in the political direction of their society.

Not all the oligarchs endorsed the government position of freezing the Radicals out of power. A more enlightened wing won out in 1911 when President Roque Sáenz Peña proposed an electoral reform. Passed in 1912, the new law called for universal male suffrage, the secret ballot, and compulsory voting. This would significantly enlarge the electorate, especially in the cities. In effect, it was a shrewd attempt at co-optation by the oligarchy. Sáenz Peña and his colleagues saw labor organizers and the working class to be the real threat—not the middle classes, who would cooperate once incorporated within the system. The Argentine political bosses were acting here much like their British counterparts in the nineteenth century, who three times adopted electoral reform bills to widen the franchise and thereby incorporate new social sectors into the established political game.

The Sáenz Peña electoral reform thus extended the vote to the frustrated middle-sector citizens. The well-organized Radicals immediately capitalized on the new rules and got their long-time leader, Hipólito Yrigoyen, elected president in 1916. Was this to be a new era?

An early test came in the government's behavior toward the workers. The Radicals began with an attitude different from that of their predecessors. They showed a genuine concern for the lot of the working class, partly because they hoped to win eventual votes in their struggle with the Conservatives. As labor-management conflicts arose, the Yrigoyen government followed an apparent pro-labor stance in its interventions. Labor organizers saw this as an improvement, but one that would depend on government action case-by-case.

The crisis came in 1918–19, when the entire Western world was shaken by strike waves. The actions, which often included general strikes, were generally aimed at both the employers and the state. It was a confluence of specific grievances and generalized hostility. In Argentina, workers were aroused over the reduced buying power of their wages, caused by the inflation of World War I. Food prices rose sharply, stimulated by European demand, but wage increases lagged. The resulting squeeze hit Buenos Aires workers hard. Union leaders called a series of strikes in late 1918, and in early 1919 syndicalist organizers decided the time was ripe for a general strike—the syndicalist instrument for bringing down the bourgeois state.

Their idea had tragic consequences. The Yrigoyen government decided it had to act firmly, and the results were a rerun of 1910. Anti-labor hysteria was promoted by a newly formed ultra-rightist civilian paramilitary movement, the Argentine Patriotic League *(Liga Patriótica*

Argentina), which effectively exploited the middle- and upper-class fear of the popular challenge. League members took to the streets to attack workers; it was class warfare with a vengeance. Hundreds of demonstrators were shot. The labor leaders were again repressed, this time by the Radicals, with the heaviest blows falling on the syndicalists and the last remnants of the anarchist leadership.

Organized labor did not disappear. Although the anarchists were beaten and the syndicalists were weakened, two other ideological currents began to make headway among Argentine workers: socialism and communism. The former stressed political action, betting on the Socialist Party as the hope for change in Argentine capitalism. The Communists, on the other hand, placed emphasis upon the labor union movement instead of the ballot box, and they soon made modest progress in gaining key union positions.

The 1920s did not bring much success to labor organizers. Real wage rates increased steadily during the decade, but not because of union pressure. On the contrary, rising wages may have undermined unionization. The relative absence of conflict (seen in Figure 3-2) also meant that the government made no effort to create a new framework for labor relations. By 1930 organized labor had become a relatively subdued actor on the Argentine stage.

At the other end of the political spectrum were the Conservatives. They had hoped the Sáenz Peña electoral reform would make it possible to co-opt the Radicals. Their hopes were soon disappointed. Instead of sustaining the tradition of agreement by *acuerdo*, electoral reform led to basic changes in the political system.

First, the electorate was steadily enlarged. All Argentine males over eighteen years old now had the right to vote, and nearly 1 million people qualified in 1912. There continued to be a sharp increase in the electorate, with the total reaching 3.4 million by 1946. Participation was high, it being mandatory, and fraud was minimized. Between 70 and 80 percent of the eligible voters cast ballots in the presidential elections. The elections were also highly competitive. Winners rarely emerged with more than 60 percent of the vote. This relatively high percentage of (male only) citizen participation distinguished Argentina from other major Latin American countries such as Mexico or Brazil, whose far more restricted electorates reflected lower literacy rates and more closed political systems.

A further consequence of expanded voter participation was the increased importance of the political parties. Almost nonexistent under the Generation of 1880, after 1912 the parties became primary vehicles for the organized pursuit of power. The parties, in turn, spawned a new kind of political elite: middle-class professionals who made careers out of politics. Their favorite forum was the national Congress.

As innovative as the Sáenz Peña electoral reform was for its day in

Latin America, it left the political system still limited in basic ways. Restricting the franchise to male citizens not only excluded all women; it also left out at least half of the adult males because so many were still foreign citizens. Few immigrants underwent naturalization, because most never intended to settle permanently in Argentina and therefore did not try. Those who did try were usually defeated by the labyrinthine procedures and deliberately obstructionist tactics of Argentine officials. Since the unnaturalized immigrants were more numerous among the working class, the reform tended to help the middle class at the expense of the lower class.

The practical result of all these changes was to leave the Conservatives far from power. The Radicals, building upon their popular base and employing machine tactics, displayed continuing electoral supremacy: Marcelo T. de Alvear became president in 1922, Yrigoyen was reelected in 1928, and the Radicals dominated both houses of Congress. Intense partisan struggles replaced fluid party allegiances. The rise of middle-class politicians threatened to end genteel, intraclass maneuvering. The political system came to represent an autonomous threat to the socioeconomic system, even in the absence of major disagreements over policy, both through the hegemony of political professionals and through the accumulation of political power within an increasingly autonomous state. For Conservatives and their allies, Argentina's experiment in limited democracy was becoming distasteful and risky.

Tension in the political sphere was exacerbated by the world economic crash of 1929, though Argentina was not hit as soon or as hard as some other countries. The prices and values of beef exports held up until 1931. The wheat market was suffering badly, but mainly because of a drought; besides, farmers exerted scant political influence, partly because so many were unnaturalized immigrants (about 70 percent in 1914). In 1930 real wages underwent a brief decline and unemployment was starting to spread, but labor agitation was still at a moderate level (see Figure 3-2). The Great Depression no doubt exposed weaknesses within the political system, but it was not enough alone to have caused a coup.

The Military Turns Back the Clock

On September 6, 1930, a coalition of military officers and civilian aristocrats ousted President Yrigoyen on the grounds that his government was illegitimate. They then set up a provisional regime. Who were these soldiers? How had they come to intervene in what seemed to be a smoothly functioning constitutional order?

The answer is partly to be found in the history of the armed forces. The liberals who came to power in 1852 believed that a professional

army was indispensable for Argentina's development. They wanted a well-trained military to crush provincial *caudillos* and to provide the order necessary for economic growth.

The professional training schools established were the *Colegio Militar* (1870) and the *Escuela Naval* (1872), which were to remain the basic training schools for Argentine officers. Like other Latin American military, the Argentines looked to Europe for their models. In 1899 General Roca and his colleagues negotiated the visit of a German mission to train staff officers in modern military technology. That collaboration with Germany was to last forty years.

The increased professionalization of the military led to a shift in the outlook of the Argentine officer corps. By 1910 there was a change in the criteria for promotion: now it became seniority and mastery of the new technology, rather than political favoritism. At the same time, there was a shift in control over promotions—from the presidency to an all-military committee made up of army division commanders and chaired by the highest-ranking general. The army was thus able to develop a new sense of its own efficiency, while acquiring a higher degree of institutional autonomy.

The increased emphasis on merit opened military careers to aspiring sons of the middle class. Not surprisingly, this soon included sons of immigrants, especially from Italy. The military offered an inviting avenue to upward social mobility. Successful recruits who made their way up the hierarchy forged a strong allegiance to the military as an institution—and a jealous regard for its independence, honor, and professional reputation. The reverse side of this loyalty was a deep suspicion of outsiders, especially politicians. Soldiers often regarded civilian officials with a mixture of scorn and apprehension. By 1930 officers concluded that the only way out of the political mess was to revise the rules of the political game.

Although they agreed on that point, they were otherwise divided. One faction, led by General Agustín P. Justo, wanted to return to the oligarchical system of the pre–Sáenz Peña reform days. These officers thought it was merely Yrigoyen and the Radicals who had abused the electoral system. If they were removed from politics, then power would revert to the aristocrats and the specter of class struggle would disappear.

Another faction, led by General José F. Uriburu, suggested a more sweeping solution: the establishment of a semifascist corporate state. They saw the problem to be not Yrigoyen, or the Radicals, but the very attempt to try democracy in Argentina. Uriburu was reflecting the antidemocratic doctrines already rampant in Europe, especially in Italy, Spain, and Portugal. In essence, Uriburu and his sympathizers wanted a hierarchical order based on social function. Voting would be in the hands of the most cultivated members of society, and Congress would no longer be in the hands of the professional politicians—"agents of

Despite its historic importance, the military coup of 1930 was a relatively gen-teel affair—here a white flag of surrender flutters from the presidential palace (see arrow) as onlookers gather in the Plaza de Mayo. (Private collection.)

political committees," as he disdainfully described them. Uriburu envisioned a "functional democracy," where the elected legislators would represent functional (or "corporate") interests, such as ranchers, workers, merchants, and industrialists. This was the corporatist model in vogue in Mediterranean Europe. The theory was that a vertical structure would reintegrate the political system with the economic system, so that the political arena would once again reflect the distribution of power in the economic arena. It was also, quite obviously, a formula for stopping class-oriented politics. The corporatist answer to the ills of the congressional system was to radically redefine the bases of representation.

Although Uriburu directed the provisional government in 1930, the Justo group eventually won out. Elections took place, but as in the days before Sáenz Peña, fraud was freely practiced. After Justo became president in 1932 he created a pro-government coalition of parties called the *Concordancia*, and in an effort to curry civilian support he replaced a number of military men in sensitive posts with well-known politicians. Clearly, Justo was hoping to form a broad, national government that would give him the authority to respond to the socioeconomic effects of the world depression.

This proved impossible. One reason for this failure was the expansion of an urban working class which, through strikes and other tactics, made repeated demands on the government. And on the elite level, political professionals—committed to partisan interests—refused to play by the old-fashioned rules. This became clear when Radical Party leader Roberto Ortiz, elected through *Concordancia* manipulation as Justo's successor in 1937, stopped electoral fraud and thereby allowed the Radicals to win control of Congress.

Ortiz' health forced him to leave office in 1940. His successor, Ramón Castillo, resorted to the technique of the embattled oligarchs when faced with elections: stuffing the ballot box. The cheating of course fooled no one. It only dramatized the illegitimacy of the incumbent civilian government.

Military officers watched the drama with increasing impatience. They saw how their counterparts in Germany and Italy had played key roles in displacing the wavering civilian governments. As the war spread in Europe in the early 1940s, and the Axis—which included Germany, Italy, and Japan after 1940—seemed to be carrying the day, the Argentine military chieftains saw the need for steady, sure leadership in their own land. The obstacle was the cabal of civilian politicians, whom the military had not sufficiently purged from power after 1930, and who had continued pursuing their petty interests and thereby rendering their country vulnerable.

Politics in Argentina was taking a path that did not appear in any other major Latin American country. The causes were several. There was first the strong Argentine antipathy to the Allies, especially the United States and Britain. In contrast to Brazil, whose government had, after a long flirtation with Nazi Germany, decided to cast its lot with the United States in 1942, Argentina wanted to preserve its "neutrality." In practice that meant it would continue to sell essential foodstuffs to besieged Britain while refusing to join the U.S.-led military effort. This reflected a strong sentiment among the Argentine elite, both civilian and military, that their country had most to gain by withholding its political and military allegiance in the world conflict.

Behind this rough consensus lay the continuing military impatience with the civilian politicians. Dissident officers mounted plots to seize power. The triumphant group was called the GOU *(Grupo Obra de Unificación* or *Grupo de Oficiales Unidos)* and justified its seizure of power as a response to popular demand: "We support our institutions and our laws, persuaded that it is not they but men [i.e., professional politicians] who have been delinquent in their application."

In fact, the ambitious officers wanted to revamp the entire political structure. They began, in 1943, by dissolving the Congress, that target of their oft-expressed scorn. The ascendant military, led by the first provisional president, General Arturo Rawson, grandly announced, "Now there are no political parties, but only Argentines." The military

set themselves to rid Argentina of politics, as well as politicians. In 1944 they decreed the end of political parties, and they excluded from the cabinet all professional politicians, aside from a few "collaborationist" Radicals.

While the military were seizing control of the political system, class consciousness was growing among the workers. By the 1940s the urban working class, especially in Buenos Aires, had changed from the days of the Sáenz Peña electoral reform in 1912. It was now about 90 percent literate and it was mobile, with many of its members having recently arrived from the countryside. In contrast to the era of the great export boom (1880–1914), most urban workers were now native Argentines, not European immigrants. Buenos Aires hosted a proletariat not unlike those which had frightened the European bourgeoisie and military into turning to corporatist and fascist solutions. Indeed, political commentators in the United States and West Europe saw Argentina as an important test case in the ability of a New World republic to adjust to the social conflicts inherent in industrialization and development.

As the drama unfolded, the principal actors in Argentine politics turned out to be the military and labor. The military had its own institutional base, but the large and growing urban working class lacked effective political representation. Why? In part because of the electoral fraud. But more important was the way the existing party system worked. All the major parties, including the Radicals and the Socialists, were geared to the electoral system as modified by the Sáenz Peña law in 1912, when over half the adult male population remained excluded from the vote. So none of the major parties, with the partial exception of the Socialists, created an authentic working-class base. As a result, party politics offered no meaningful outlet for urban workers.

Enter Juan Perón. A man of middle-class origin, he had risen to the rank of colonel through a career in the Argentine army. Ambitious and outgoing, nearing fifty, he had taken an active part in the GOU movement that ousted Ramón Castillo from the presidency in 1943. In recognition of his role he became secretary of labor, a fairly minor post, but one that he transformed into a bastion of strength. Using both carrots and sticks, Perón courted the support of industrial workers: old laboring groups as well as new, lifelong urban residents as well as recent migrants from the countryside. Thus he made the union movement a resource of his own, and partly because of this influence he later became minister of war and vice president of the nation. A hero to the dispossessed, he won the presidential election of 1946 with a solid 54 percent majority—over the indiscreet resistance of the U.S. State Department, which denounced him for pro-fascist sympathies, and over the combined opposition of all the national political parties. The Argentine Communist Party was especially vehement in denouncing Perón as a fascist demagogue.

During this period Perón grew to rely on the political instincts of his mistress, later his wife, Eva Duarte. A former radio actress not long arrived from the interior, she was determined to make her mark in the world of Buenos Aires.

Peronism and Perón

Once installed as president, Perón proceeded to put into practice the corporatist principles the GOU officers had espoused. Argentina would now be organized according to functional groups: industrialists, farmers, workers. At the peak of this hierarchy would be the state. The government would act as the final arbiter in case of conflict among groups. The new government quickly issued a host of statutes regulating functional organizations, with Perón rapidly asserting an interventionist role for the state in the economy. A Five-Year Economic Plan was issued and a powerful new foreign trade institute (*Instituto Argentino de Promoción del Intercambio,* or IAPI) was given a state monopoly over the export of key agricultural crops. Argentina was now positioned to undertake the most state-directed economic policy thus far seen in twentieth-century Latin America.

Perón was carrying out the 1930s corporatist vision of General Uriburu, but with a vital difference: Perón made urban workers his most important political ally, flanked by industrialists and the armed forces. Uriburu would have reduced labor to a minor role. Perón created a political alliance completely unprecedented in Argentine history: a union of workers, managers, and the military.

Perón had campaigned for the presidency on a nationalist and populist note. "Argentina was a country of fat bulls and undernourished peons," as Perón described Argentina in 1946. He promised to promote truly Argentine solutions while channeling to the workers the material and psychological rewards they had been unjustly denied. His government lost no time in rewarding the workers. Perón continued the tactics he had been perfecting since he took over the labor secretariat in 1943: the encouragement of strikes which the government then settled in favor of the workers. (Notice the upsurge in the number of strikers in the late 1940s in Figure 3-2, though the number of strikes was not especially large: under Perón the average *size* of strikes went up as well.) Real hourly wage rates jumped 25 percent in 1947 and 24 percent in 1948. Labor's share of the national income increased by 25 percent between 1946 and 1950. Argentine urban workers received a sharp rise in their standard of living. The losers in this populist drama were the owners of capital, especially the landowners, since the government trade monopoly (IAPI) bought most of their products at low, fixed prices.

At first this bold strategy seemed to work well. The GDP grew by 8.6 percent in 1946 and at the startling rate of 12.6 percent in 1947. Even

the lesser rate of 5.1 percent in 1948 was still very respectable by world standards. This growth was fueled in part by Argentina's booming exports, which produced healthy trade surpluses from 1946 through 1948.

Perón also made good on his promise to reduce foreign influence in the economy. In 1946 the central bank was reorganized so as to increase control over all foreign-owned monetary assets. In 1948 Argentina nationalized the British-owned railways, still the heart of the national transportation system. Also nationalized was the leading telephone company (from U.S.-controlled ITT) and the French-owned dock facilities. In every case the Argentines compensated the owners, at prices Argentine nationalists later claimed were too high. And in July 1947 Perón paid off Argentina's entire foreign debt, marking the occasion with a huge ceremony where he issued a "Declaration of Economic Independence."

Eva Perón also emerged as a political power in her own right. Snubbed by the society matrons *(señoras gordas)* who had always monopolized the political careers of past first ladies, "Evita" set up her own foundation in 1948. Dispensing cash and benefits personally from within her monumental marble building, Evita rapidly built up a fanatically loyal following. Her charisma complemented her husband's, and they together succeeded in building an imposing political machine. They presided over an elected government, but one that was steadily choking off any open political dissent.

By 1948 it seemed clear sailing for the Peronists. Social justice was rapidly being accomplished, and the regime would soon pronounce its "Justicialist" doctrine, the label used to describe the social welfare measures aimed at helping the urban working class. The economy continued to hum. The political opposition had been demoralized and humiliated. The streets were continuously full of the faithful. It was the realization of the "New Argentina" Perón had promised.

This success was soon clouded by economic problems. 1949 brought the first foreign trade deficit since the war, reducing foreign exchange reserves to a dangerously low level. Equally important was the sudden jump in inflation to 31 percent, double the previous year. A severe drought curtailed the production of exportable goods.

Perón ran into economic realities that had remained hidden during the first few postwar years. Profiting from the world economy seemed so simple in 1946, but now became problematic. World prices for Argentina's exports were dropping; prices for imports, especially manufactured goods, were rising. Peronist policies also compounded the problem. IAPI, the government foreign trade institute, had set unrealistically low prices for agricultural goods. The objective was to keep down food prices in the cities, but the effect was also to discourage production, thus hurting exports.

Perón reacted to the economic crisis in 1949 by appointing a new

finance minister who launched an orthodox stabilization program: tight credit, reduced government expenditure, and tough limits on wage and price increases. Perón was determined to get the economy under control and resume as soon as possible his ambitious social politics.

The onset of the economic crisis coincided with Perón's decision to strengthen his political grip. His first problem was the Argentine Constitution of 1853, which prohibited reelection of the president. Could there be any doubt that Perón wanted another six-year term? The Peronists had their way. The constitution was amended not only to allow reelection of the president but also to reiterate a 1947 law which gave women the vote. In 1951 Perón was reelected with 67 percent of the 6.9 million votes cast, drawing especially heavily among women voters. A Peronist party was now founded, with a Tribunal of Party Discipline as one of its central organs. The government now reverted more frequently to authoritarian measures, such as the expropriation in 1951 of *La Prensa*, the leading opposition newspaper.

On one political front, however, Perón was defeated. In running for reelection in 1951 he wanted Evita to be his vice presidential candidate. Her political influence had grown enormously, as many workers had come to identify her as the heart of Peronism. She was brilliant at promoting this image, aided by huge (and largely unaccounted for) government funds. But the military put their foot down: they refused to accept the possibility that a woman might succeed to the presidency and therefore become their commander-in-chief. Evita was bitter about the decision, which suggested that there were limits to Peronist power.

The military veto of Evita's candidacy foreshadowed a far greater blow. Evita fell ill and eventually could not hide the fact that she was dying of cancer. She grew hauntingly thin but fought the disease ferociously and continued her exhausting vigil at the foundation, facing the endless lines of unfortunates begging for her help. In July 1952 she finally died, depriving Perón of a political partner who had become fully as important as he.

Evita now became larger in death than she had ever been in life. The government suspended all functions for two days and the labor union confederation, the CGT *(Confederación General del Trabajo)*, ordered its members to observe a month's mourning. The outpouring of grief was astounding. Merely transferring her body from the presidential residence to the ministry of labor, where she was to lie in state, provoked such a crush of onlookers that eight people were killed. Over 2000 were treated for injuries in the following twenty-four hours. There were immediate plans to build a mausoleum 150 feet taller than the Statue of Liberty. Dead at the age of thirty-three, Evita became a powerful myth binding together the Peronist faithful.

Meanwhile the tough austerity plan of Finance Minister Alfredo Gómez Morales was beginning to produce results by 1952. Perón and his advisers now opted for a second Five-Year Plan, far less populist and

nationalist than the policies of the late 1940s. There was a direct appeal for foreign capital, resulting in a contract with Standard Oil of California in 1954. There were new incentives to agriculture, previously a prime target for exploitation under the cheap-food strategy. Workers were asked to accept a two-year wage freeze, a sacrifice in the name of financing much-needed investment.

The political implications of this shift in economic strategy were obvious. In order to regain economic growth Perón believed he had to reverse, at least in part, some of his nationalist and redistributionist policies. As long as the economy was expanding it was easy to favor one social sector; with a stagnant economy, however, the workers could only gain at the direct expense of the middle or upper sectors. Class conflict threatened to tear apart Perón's carefully constructed populist coalition.

Perhaps for this reason, the Peronist political strategy seemed to become more radical. After 1949 the few aristocrats once found in the Peronist ranks disappeared. Perón moved to win control over the army by giving preferential promotion to political favorites. There was also a new program to indoctrinate cadets with Peronist teachings, and to dress up the lower ranks with flashy uniforms. Perón knew he had opponents within the army, and in 1951 they attempted a coup against him. He easily suppressed them, but the germ of discontent remained alive.

After Evita's death in 1952 Perón shifted his attention from the army to the labor unions, led by loyalists. As the economic policy became more orthodox, Peronist rhetoric became more strident. A militantly Justicialist working-class tone became evident. In 1953 a Peronist street crowd pillaged the Jockey Club, the bastion of the Argentine aristocracy.

In 1954 the Peronist radicals took on another pillar of the traditional order: the church. Divorce was legalized and all parochial schools were placed under the control of the ministry of education. 1955 brought mass demonstrations against the church, orchestrated by the Peronists. Several famous cathedrals in Buenos Aires were burned by Peronist crowds. The Vatican retaliated by excommunicating the entire governmental cabinet, including Perón. The president vowed to mobilize his masses against the "conspirators" who menaced Argentine independence, threatening vengeance on five opponents for every Peronist who fell in political combat.

In fact, the Peronist government was out of control. The anticlerical crusade could not possibly bear enough political fruit to justify its disruptive effect on the public, including some faithful Peronists. Most important, the ugly battle with the church had now given Perón's enemies within the military their chance. Many officers who had been disturbed by varying aspects of Peronism became convinced that he was bent on destroying the country. In September 1955 the military con-

spirators mobilized their forces and presented their former fellow officer with an ultimatum: resign or face civil war. Perón, so often given to extreme rhetoric, had no stomach for a bloodbath. Unlike Evita, he had never wanted to arm the workers. Now it was too late. He accepted the ultimatum and retreated to the refuge of a Paraguayan gunboat that took him to an ignominious asylum.

The Military Stewardship

As General Perón fled across the wide river, the mood changed in Argentina. Where were the workers, anxious to protect their leader? What had happened to the huge political machine? How could a relative handful of military officers so easily thwart the working class?

Perón had not really been defeated. He had left. He departed under duress, but he had made no effort to mobilize his followers against the men in uniform. The sudden vacuum created by his departure was indicative: neither Perón nor Peronism was finished.

The general who now became president was Eduardo Lonardi, a moderate who wanted to avoid a vindictive policy that would keep the Peronists united. But the hard-line military grew impatient with his conciliatory approach; they demanded tougher measures. In November these hardliners deposed Lonardi and installed General Pedro Aramburu as provisional president. The anti-Peronist zealots now got their chance to purge everything Peronist. The party was outlawed, Peronist functionaries were fired, and every scrap of Peronist propaganda became contraband.

The hard-line military seemed to believe that Peronism could be eradicated in a relatively brief interval of military rule. Familiar pillars of pre-Peronist Argentina were quickly restored. The vehemently anti-Peronist *La Prensa* was returned to its owners, the Gainza Paz family. Former property owners hit by Peronist expropriations had their holdings restored. The Aramburu government pushed the crackdown on Peronist leaders, especially in the unions, on the assumption that repression could reverse their influence. In June 1956 the Peronists struck back. A revolt of pro-Peronist military took place in several provinces, and the government responded with force. In the follow-up some forty leaders, including some military officers, were executed. Spilling of blood on this scale was ominous; however authoritarian Perón's government, it had never resorted to such a level of official killing.

The Aramburu political strategists believed that they could reshape the political system for a transition to a post-Peronist era. In October 1956 they created a "Board for the Defense of Democracy," which was to monitor all "nondemocratic" parties and movements. All parties were now required to commit themselves to democracy. Having to es-

tablish such a "requirement" was itself dramatic proof of the fragility of Argentine democracy. Early 1957 saw a recrudescence of Peronist violence, met by government repression. Despite this conflict, the military still expected to transfer power to a civilian government capable of handling the Peronists.

On the economic front the military governments of 1955–58 found a sluggish economy that had fallen far short of its ability to extract gains from foreign trade. In fact, the economic policymakers summoned by the military took few bold steps. There was a devaluation, and the renegotiation of Argentina's many bilateral foreign debts. The years 1955–57 saw a healthy growth rate in the GDP, although agricultural output lagged. The military governments had not found a way to create convincing incentives for the landowners, still traumatized by the discriminatory Peronist policies. Notwithstanding their rhetoric, the military policymakers followed wage policies that resulted in virtually unchanged real hourly wage rates for 1955 and 1956 and in a 7.2 percent increase in 1957. Any attack on the major economic problems awaited a government with more legitimacy.

Unfortunately for Argentina, the anti-Peronist civilian politicians were deeply divided. The largest party was still the Radicals, the venerable party of Yrigoyen and his acolytes. At a party convention in 1956, the Radicals (UCR) split in two. One faction was the "Popular Radicals" (UCR del Pueblo, or UCRP), led by Ricardo Balbín, the party's elder statesman who had run for president in 1951. The other was the "Intransigent Radicals" (UCRI), led by Arturo Frondizi, an economics professor. The Balbín faction was more fanatically anti-Perón, while the Frondizi faction advocated flexibility in dealing with the Peronists. This paralleled a similar split of opinion within the military.

In July 1957 Argentina held its first elections since Perón's overthrow. The two Radical factions won an almost equal number of seats in the constituent assembly, which promptly restored the Constitution of 1853. But the delegates wrangled and walked out so often that the assembly was finally disbanded.

The military were nonetheless determined to move to a civilian government. They held a presidential election in February 1958 and the victor was Frondizi, who had mounted an aggressively nationalist campaign. His wing of the Radicals, the UCRI, was still bitterly opposed by the Balbín wing (UCRP), so Frondizi had needed votes elsewhere. To get them he struck a deal with the Peronists, promising some Peronist-type policies and a willingness to work to restore their party to legality. Frondizi's wing won not only the presidency and the national Congress, but most provincial governments as well. It looked as if Argentina now had a strong centrist government that could take up the difficult tasks left by Perón and the inconclusive military government of 1955–58. The major cloud hanging over Frondizi was the degree of his debt to the Peronists, and how he might have to repay them.

The Failure of Developmental Reformism

The Frondizi government (1958–62) was an all-important test of Argentina's ability to rejoin the ranks of elected civilian regimes. The new president seemed to be from the same mold as other democratic reformers then making their mark in Latin America, such as Eduardo Frei of Chile and Juscelino Kubitschek of Brazil. On both the economic and political fronts, Frondizi decided to take major gambles. The risks were great, but so were the potential rewards.

Frondizi had an ambitious economic program designed to accelerate industrialization while also stimulating agricultural production, thereby boosting export earnings. Much of the financing for new industry was to come from abroad, while at home the extensive state intervention in the economy was to be reduced, starting with the recognition of free collective bargaining in the labor sphere. There was one unfortunate catch to this plan: it could succeed only if there was a shift from consumption to investment. That meant that consumers, once so favored under Perón, must be prepared to make short-term sacrifices for the sake of long-term national development.

Frondizi disappointed the nationalists by one of his early investment decisions: to sign oil exploration and production contracts with foreign oil companies in an effort to correct Argentina's 50 percent dependence on imported oil. Argentina was potentially self-sufficient in petroleum and badly needed to save the foreign exchange spent on imported oil. Nonetheless the nationalists bitterly attacked Frondizi for his "sell-out" to the imperialists, especially since he had struck a dramatically nationalist note in his campaign.

The rest of the new president's economic plan soon came under a more ominous shadow. Only a few months into his presidency, Frondizi was faced with an acute balance of payments crisis. Since 1955 the military government had run consistent deficits on the commercial account, leaving the new civilian regime in 1958 with fewer foreign exchange reserves than any government since the war. Perón had faced a similar problem in 1949 and again in 1952 by undertaking a tough stabilization program without any commitments to foreign creditors. Most specifically, he had avoided the International Monetary Fund (IMF), which Peronist Argentina had chosen not to join. But Frondizi chose a different tack. He wanted to impress Argentina's foreign creditors, from whom he sought new investment. With Argentina now in the IMF (the military government had made a point of joining), Frondizi decided to accept the IMF's extreme medicine: a huge devaluation, stiff controls on credit, cuts in public spending, tough wage limits, elimination of subsidies on public services, and dismissal of redundant public employees.

Frondizi was now caught in a contradiction: he was trying to launch a major economic development program while at the same time cutting

back in order to satisfy foreign creditors. His truncated presidency became a classic example of the political costs of economic stabilization.

The contradiction in Frondizi's economic policy was matched by a contradiction in his political strategy. He owed his election to Peronist support, and he was clearly hoping to coax Peronist voters over to his side. Yet the military, whose approval was essential for any government, was deeply suspicious of this conciliatory policy. Within a year Frondizi was forced by the military to fire his economic team and replace them with a dogmatic free-enterprise group headed by Alvaro Alsogaray, a rigid advocate of IMF-style monetarism.

Frondizi was now committed to carrying out the IMF-prescribed "shock treatment," beginning with a 200-percent devaluation and a sudden removal of price controls and subsidies. The objective was to jolt the economy into adjusting domestic prices to international prices. The shock-treatment advocates argued that this process would be painful in any case, so it should be done quickly. An important sectoral goal was to increase the real price of agricultural products, thereby presumably calling forth the increased production to satisfy both the external and the domestic market.

The inevitable effect of these policies was a sharp shift in income. The real income (or buying power) of industrial workers dropped by 25.8 percent in 1959, while the real income accruing to beef production rose 97 percent in the same year. Here was an exact reversal of Perón's *Justicialismo.* Even before their real wages had declined significantly, urban workers went into battle against the new government. There were general strikes in April, May, and September of 1959, and an extended railway strike in November. The latter was the most damaging, because the swollen employment rolls of the deficit-ridden nationalized railways were a prime target in the government's campaign to reduce the government deficit. Due to apparent public support for the strikers, Frondizi was forced to accept a compromise settlement in which labor was the effective victor.

The stabilization policy also came under attack from Argentine businessmen, especially those from smaller firms. They denounced the credit restrictions and the huge increase in import prices resulting from the massive devaluation. Businessmen from the larger firms, especially those linked to foreign capital, were happier, since government policy helped them. Farmers proved to be one of Frondizi's greatest disappointments. Having been guaranteed higher prices, they were expected to expand production, but they proved suspicious and refused to make the long-term commitment needed to boost output.

Despite the fierce public opposition, some of the Frondizi policies began to bear fruit. The stagnation of 1958–59 was followed by growth rates of 8 percent in 1960 and 7.1 percent in 1961. The rate of inflation, which rocketed up to 113.7 percent in 1959, declined to 27.3 percent in 1960 and only 13.5 in 1961. Industrial production was up

sharply, and two key sectors showed success: the country's first integrated steel complex was finished in 1960 and domestic oil production was trebled, attaining virtual self-sufficiency.

Yet the fate of Frondizi's presidency would not depend on economic indicators. It rested on the strength of his political support. Here the odds were stacked against him. Labor and the nationalist left never forgave him for the orthodox stabilization policy, with its cut in real wages and its embrace of foreign capital. The congressional elections of March 1960 dramatized the problem. Frondizi's Radicals got fewer votes than the Balbín faction; Peronists cast blank ballots on instruction from their exiled leader. Frondizi was already losing his gamble that he could woo the Peronists to his side, and this increasingly obvious failure aroused the military.

The climax came in the congressional and local elections of March 1962. Now the Peronists were allowed to run candidates under their own banner (for the first time since 1955), as Frondizi fulfilled his commitment to restore the Peronist party to legality. The result was a disaster for the government. The Peronists led all parties in total votes, with 35 percent. They won a string of provincial governorships and a large bloc of congressional seats. The Frondizi Radicals got 28 percent and the Balbín Radicals 22 percent, the rest going to smaller parties.

The furious military quickly forced the president to annul the Peronist election victories in the provinces. The Frondizi Radicals then tried for a coalition with the Balbín faction, an obvious way out. Together, the Radicals represented half the electorate. But the Balbín followers rejected the Frondizi overtures, dismissing the gravely weakened president for his "antinationalist" economic policies. Once again the middle-class party, the Radicals, proved unequal to the task of governing Argentina, the most middle-class country in Latin America.

The military now saw Frondizi as discredited. He had gambled on converting the Peronists and lost; yet he stubbornly refused to resign. On March 29, 1962, the army tanks rolled onto the streets and removed Frondizi from the Casa Rosada. Into the presidency stepped the constitutional successor, Senate President José María Guido.

Guido served as acting president for a year and a half, but the real power was held by the military, still deeply divided on how to deal with the civilian politicians. These splits led to intramilitary revolts and repeated small-scale bloodshed among the men in uniform. The fact was that the military were by no means united over the advisability or possibility of trying to "reintegrate" the Peronist masses into the political system. And complicating this was the exiled Perón himself, continuously sending instructions to his lieutenants in Argentina.

The military finally decided to annul entirely the election results of 1962, and to hold a new round of elections in July 1963. The voting followed a familiar pattern, although this time the Balbín Radicals won the largest total, with 27 percent of the ballots. The new president was

Arturo Illia, a colorless provincial physician who was to lead the second Radical attempt at governing post-Peronist Argentina.

Illia's political style was decidedly low-key. This seemed suitable, since he had gained only slightly over a quarter of the popular vote, and would face a Congress in which the opposition held almost two-thirds of the seats. Unlike Frondizi, Illia had made no overtures to the Peronists. Nonetheless, the hardline military were ever vigilant to find any signs of softness toward Peronism or the left.

Illia was relatively fortunate in the economic situation he found. 1962 had seen a recession, but good agricultural harvests would soon improve the trade balance and increase foreign exchange reserves. The Illia government began very cautiously on the economy, announcing no overall program. It soon became evident, however, that the policy-makers were set on expansion, granting generous wage increases and imposing price controls. These measures helped to swing Argentina into the "go" phase of the "stop and go" economic pattern (alternately stimulating and contracting the economy) it had exhibited since the war. The GNP showed small declines in 1962 and 1963, but spurted to gains of 10.4 percent in 1964 and 9.1 percent in 1965.

On the agricultural front the Illia government suffered through a downswing in the "beef cycle," when the depleted herds were withheld for breeding. The resulting shortage irritated urban consumers—always voracious beef eaters—and reduced the production available for export. Cattlemen were angry because the government did not let prices rise to the levels indicated by market demand. Illia, like virtually every other president since 1945, found the rural sector virtually impossible to harness for the national interest.

The Peronist unions were opposed to Illia from the moment he entered office, in part because the Peronists were barred from the 1963 elections. Despite Illia's initially large wage settlements, the Peronist-dominated CGT drew up a "battle plan" *(plan de lucha),* which included strikes and workplace takeovers. Like Frondizi, Illia hoped his policies would attract some Peronist votes. That illusion was shattered in the congressional elections of March 1965, when the now legalized Peronist party won 30.3 percent of the vote, as against 28.9 percent for the Illia Radicals. As in 1962, the Peronists showed their ability to get out the vote in demonstration of loyalty to the movement.

Perón, in his Spanish exile, was encouraged by the vote and sent his third wife, Isabel, to Argentina to negotiate directly with the feuding Peronist groups. The hard-line military grew more worried over the apparent Peronist comeback. Illia had taken the same political gamble as Frondizi, with similar results. The economic scene had also taken a disquieting turn. Inflation had erupted anew, the government deficit was out of control, and investor confidence plummeted, with the Argentine stock exchange virtually closing down. In June 1966 the military intervened again. Illia was unceremoniously ejected from the Casa

Rosada. Once again the officers had removed a Radical government unable either to court or to repress the Peronist masses.

The Bureaucratic-Authoritarian Solution

The military coup of 1966 appeared to mark a sharper break with the past than any coup since 1943. At a minimum, it was the most repressive, at least in the initial stages. Proclaiming the advent of "the Argentine Revolution," General Juan Carlos Onganía sought to implant a new kind of regime—a bureaucratic-authoritarian state. The goal was to attack the root causes of Argentina's problems, rather than deal with the symptoms: society must be transformed. The Onganía government shut down the recalcitrant Congress, ousted opponents from the universities, and set out to control (and purportedly "uplift") the tone of social life. Dismissing politicians from positions of authority, the military leaders forged alliances with technocrats and foreign investors, whose capital was sought as a means of spurring economic growth. A key part of the overall plan called for the suppression of the labor movement, since the increase in investment was to be provided in part by a decline in real wages.

The Onganía government attempted yet another economic stabilization program. Perón had been forced to adopt one in 1949 and again in 1952; he controlled inflation and improved the balance of payments, but the cost was economic stagnation. Frondizi tried in 1958, gaining limited success with inflation and the balance of payments, but only at the cost of stunting investment and sacrificing his longer-term developmentalist plan. Illia had simply ridden with the expansionary phase of the economic cycle, imposing quasi-Peronist wage and price policies. None of these governments had succeeded in getting at the root of Argentina's problem: the lack of sustained growth, based on a productive rural sector able to satisfy both export and domestic demand. Frondizi had put forward the most coherent vision, but it was immediately compromised by the need for painful short-term stabilization measures.

The Onganía government was determined to undertake a more profound economic program. After an initial pause the economics minister, Adalberto Krieger Vasena, announced a wide-ranging plan which one expert foreign analyst later called "the most balanced anti-inflationary program of the postwar era," one which "combined the expansion of production, price stability, and the improvement of the balance-of-payments objectives with an effort to increase the level of aggregate consumption, albeit very gradually." A key feature was a two-year wage freeze in 1967, which the government was able to enforce because of its authoritarian methods and its relative success in holding down price increases. (Note also the sharp reduction in strikes shown in Figure 3-2.)

The government had another factor working in its favor. A significant wing of organized labor, led by CGT executive Augusto Vandor, welcomed the coup against Illia and wanted to collaborate with the new military government. The Onganía-led officers in turn relished the prospect of dividing labor and thereby breaking the Peronist stranglehold. This tactic proved partially successful in 1967 and 1968. But in 1969 it was engulfed by explosive labor opposition.

In that year an opposition movement developed in the provincial city of Córdoba, the heart of the newly created Argentine automobile industry. There had been a series of antigovernment protests and labor stoppages. The local army commander became nervous, his troops jumpy, and during a street protest the troops opened fire, killing some tens of protestors and bystanders. A howl of protest went up in the country, despite the authoritarian government's control of the media. The many enemies of the government's economic program, including some military who opposed the wage freeze, seized the occasion to lobby for Krieger Vasena's ouster. Onganía hesitated, the pressure grew, and Krieger Vasena was finally forced out in June 1969. The Onganía government lasted another year, but its credibility was shattered.

It was not only the labor opposition that doomed Onganía's regime. There was also a shocking rise in political violence, such as clandestine torture and execution by the military government and kidnapping and assassination by the revolutionary left. The Onganía coup began in violence and the victorious military immediately made clear that all normal legal guarantees were suspended. The labor policy, tinged with a conciliatory aspect at the outset, soon came to depend on coercion. This had happened before, as under Aramburu, and even under Illia, but now there was a difference. The left decided to reply with its own violence. Splinter revolutionary groups sprang up, kidnapping prominent businessmen and ransoming them for huge sums. Foreign multinational executives became targets. The firms countered by locating their people across the river in Montevideo, from where they would commute daily under armed guard. In 1970 leftist terrorists kidnapped ex-President Aramburu, who had ordered the execution of Peronist conspirators in 1956. Aramburu was later found murdered.

A deadly toxin had entered the Argentine body politic. There was now a revolutionary left, committed to traumatizing the nation by violence against those they identified as the oppressors: the military and the police, along with their collaborators, the well-tailored executives of the multinationals. And the government struck back with violence of its own. Civil war had broken out.

The Onganía government was by all standards a political failure. Although it brought off postwar Argentina's most successful economic stabilization program, it failed to meet a more difficult challenge: creating a broad-based political coalition which could give continuity to pol-

icy and make possible genuine planning for the future. The Onganía military regime could never have achieved such a coalition because it seized power against both the Radicals, chief voice of the middle class, and the Peronists, still the strongest voice of the working class. Unlike the Brazilian generals, whose apparent success the Argentines envied, the Onganía officers could not forge a military-civilian alliance capable of holding power long enough to carry out a policy that would bring sustained economic growth. No less important, the Argentine political scene had grown far more polarized than in Brazil. Onganía's failure left Argentina with few alternatives.

The new president was another general, Roberto Levingston, a little-known intelligence officer stationed in Washington, D.C., since 1968. Returning to Buenos Aires, Levingston faced a treacherous economic problem: inflation, which Krieger Vasena had gotten down to 7.6 percent in 1969, was on the rise, reaching 34.7 percent in 1971. Levingston pursued a moderately expansionary course, which was soon doomed by a downswing in the beef cycle that caused shortages and high prices. Never well endowed with military prestige, Levingston found himself isolated. Recurrent rioting in Córdoba in March 1971 sealed his doom. Yet another military coup removed Levingston, installing General Alejandro Lanusse, who had been the brains behind the ouster of Onganía eight months earlier.

Lanusse followed a moderate nationalist course in economic policy. Disinclined to take bold steps, he decided to ride along with rising budget deficits. Predictably, inflation increased, hitting 58.5 percent in 1972. The Lanusse government made no pretense at having an answer for the economy.

Lanusse's real ambition was to achieve a new political accord. He opted for a relegalization of the Peronists, along with a new electoral system which he hoped would favor pro-government parties. Lanusse took an even greater gamble: he decided to allow Perón to return. Elections were announced for March 1973. Perón briefly returned to Argentina in late 1972 and lobbied intensively for his stand-in, Dr. Héctor Cámpora, as his own presidential candidate. Meanwhile the violence continued, with business executives being kidnapped, ransomed, or murdered. The guerrillas became bolder and were directly striking at high-ranking military officers, as well as at prisons and barracks.

Cámpora received 49 percent of the popular vote, not an absolute majority, but far ahead of Balbín's 22 percent. The president and like-minded officers began to see Perón as the only hope against the left. They saw him as an essentially conservative figure, having nothing in common with the leftist guerrillas, some of whom claimed to fight in his name. When Héctor Cámpora was inaugurated in May 1973, more than a few officers felt that the first step toward a solution to the leftist threat might be at hand.

The Peronists Back in Power

Cámpora had left no doubt that he was only a stand-in until Perón could return and run in a new election. Notwithstanding Cámpora's thoroughly weak personality, his government launched a bold new economic policy. It had been largely drawn up by the CGE *(Confederación General Económica)*, an Argentine businessmen's association; it was aimed at first stabilizing prices and then boosting workers' earnings back to the share of national income they had reached in the earlier Peronist era. It was to be done by (1) a one-time wage increase; (2) a price freeze agreed to by government, labor, and industry; (3) a two-year wage freeze; and then (4) a set of policies designed to improve the workers' real incomes. Obviously this would require extraordinary cooperation from all interest groups. The Cámpora government seemed to have negotiated agreement to that in their proposed "Social Contract" *(Pacto Social)*, which was formally ratified by both the CGT and the CGE. They drafted a parallel compact with rural producers (except for the rabidly anti-Peronist cattle breeders) which promised price, tax, and credit incentives in return for a promise to double farm production by 1980. Surprisingly, the new Peronist regime had constructed a coalition that included almost every interest group in Argentine society. How was it possible? In part, because both exhaustion and realism had taken hold of Argentines. Indeed, more than a few long-time anti-Peronists looked to the new Perón government as perhaps their country's last chance to solve their problems by something short of naked force.

As every Argentine and foreign observer could see, the odds for success were not high. Political violence was rising steadily, as guerrilla forces scornfully rejected the new Peronist regime and tried through kidnappings and assassinations to destabilize the fragile political balance. A further liability was the age and health of the once charismatic figure around whom the new social consensus had to be built: Perón was seventy-seven and in failing health.

New presidential elections were scheduled for September. Perón now succeeded in a political tactic which had failed in 1951: he got his wife, Isabel, nominated for the vice presidency. They swept the election with 62 percent of the vote. Perón now began to turn on the revolutionary left, whom he had often encouraged in his comments from exile. The People's Revolutionary Army (ERP) was outlawed, in good part because of its repeated assassination of military officers. Perón gave his blessing to crackdowns on leftist rallies and publications. Lanusse's judgment seemed vindicated; Perón was proving the perfect sponsor to preside over a military and police counteroffensive against the revolutionary left.

On the economic front, the ingenious program launched by Cám-

pora seemed to be working. Early 1973 had seen an economic boom, fueled by high export earnings, as world meat prices rose and Argentina's grain production swelled. During 1973 inflation dropped dramatically and real wages rose 13.3 percent in the second half of the year. But 1974 brought trouble. The OPEC oil price increase hurt Argentina's balance of payments, although it was importing only 16 percent of its oil. Furthermore, some non-CGT unions won new wage agreements, in violation of the Social Contract. Several CGT unions soon followed suit. Now under growing pressure from union leaders, Perón agreed to large year-end bonuses for all CGT unions, thereby undermining his own anti-inflation program.

Whether Perón could have yet worked his magic with the workers again was not to be known. In July 1974 he died; the president now was Isabel. Perón had met her when she was a nightclub dancer in Panama, during his leisurely journey after his 1955 overthrow. Isabel was no Evita, as her insecurity and indecision had already made clear. *La Presidente* assumed office as the Peronists were bitterly squabbling.[1] There was an immediate scramble to gain influence over the frightened woman who had succeeded to the presidential duties.

The adviser with greatest influence was Isabel's minister of social welfare, José López Rega, an ambitious and bizarre figure as well known for his fascination with astrology as for his militantly right-wing Peronist views. López Rega first helped convince Isabel to purge her cabinet of the more moderate ministers in October 1974, then persuaded her to crack down on the left—including left-wing Peronists. This became the direction of policy in 1975, as unions began negotiating new contracts with 100 percent wage increases or more. Isabel mounted a counter-campaign, annulling the huge wage settlements and later, after a series of massive strikes, reinstating them. This dramatic reversal, on top of the increasingly bloody battle between leftist guerrillas and the military, led to López Rega's resignation. The president also lost her congressional majority as the Peronist delegation split apart.

The economy now careened out of control. Inflation rocketed to 335 percent in 1975, as the wage-price spiral picked up speed. 1975 also proved to be disastrous for exports, with a disappointing harvest compounded by a new European Common Market ban on imported beef. Foreign exchange reserves dropped by more than 50 percent. By early March 1976 Isabel's government was reduced to adopting a stringent stabilization plan in return for help from the IMF.

Deterioration of the economy was accompanied by increasing violence in politics. The guerrillas continued their deliberately provocative

1. Isabel should normally have been referred to by the feminine form *la presidenta;* but the Argentine Constitution spoke only of *el presidente,* and her supporters did not dare take any chances. So legal nicety triumphed over grammatical logic, and she came to be known as *La Presidente.*

attacks on the police and military, bringing off some dramatic assassinations. The right answered through equally violent organizations such as the Argentine Anti-Communist Alliance. The "great accord," on which the new Peronist era was to have been based, now seemed a cruel hoax. The value of money shrank daily, almost hourly. Fear of terrorists, whether of the right or left, took hold of the populace, especially the urban middle class. The president was terrified, utterly unable to wield command, and increasingly bewildered. Only one question was in the air: when would the military remove her?

Isabel's term ran to 1977, and the military seemed determined to let her serve it out. If they siezed power, then they would have formal responsibility for dealing with the economic mess. Better to leave her in office, especially since she was giving carte blanche to the security forces in their war on the guerrillas. By retaining her as president they had the semblance of civilian legitimacy. And there may have been another motivation at work. The military may have decided to let the national situation become so violent and the economy so chaotic that no one could doubt the need for the military to step in. If so, they had succeeded by March 1976. In Argentina's best predicted coup, the men in uniform placed *La Presidente* under house arrest (she would be investigated for corruption) and once again an elected government disappeared from the Casa Rosada.

The Military Returns

When the armed forces finally moved against Isabel, they were determined to impose a bureaucratic-authoritarian solution that would last. Under General Jorge Rafael Videla, the regime launched a vicious campaign, alternatively known as a "dirty war" or "holy war" against the opposition. The government began arresting "subversives" at will, at one point acknowledging that it held nearly 3500 prisoners. And then there were the *desaparecidos*, those who simply "disappeared," perhaps 10,000 or 20,000 in all. These people were abducted by heavily armed men who refused to identify themselves but who were undoubtedly "off-duty" security men or paramilitary operating with the military government's knowledge. Virtually none of the abducted were ever heard of again. The secrecy for this ghastly slaughter may have come from the Argentine military's study of how the Brazilian security forces in the early 1970s were highly criticized for mistreatment of political prisoners. The Brazilian mistake, according to the Argentines, was to place suspects under official arrest, thus leaving a legal trail.

We shall never know how many of the "disappeared" were totally innocent and how many actively supported the guerrilla movements. Thousands of Argentines were no doubt involved in one way or another, including couriers, gunmen, infiltrators, arms smugglers, and lookouts, while broad sectors of the populace shared the rebels' antipa-

thy to right-wing domination. From bank robberies and ransomings the guerrillas built a war chest of at least $150 million, and they proved highly adept at paramilitary strikes (one bomb went off in an army mess, killing fifteen and wounding many more). The army and police faced a formidable challenge.

Driven by a fanatical adherence to doctrines of "national security," the infuriated generals decided to pursue an all-out offensive without any legal constraints. The "disappeared" were victims in a tactic consciously designed to terrorize the country. Combat with the guerrillas was brutal, with even a conventional battle in the province of Tucumán. In the end the generals won, but at a terrible price. They thought they had had no choice, and savagely rejected criticism from every quarter. One of the most widely read exposés was by Jacobo Timerman, a former Buenos Aires newspaper publisher, who described his tortures as abusively anti-Semitic and even pro-Nazi. Once proud Argentina became an international pariah along with Chile and South Africa, and its people, by habit articulate and argumentative, suffered the ignominy of silence and intimidation.

What had the guerrillas wanted? There were several groups, but all sought the violent overthrow of the government and the installation of a revolutionary socialist regime along Marxist-Leninist lines. They were predominantly middle-class and many were university students or recent graduates. Desperately idealistic and deeply alienated by the merry-go-round of Argentine politics, they were caught up in a passionate rebellion against a socioeconomic structure that was, ironically, one of the most "modern" in Latin America. Once locked in battle, there was no exit for the guerrillas. The military were so set on liquidating their opponents that amnesty was never a remote possibility. It was a war to the death.

The war showed, as have others in Latin America, that a well-equipped and determined government can, barring any major split among the society's ruling elites, normally defeat a guerrilla movement. A key factor was the tacit (and often explicit) support of the middle class for the antiguerrilla campaign. As a proportion of the society, the Argentine middle class was the largest in Latin America, and therefore a crucial actor in the political drama. It had watched with dismay the decay of order under Isabel from 1974 to 1976, and most of it supported the coup of 1976.

The March 1976 takeover was intended as a coup to end all coups. Videla and his colleagues proclaimed that their goal was not merely to terminate the chaos of the Peronist years, but also to restructure Argentine society. The junta promised to eradicate terrorism and thereby remove some potent actors from the political scene. They planned to revitalize the public sector, including foreign as well as domestic capital, to reduce the public sector, and consequently to rearrange relationships among business, labor, and the state. They affirmed Argentina's

alignment with the "Western and Christian world," and in keeping with these lofty principles, they promised to "reeducate" the populace by emphasizing values of "morality, uprightness, and efficiency." Thus the soldiers would lay the foundations for an eventual "democracy" that would be, in carefully chosen words, "suited to the reality and needs and progress of the Argentine people."

In pursuit of these ideas the military not only embarked on all-out war against the opposition. They also penetrated Argentine society more deeply than ever before: in addition to abolishing the General Confederation of Labor, military officers also took over other institutions, such as sports and charitable organizations.

In 1978 the generals got a heaven-sent propaganda opportunity when Argentina hosted the World Cup soccer matches. Argentina won the cup, to the ecstatic cheers of the home crowd and the obvious pleasure of the heavy-handed generals. For a few weeks at least, ordinary Argentines could take pride in their country. But the euphoria was soon dissipated by the realities of Argentina's plight.

Among the gravest worries was the economy. Economics Minister José Martínez de Hoz, an outspoken representative of the "neo-liberal" view, immediately imposed a stabilization program designed to reduce inflation and win back the confidence of foreign creditors. Labor faced declining real wages, while businessmen found credit increasingly hard to obtain. The most favored sector was the banks and financial institutions, which earned huge profits because of record-high real interest rates (often 20–40 percent). Foreign capital inflow rose dramatically, but much of it was speculative. Martínez de Hoz also moved to denationalize a number of state enterprises, while slashing tariffs on almost all industrial goods.

These policies succeeded in bringing inflation down to 88 percent in 1980, and in achieving a surplus on the balance of payments for four successive years (1976–79). By 1981, however, the picture had darkened. A wave of bank failures provoked a huge outflow of funds. Inflation again exceeded 100 percent, and a recession set in. In 1981 industry operated at only half capacity, and real income was less than it was in 1970.

Despite these economic troubles, the armed forces demonstrated a notable degree of coherence and unity. This was an institutional regime, not a one-man show, and when Videla turned the presidency over to General Roberto Viola in March 1981 it was merely confirmation of the fact. Viola lacked the stamina needed in the pressured position and passed the presidency in early 1982 to General Leopoldo Galtieri, the commander-in-chief of the army.

In March Galtieri chose to stake his government's fate on the Falkland Islands controlled by the British but long claimed by the Argentines, who called them the Malvinas Islands. During his 1946–55 rule Perón reawakened Argentine passions over the islands, but Britain sim-

ply ignored him. In 1965 the United Nations invited Britain and Argentina to begin talks leading to a peaceful solution of the dispute. The sovereignty issue was not formally discussed until 1977, the same year in which British intelligence warned of an impending invasion. Britain sent a small war fleet, without publicity, and the threat faded.

In 1982 the Argentines thought the British had changed signals and would not bother to defend the desolate islands—8000 miles away from Britain, populated by only 1800 British subjects and 600,000 sheep. On April 2 a large Argentine force invaded the islands and quickly overwhelmed the badly outgunned royal marine garrison.

British Conservative Prime Minister Margaret Thatcher was not about to see Britain's sovereignty and subjects snatched away by a South American military dictatorship. The British denounced the invasion and mobilized a major task force. Britain justified its impending military intervention as needed to prevent the aggressor from usurping the islanders' right to self-determination. In late May the British task force landed thousands of troops onto Falklands/Malvinas beachheads. All but three other Latin American countries backed Argentina in an Organization of American States vote condemning Britain as the aggressor.

Why had the Galtieri government decided to invade? Clearly the Argentine economy was on the rocks again, with inflation and the foreign debt rocketing upward. Only days before the April 2 invasion there had been the largest antigovernment demonstration since the military seized power in 1976. Galtieri and the few fellow officers he consulted undoubtedly saw the lure of a quick military victory in the Falklands/Malvinas as a boost to the government's sagging popularity. Furthermore, Galtieri felt certain that he would have at least the tacit support of the Reagan administration, with which the Argentine generals had developed a warm relationship.

In the short run Galtieri was right about the Argentine reaction. The invasion brought an outpouring of patriotic sentiment in Argentina. In part that was due to hyperbolic, government-controlled reporting that told of nothing but Argentine victories. Had long-fragmented Argentina finally found a way to come together?

The Argentine public soon suffered a rude return to reality. The better trained and more experienced British troops swept across East Falkland and laid siege to the 7500 Argentine troops holed up in the capital, Port Stanley. After nervous consultation with Buenos Aires and sporadic resistance, the Argentine commander surrendered promptly—the only sensible option given the poor morale, condition, and positioning of his troops. Yet the sudden surrender hit Buenos Aires hard. The government's propaganda now turned to ashes. Britain, supposedly enfeebled and unable to defend these distant islands, had decisively defeated numerically superior Argentine forces. Only

The seizure of the Falkland/Malvinas islands in 1982 led to a massive out-pouring of national pride and defiance. (C. Carrión/SYGMA.)

the Argentine Air Force emerged as having had both the skill and the courage to fight effectively.

Transition to Democracy

The Galtieri-led junta had made a mortal error: as a military government it began a military adventure that it failed to win. The public reaction was inevitable: what good are generals if they can't even win a small war? Patriotic fervor turned into ugly demonstrations outside the Casa Rosada. Galtieri came under intense fire from his fellow officers. He resigned, as military unity began to unravel. The navy and air force withdrew from the junta, leaving the army holding power alone. The new president was an obscure retired general, Reynaldo Bignone, an officer in the engineers corps. Upon assuming office in July 1982, Bignone bravely repeated Argentina's claim to the Falklands/Malvinas. He promised an election in 1983 and a return to civilian government by 1984. It was as if the Argentine generals had by their incompetence restored legitimacy to the civilian politicians.

The Argentine economy went from bad to worse in 1982. Inflation shot up to 200 percent, workers lost about one-quarter of their real income, and the country went into de facto default on its private for-

eign debt. By early 1983 the nominal take-home value of the minimum wage was one million pesos a week, worth only about $19. The Brazilians (their inflation rate was only 100 percent) thought it amusing when the Argentines printed their first one-million-peso note. The Argentines themselves grew frustrated. The 100-peso coin—worth far less than one U.S. cent—dropped out of circulation and proved useful only to irate soccer fans who threw them at referees, and to dissident activists who hurled them at police.

In order to obtain desperately needed external financing, the government agreed to an austerity plan outline in collaboration with the IMF. But there was real doubt as to whether the military government could manage the transition to a civilian government. In December 1982 the opposition staged the largest antigovernment protest in seven years of military rule.

To virtually everyone's surprise, Radical Party leader Raúl Alfonsín won 52 percent of the vote in the presidential election of 1983. The Radicals also gained majority control of the Chamber of Deputies. The Peronists, who had not lost a free presidential election since 1945, got only 42 percent. Alfonsín had been a courageous battler for human rights during military rule. Also, his party was the only non-Peronist group capable of forming a viable government.

The new regime faced formidable problems. First was the commitment to prosecute the military personnel and police who had killed or "disappeared" more than 10,000 suspects. The public revulsion against the perpetrators was deep and had helped give Alfonsín his vote. Argentina, however, would be the first country to try its own military for domestic crimes—the Nuremberg trials of the Nazis were imposed after all by the victorious foreign powers. This was untrod ground. How many should be tried? Where did criminal responsibility end? And how would this prosecution affect the effort to build a new democratic military?

The second major problem was the economy. Inflation had reached 400 percent in 1983, and Argentina could not make the payments on its huge foreign debt. Argentina had also failed to modernize its economy for survival in world trade. Finally, Alfonsín faced the ever-present struggle for income among competing classes and sectors, with the huge labor unions bloodied but unvanquished by military repression.

The third problem was finding a viable political base. Could the Radicals, a minority party since 1945, retain the majority Alfonsín had won? If not, was an effective coalition feasible?

Alfonsín struggled valiantly with all these challenges. Prosecuting the torturers proved almost a no-win situation. A presidentially appointed commission documented the death or disappearance of 8906 Argentines. The government charged the nine military commanders-in-chief for crimes ranging from murder to rape. Five were convicted and

given prison terms, while three of the four acquitted were later tried by military justice and sentenced to prison. But how far down should the prosecutions go? A 1987 military revolt protesting the impending prosecutions forced Congress to exempt all officers below the rank of general. Even the ongoing prosecutions bogged down, spurring human rights advocates and relatives of the "disappeared" to denounce the failure to pursue the hundreds, if not thousands, of other cases. Alfonsín supporters replied that no other Latin American government had ever dared to prosecute its officers for crimes committed during a military government. Meanwhile, the Argentine officer corps was largely unrepentant. Clearly, the repression and subsequent search for justice would leave a deep scar in Argentine society.

Meeting payments on the massive $50 billion foreign debt was an immediate economic problem for the Alfonsín government. Since Argentina's exports did not earn a sufficient surplus to service the debt, Alfonsín had to seek new loans. The price for this money was an IMF-designed austerity policy at home. Nonetheless, inflation roared up to 627 percent in 1984, and approached 700 percent in 1985. With its back against the wall, the Alfonsín government unveiled the "Austral Plan," a wage-price freeze that created a new currency (the "austral") in the hope of breaking the inflationary psychology. Inflation dropped to less than 100 percent (a victory by Latin American standards), but a recession and a sharp fall in real wages also occurred. Only makeshift measures allowed the government to avoid defaulting on the foreign debt, and few believed that Argentina could continue payments that totaled 6 percent of its GDP.

On the political scene, Alfonsín managed to hold his ground against uphill odds. In the 1985 congressional elections the Radicals retained their lower house majority, as bitterly divided Peronists continued to lose middle-class support. Labor unions were also on the defensive. In the 1987 elections, however, Peronist congressional candidates outpolled Radicals by 41 to 37 percent and extended their grip on provincial governorships to sixteen of twenty-two.

Then the Austral Plan unraveled, partly as a result of Alfonsín's propensity for deal making and compromise. Inflation mounted and the exchange rate ran out of control. By early 1989 prices were rising at more than 30 percent *a month;* they would reach more than 100 percent a month by midyear. Output plummeted and income declined. The gross domestic product shrank by 3 percent in 1988 and 6 percent in 1989 (overall, per capita income for Argentines declined by nearly 25 percent during the 1980s).

Peronists seized the opportunity. In the presidential elections of May 1989 the party's candidate, Carlos Saúl Menem, governor of the interior province of La Rioja, took approximately 47 percent of the popular vote—and a clear majority in the electoral college—winning handily over the UCR candidate Eduardo Angeloz and two other also-rans.

This marked a potential watershed in Argentine politics: it was the first time that an opposition party had triumphed in a presidential election in over seventy years, and the first time ever that Peronists would come to power without the specter of Perón himself. If the country could take these steps, some analysts reasoned, Argentina might have a realistic chance of achieving genuine democracy.

It would not be an easy task. The economic crisis intensified. Looting erupted in May 1989, the month of the election. Argentina, the proverbial breadbasket of the continent, suffered the humiliation of food riots. A stunned President Alfonsín declared a state of siege, then announced that he would resign from office six months ahead of schedule. "No president has the right to endlessly demand the sacrifice of his people if his conscience tells him he can lessen it with his own." Chastened and disheartened, Alfonsín surrendered his ambition of being the first freely elected Argentine president to complete a full term in office since the 1920s.

Taking power amid these somber circumstances, Menem had his hands full. Inflation was running at 150 percent per month. The country was nearly $4 billion in arrears in payments on the external debt, which had by now increased to $64 billion. Near the end of the year Menem installed a new economics minister, Antonio Ermán González, who immediately imposed a strict austerity program: he lifted price controls, allowed the exchange rate to float freely, slashed export and import taxes, and removed a host of restrictions on foreign trade. In January 1990 he shocked the public by transferring interest-bearing bank certificates into ten-year bonds—in effect, confiscating savings of the middle class. These and other hard-nosed policies eventually provoked a recession that brought an end to hyperinflation.

Violating cherished principles of Peronism, Menem and his ministers embarked on a program to "privatize" state-owned companies by selling them off to private investors. In June 1990 the government auctioned off Entel, the national telephone company, to consortia of Spanish, Italian, and French investors. In July 1990 the government sold the national airlines, Aerolíneas Argentinas, to Iberia of Spain. Not content with these bold strokes, Menem announced his intent to proceed with privatization of electricity, coal and natural gas, subways, and shipping. Neo-liberal economic doctrine seemed to be triumphant.

In early 1991 Menem named as economics minister Domingo Cavallo, a powerful personality and a firm believer in strict market-oriented reforms. Cavallo extended the privatization campaign, which earned over $9 billion for the government by 1994, and centered his program around a "convertibility law"—which restricted public expenditures to revenues, forbade the printing of new money, and, most important, established a one-to-one exchange rate between the Argentine peso and the U.S. dollar. Adherence to this exchange rate became the key to economic credibility, and stimulated a substantial inflow of pri-

vate capital. Cavallo also fashioned a restructuring of Argentina's foreign debt in 1993, while the World Bank and the International Monetary Fund continued to support his hard-line policies. Inflation plummeted from 4900 percent in 1989 to less than 4 percent in 1994, and economic growth climbed to around 6 percent per year. Astonished onlookers at home and abroad hailed the achievement of an "economic miracle."

There were negative features as well. One was overvaluation of the peso, which encouraged imports and discouraged exports, leading to a trade deficit of more than $6 billion in 1994. Another was unemployment and impoverishment of the middle class. According to one study, nearly half the country's middle class slipped down into the lower class during the early 1990s. In the meantime, open unemployment increased from 6.5 percent in 1991 to 12.2 percent in 1994.

Not surprisingly, the Menem initiatives caused disruption and discord within the labor movement. The CGT split into two wings, one headed by Saúl Ubaldini, a steadfast critic of the Menem policies, the other headed by presidential supporter Guerino Andreoni. In September 1990 the administration defeated a strike movement by telephone workers in Buenos Aires who were seeking a pay increase that was judged to be inflationary. Accelerating unemployment and government layoffs also sparked protests in the interior provinces, especially in La Rioja and Santiago del Estero in 1993, and dissidents organized a major rally in Buenos Aires in mid-1994. It seemed ironic to many, and grievous to some, that a Peronist government was breaking strikes by organized labor and confronting protests from the working class.

The armed forces at first presented Menem with a vexing challenge. Several months after taking office he issued sweeping pardons for, among others, participants in minor military revolts in 1987 and 1989 whose purpose was political protest rather than governmental takeover. In December 1990 there erupted yet another military rebellion, undertaken by *carapintadas* ("painted faces") in the name of strident nationalist Colonel Mohamed Ali Seineldín, only days before the expected arrival in Buenos Aires of U.S. President George Bush. The uprising was ultimately crushed but it represented a serious challenge to Menem's authority. Claiming that he had made no bargain with the rebels, Menem issued—on December 29, in the midst of the holiday season—a new round of pardons in favor of former leaders of the military government and of its campaign of political repression. The decision prompted protest rallies and some principled resignations, but the military won their case: there would be no continuing sentences or prosecutions for human-rights offenses committed in the dirty war.

As the armed forces continued to show signs of restlessness, especially over modest cuts, Menem apparently sought to mollify the generals in late 1994 by expressing gratitude for their grisly campaign: "We triumphed in that dirty war, which had placed our society on the verge

of dissolution." Human-rights advocates vigorously denounced Menem's declaration, and the president would be subsequently embarrassed by public confessions of former military officers who recounted the routine practice of throwing political prisoners into the sea during naval flights in the 1970s. Legacies of the dirty war stubbornly persisted.

In 1994 the administration gained congressional approval for reform of the country's 140-year-old constitution. According to an agreement between Menem and Raúl Alfonsín, now leader of the Radical Party, the amendments would reduce presidential terms from six to four years, but permit one reelection; reduce the president's authority to rule through emergency decree; create the post of cabinet chief, who would be subject to removal by majority vote in Congress (thus injecting a dose of parliamentary authority); strengthen the judiciary; and provide a measure of autonomy for the city of Buenos Aires. Proponents insisted that the reforms would improve governmental accountability, decentralize power, and institutionalize a system of checks and balances. Opponents, including many Radicals, regarded the reform as a maneuver by Menem to perpetuate himself in power.

Menem promptly declared himself a candidate for the presidential elections of May 1995. Despite continuing rumors of high-level corruption and widespread resentment of the president's authoritarian style, Menem won a solid victory with 49.8 percent of the vote (under the new constitution he needed only 45 percent to avoid a second round). Divided, demoralized, represented by a lackluster candidate, the once-proud UCR earned only 17.1 percent. The strongest opposition came from José Bordón, a dissident Peronist who headed a center-left coalition known as FREPASO *(Frente País Solidario)* and took 29.2 percent of the vote. It appeared, to some, that Argentina's long tradition of two-party politics was nearing an end. It seemed uncertain, as well, whether Bordón would be able to develop FREPASO into a durable party.

In the international arena, Argentina took two novel and decisive steps. One was to promote the continued development of MERCOSUR (the "Common Market of the South"), a four-partner association that included Argentina, Brazil, Uruguay, and Paraguay. Established by the Treaty of Asunción in 1991, the scheme envisioned the creation of a free-trade zone that would eventually evolve into a customs union and, finally, into a full-fledged "common market" along the lines of the European Union. Despite occasional tension among the members, the volume of trade and investment within MERCOSUR grew rapidly throughout the early 1990s. Its apparent success bolstered Argentina's claims to leadership in South America, although Brazil would claim this mantle as well.

Second, Argentina under Menem adopted a foreign policy in line with the United States (the foreign minister, indeed, was reported to

have quipped that Buenos Aires was seeking "carnal relations" with Washington). Menem actively supported U.S. military actions in the Persian Gulf in 1991, and in Haiti in 1994. Menem normalized relations with the United Kingdom, proclaiming his dedication to a peaceful settlement of the Falklands/Malvinas issue, and sought to strengthen contacts with the European Union. He also became a vociferous critic of Fidel Castro's Cuba, and abandoned political links with developing countries in Asia and Africa. "I don't want to belong to the Third World," he said on one occasion. "Argentina has to be in the First World, which is the only world that should exist."

FOUR

CHILE
Socialism, Repression, and Democracy

The territory we now call Chile was one of the most distant realms of the Spanish empire in America. It evolved into a secondary center valued largely for its agricultural and mining production. The Spaniards found a native Indian population, but many perished under the onslaught of diseases brought by the Europeans. A relatively homogenous population emerged from the colonial era, *mestizo,* although few of the "European" inhabitants wished to admit the extent to which their Spanish forebearers had mixed with the Indians.

When Napoleon invaded Spain, the colonists in Chile reacted much as their counterparts elsewhere, showing strong loyalty to the crown. They were indignant over Napoleon's cavalier treatment of Spain and its colonies: turning them over to his brother Joseph. As the French control of Spain dragged on after the conquest in 1808, the Chileans took matters into their own hands, holding a Congress in 1811. They seemed headed for independence, but the royalist forces regained the initiative and by the end of 1814 won control of Chile. It was against this royalist "reconquest" that Bernardo O'Higgins helped lead a revolutionary army from Mendoza. The rebels won Chilean independence from the Spaniards and their royalist supporters in 1818. As the new republic's supreme director, O'Higgins proved a decisive but autocratic leader. He created a navy (to be one of Latin America's finest), promoted education, and won recognition of Chile's independence from the United States, Brazil, and Mexico. The constitutional congress he had promised was rigged, however, and in 1823 the discontented Chilean aristocrats forced him to resign.

The following years saw political instability, as Liberals and Conservatives struggled for control. The latter won in 1830, beginning the three decades of the "Conservative Republic." The key figure was Diego Portales, who became the strong man of the regime, although never president in name. A Constituent Assembly was held in 1831, producing a constitution in 1833. It created a strong central government, with economic power in the hands of the landowners.

Portales ruled unchallenged because the government controlled the electoral machinery and the landowners were happy to let Portales exercise power (including repression when deemed necessary) for their benefit.

Portales' undoing was a war with Peru (1836–39), which provoked a military rebellion at home and brought the dictator's assassination. Chile then went on to defeat the Peruvians. Chile's principal war hero was General Manuel Bulnes, who served as president for a decade after 1841, presiding over an era of ferment and creativity. Cultural life was enlivened by the presence of exiles from elsewhere in South America, especially Argentina, suffering under the dictator Rosas. In foreign policy, the Bulnes government took possession of the Straits of Magellan, initiating a territorial struggle with Argentina that was settled only in 1984.

The 1850s brought another decade of fruitful consolidation for the new nation. The status of the church proved a key political question. Of all the legacies from the Spanish colonial era, none was to cause more controversy than this issue. One wing of the landowning elite wanted greater state control over the church, especially in education and finances. Their opponents defended all of the church's privileges. When the normally anticlerical Liberal Party softened its stance in the late 1850s, dissident Liberals founded the Radical Party, an organization that would come to play an enduring role in the political life of the nation.

Overview: Economic Growth and Social Change

For Chile, as for many Latin American countries, the nineteenth century marked a period of far-reaching economic and social transformation. During the colonial era Chile played a relatively minor role in the Spanish American economy. Land in the fertile central valley was concentrated in the hands of a small number of powerful landlords. Their vast estates produced agricultural goods, especially fruit and grain, some bound for such cities as Santiago or Valparaíso but most destined for export to Lima and other urban markets in Peru. Maritime trade along the west coast of South America thus connected Chile to the centers of the Spanish empire.

The Wars for Independence interrupted this coastal trade, and Chilean agriculture promptly entered a period of relative stagnation. The situation was further affected by protectionist policies in Peru, which sought to encourage its own agrarian development by placing tight restrictions on imports from Chile. In the 1840s the California gold rush provided temporary stimulus for a boom in agricultural exports, which jumped from $6.1 million in 1844 to $12.4 million in 1850 and reached $25 million by 1860. But there they leveled off and then declined again. Completion of the U.S. transcontinental railroad helped to take

Though mining became the most dynamic sector of the Chilean economy, agriculture continued to play a significant role; here sacks of beans are loaded for export at the port of Valparaíso sometime after 1900. (Courtesy of the Library of Congress.)

away the California market, although export to England continued. With its advantageous location and fertile pampas, Argentina had better access to Europe. Agricultural production and commerce in Chile continued, of course, but they did not become the leading forces for economic growth.

Mining played that role. Between the mid-1840s and mid-1850s the production of silver grew by four or five times. Copper mining accelerated during this same period, and by 1870 Chile controlled about one-quarter of the world market for this product. Thereafter came a sharp decline. Chile would regain its preeminent position as a copper producer only after the turn of the century.

In the meantime it was nitrates, used for fertilizer and explosives, that became the country's leading export. This development was made possible by the acquisition of northern territory from Peru as a result of the War of the Pacific (1879–83). Foreign investors (especially British) quickly rushed in, and Europeans owned about two-thirds of the nitrate fields by 1884. But Chilean investors retained a hold in this area, reaping over half the total earnings by 1920. Eventually, however, the nitrate market declined. An increase in exports during World War I was followed by a cutback in the early 1920s, then a brief recovery, then a steep and final reduction in the 1930s. Synthetic nitrates took over after that.

The nineteenth-century growth of Chilean mining—in silver, copper, and nitrates—led to important changes in the country's social structure.

One was the appearance of new elements within the elite, consisting of mine owners in the north and merchants from the growing towns and cities. Yet these elements did not truly rival the traditional landowners. For in Chile, more than in most Latin American countries, the landowning elite did not remain isolated and apart from the manufacturing and mining elites. There was, instead, a kind of merger, often achieved through family ties, so landowners frequently had relatives in upper levels of the other sectors if they did not take part themselves. Brothers, cousins, and brothers-in-law provided important links, and these connections tended to minimize conflict between the city and the countryside.

The growth of towns and cities led to a higher level of urbanization than in most of Latin America. In 1850 only 6 percent of the Chilean population was living in urban areas, but by 1900 the figure climbed to 20 percent. It would remain around this level, between 25 and 30 percent, until the 1930s. (By 1970 the proportion was over 60 percent, surpassed only by Argentina and possibly Uruguay.) Santiago retained its position as the nation's most important city, and bustling ports like Valparaíso became vital centers of commercial activity.

There also appeared a working class, first unionized in the nitrate fields of the north. Chile's economic development in the late nineteenth and early twentieth centuries did not require the massive importation of labor, however, and this fact points to a central feature of the country's working class: it was native-born. This stands in clear contrast to Argentina, where 25 percent of the population was foreign-born in 1895; for Chile this proportion was less than 3 percent. From the outset, Chilean workers had direct access to the political scene.

Copper production underwent a technological revolution just after 1900, due to the invention of a new smelting process, and this led to a major transformation in Chile. Investments required large amounts of capital, and these came from abroad. In 1904 the Braden Copper Company began exploiting the El Teniente mine near Santiago. British interests were soon taken over by the Guggenheims, and by 1920 the industry was dominated by only three companies, known from their initials as "the ABC": Andes Copper, Braden Copper, and the Chile Exploration Company–Chuquicamata. The first and third belonged to Anaconda, while Braden was a subsidiary of the Kennecott Corporation.

In less than twenty years the Chilean copper industry thus acquired characteristics that would affect the shape of national life for some time to come. It was concentrated in a few hands, and these hands were American. It came to constitute a foreign enclave, one that would provide relatively little stimulus throughout the rest of the economy. The heavy reliance on capital and technology meant modest levels of employment for Chilean workers. The importation of equipment and

parts did not give much business to Chilean manufacturers. And most of the profits, often large, were returned to parent companies in the United States instead of being invested in Chile. It is little wonder that resentment grew.

An additional problem came from the great instability of copper prices on the world market. Indeed, copper prices could fluctuate as much as 500 or 1000 percent within a single year. This made it extremely difficult for Chile to anticipate the dollar amount of foreign exchange earnings, and this posed a serious problem for economic planning. Unpredictable gyrations in the world copper market could wreak havoc with the most carefully laid-out plans. As dependent on copper as it was, Chile could only accept the consequences.

And copper came to dominate the Chilean economy (see Figure 4-1). By 1956 copper production accounted for half of all the country's exports, and taxes on the companies' profits yielded one-fifth of the government's entire revenue. As copper went, it was often said, so went Chile's economy.

In summary, these developments formed a complex social structure. The rural sector contained a traditional landowning elite, a peasantry tied by labor obligations to the estate where they lived, and a small but mobile work force that provided wage labor for the large commercial estates. There was a mining and industrial elite, many of whose members had kinship ties to the landed aristocracy. There were middle classes as well, and a growing, native-born, urban working class. Foreign investors were conspicuous from independence onward, but by the twentieth century their presence was epitomized by the preeminence of U.S. copper companies.

Tensions would appear between these various groups, but Chile has not had to face one problem that has beset so many other countries of Latin America: excessive population growth. Indeed, Chile has consistently had one of the lowest annual rates of population growth in the hemisphere: in 1900–1910 it was just 1.2 percent, and in 1970–80 it was only 2.1 percent (compared to 2.8 percent for Latin America as a whole). Birth control and family planning have kept the country's population size within manageable limits, around 12.1 million by 1985, although there have never been enough jobs for even this limited population.

As cause or effect of this situation, women in Chile have enjoyed more opportunities than in many other countries. Females entered the work force with relative ease, and by 1970, for instance, nearly 16 percent of Chile's employed females held professional or technical jobs (higher than the U.S. rate of 14.7 percent). Social customs also reflected fairly open and egalitarian standards in the relative treatment of the sexes.

FIGURE 4-1   Chilean Copper Production, 1912–1987
(Major Companies)

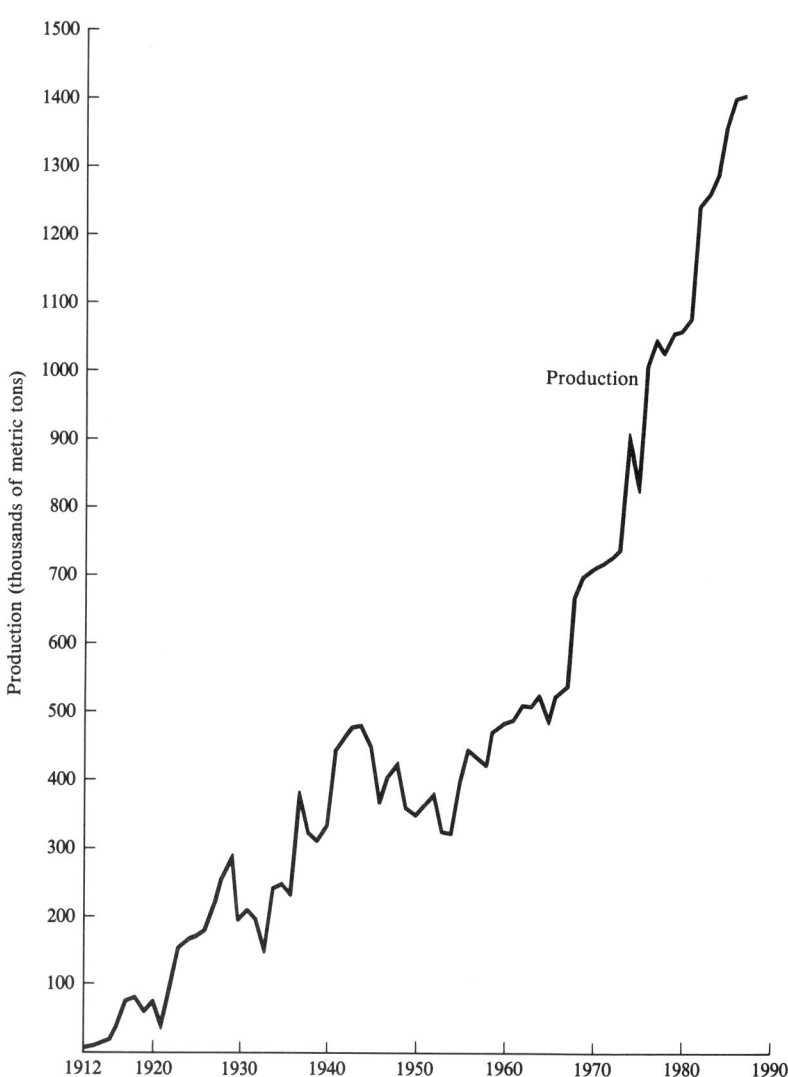

Sources: Markos Mamalakis and Clark W. Reynolds, *Essays on the Chilean Economy* (Homewood, Ill.: Richard D. Irwin, 1965), pp. 371–72; Ricardo Ffrench-Davis, "La importancia del cobre en la economía chilena," in Ffrench-Davis and Ernesto Tironi, eds., *El cobre en el desarrollo nacional* (Santiago: Universidad Católica de Chile, 1974), Cuadros 2, 7; Manual Lasaga, *The Copper Industry in the Chilean Economy: An Econometric Analysis* (Lexington, Mass.: D. C. Heath, 1981), p. 10; International Monetary Fund, *International Financial Statistics*, various years.

Politics and Parliament

As nineteenth-century Chile began to consolidate its place in the international economy, political crisis ensued. A civil war in 1859 had convinced the elite it was a time for quiet consolidation. They got it from José Joaquín Pérez, who began a ten-year presidential term.

The two most important political issues of this era were the structure of the constitution and the status of the church. On the second issue, the Liberals continued their campaign for equality of religion, while the Conservatives fought to protect the state-favored position of the Catholic Church. Slowly the Liberals won concessions, as non-Catholics received the right to have churches and religious schools. In effect, this was a modest opening of the elite, making it more pluralistic.

As for the constitution, the elite was struggling with the perennial problem of how to achieve effective government while avoiding despotism. In 1871 the constitution was amended to prohibit presidents from serving two consecutive terms. In 1874 further changes made government ministers more accountable to Congress, thereby strengthening the legislative powers. This diminution of the power of the church and president led to labeling the years between 1861 and 1891 the "Liberal Republic."

The mid-1870s brought a severe economic depression, as mining output went into decline. They also brought Chile's most famous foreign conflict: the War of the Pacific (1879–83), when Chile fought both Peru and Bolivia. The ostensible issue was the treatment of Chilean investors in the desert territories governed by Peru. After extended fighting the Chileans won an overwhelming military triumph. As victors, the Chileans took control of the mineral-rich coastal strip that had belonged to Bolivia and Peru, justifying the war as the only honorable response to the ill-treatment of their countrymen by Peru and Bolivia. This conclusion had two important effects: to increase the self-confidence of the Chileans and to arouse deep resentment among the Peruvians and Bolivians. It also led to Chile's nitrate boom.

The 1880s saw much activity on the church-state issue. The Liberal reformers made new gains. Civil registration of marriage, birth, and death was made compulsory, further eroding church control over daily life. In these same years the Congress extended the vote to all literate males over twenty-five, eliminating the previous income test.

The second half of the 1880s brought the presidency of José Balmaceda (1886–91), the most controversial leader of late nineteenth-century Chile. Although Balmaceda was a Liberal, the political lines fragmented so badly that disciplined political activity was endangered. The new president was soon engulfed in a bitter battle over food policy.

The issue arose when Chilean cattle raisers proposed a tariff on Ar-

gentine beef, which would have meant less meat and higher prices for Chileans. They were countered by the new middle-class-led Democratic Party (founded 1887), which helped mobilize Santiago's artisans, small merchants, and skilled workers against the tariff. The opposition carried the day. Balmaceda convinced the bill's proponents to withdraw it.

This early triumph of the Democratic Party signaled the start of an important trend. It was a forerunner of populist politics, i.e., a direct appeal on economic grounds to the middle and lower sectors of the cities. Seeking a broad electorate, the Democrats argued for laws that would help workers, while they also presented classic liberal demands for such items as compulsory free education and democratic procedures in electing governments. The party's articulation of mass demands showed how far Chile had already come on the road toward modern politics.

The fate of Balmaceda's presidency was sealed by the civil war of 1891. To this day Chileans argue passionately over the war, its causes, and its meaning. The origins lay in Balmaceda's efforts to push economic development faster than most of the oligarchy was ready to accept. The president wanted to increase government intervention in the economy. In order to pay for the building of new railways, roads, and urban infrastructure (water and sanitation), Balmaceda needed growing tax revenues from the nitrate industry in the northern province of Tarapacá. The obstacle was the primarily foreign ownership, especially in the railways, where Balmaceda proposed to break a foreign monopoly.

His plans met with stiff opposition in the Congress. In fact, Balmaceda's boldness masked a deeper constitutional struggle, that of congressional versus presidential authority. Congress had been fighting to establish its supremacy in the constitutional structure. But Balmaceda was determined to impose his will. The result was extreme ministerial instability. Between 1886 and 1890 Balmaceda was saddled with thirteen different cabinets, and the battle between the two branches of government led to an impasse.

In 1890 the Congress failed to produce a budget, whereupon the president ruled that the previous year's appropriations would apply. Balmaceda had earlier ventured into an area which had always proved sensitive: the choice of a presidential successor, whom he tried to nominate himself. The Congress passed a law voiding any such nomination, which Balmaceda then refused to sign.

Congressional opposition was now ready to seek a remedy by force of arms. Its leaders sought out possible military conspirators but with no success. The emerging conflict had complex implications. First, Balmaceda had alarmed conservative Chilean interests with his economic plans. He wanted a national bank, an obvious threat to interests of the established oligarchy, which dominated the private banking system.

Above all, Balmaceda was asserting the power of the presidency against a parliamentary system. Given his unorthodox economic ideas, this endangered the ruling landowner-merchant network.

The rebels against Balmaceda had the support of the navy, but not the army, which supported the beleaguered president. The mining region in the north proved a rebel stronghold, where the mine owners were happy to support a force that promised to depose the president who threatened to savage their economic interests. The owners also cut off export tax revenues which were vital to the Santiago government. A northern rebel army mobilized to sail south and depose the president.

The resulting combat at Con Con and La Placilla produced the bloodiest battles in Chilean history, with Balmaceda's forces going down in defeat. The president took refuge in the Argentine embassy, where he committed suicide one day after his presidential term ended.

In less than a month a new president was elected: Jorge Montt. But now the president's power was seriously circumscribed, as Chile now embraced the parliamentary system.

One of the key issues of the war—the status of foreign investors—had been settled. Nationalization was out, but the parliamentary victors continued pressure on the European investors. Nationalistic sentiment had penetrated all sectors of the Chilean elite.

The trauma of the 1891 civil war might at first seem a turning point in Chilean history. In fact, it only underlined the relative stability of the Chilean political system. The elite weathered the crisis of the Balmaceda challenge in a manner that promised stability at least equal to that in neighboring Argentina. Chile was poised to participate in the export boom that was carrying Latin America ever deeper into the North Atlantic economy.

Balmaceda's fall at the hands of the congressional rebels changed Chile's constitutional structure. The strong presidency gave way to a parliamentary system, but it proved impossible for any ministry to last long. Cabinets came and went at an average of four every year between 1895 and 1925. This instability was reinforced by the fragmentation of major parties, which proliferated to five by 1900.

Political control continued in the hands of an oligarchy representing primarily agricultural interests. It was dubbed *la fronda aristocrática* ("bourgeois aristocracy") in Alberto Edwards's classic book (1928) of that name. Occasionally they were challenged by urban groups, such as merchants. The workers, although not yet organized into political parties, were already beginning to make their weight felt. The issue that aroused them was rising prices. In 1905 workers staged a series of protests that led to direct confrontation with armed members of the oligarchy, and a miners' strike in 1907 in the northern city of Iquique similarly erupted into violence and bloodshed.

After 1910 workers became even more militant. The leading organizers were the anarcho-syndicalists, indefatigable activists who excelled in

organizing the many small firms. Although their long-term goal was the abolition of all government, they concentrated in the short run on winning immediate concessions for their followers. Centered in Santiago, the anarcho-syndicalist unions won significant improvements in wages and working conditions. But they were vulnerable to reprisals against their leaders, who were subject to firings or arrests or imprisonment.

These unions did not, however, represent a basic threat to the political system. In their wage demands, the workers sought to hold their own against inflation. As for working conditions and fringe benefits, the employers could undercut much of the militancy by granting social welfare benefits. The Congress did exactly this, legislating workmen's compensation in 1916, employer's liability in 1917, and a retirement system for railway workers in 1919.

After a slowdown in strike activity and a loss in bargaining power, organized labor began to revive in 1917. Economic recovery strengthened labor's hand, as World War I had greatly stimulated the demand for nitrates, a key ingredient in explosives. But inflation was again eating up wages, making workers receptive to the organizers' appeals. For the next three years unions grew steadily, despite the fact that Chilean law did not recognize labor unions.

This growth of labor strength worried the incumbent political oligarchy as well as the middle sectors. As in Argentina and Brazil, the elite presumed that discontent must be the work of foreign agitators. In 1918 the Congress passed a *ley de residencia*, much like the Argentine and Brazilian laws, designed to facilitate the deportation of aliens who were active labor organizers. But neither the politicians nor the elite had done their homework, because Chile had almost no such aliens! Since immigration from Europe had been minimal in Chile, the deportation strategy simply could not work.

The year 1919 marked the peak of labor mobilization. In January-February union leaders called huge rallies in Santiago to protest the high prices of wartime inflation. In August came another mammoth demonstration, with 100,000 participants marching past the presidential palace. The next month a general strike in Santiago failed, however, and worker morale was shaken. Thereafter the rate of strikes declined.

Surprisingly enough, the government response to strikers had been moderate since the labor upsurge began in 1917. In December an executive edict (the Yáñez decree) made the government a mediator in stalemated labor conflicts. Although rejected by anarchist and syndicalist leaders, it was heavily used, often for labor's benefit, in 1918 and 1919. That pattern continued into 1920, in part because the government was worried about the presidential elections scheduled for June.

Chile, like Argentina, had opened the door to middle-sector political participation, a process far less advanced in Brazil. The number of

working-class voters, although still small, had begun to attract the attention of Chilean bourgeois politicians, especially in Santiago. Their support could be crucial, especially when the vote was divided among many parties, as in Chile. The political leader who saw this most clearly was Arturo Alessandri, who campaigned for the presidency by making a passionate appeal to urban voters, including workers. Alessandri represented an "enlightened" middle-sector view—accepting the inevitability of working-class participation, while also hoping to channel it into controllable lines of action. He proposed to legitimize unions, but also put them into an intricate legal framework determined by the government.

Alessandri won the election, if narrowly. With this democratic exercise past, the incumbent President Sanfuentes felt free in his few remaining days to answer the labor challenge. In July 1920 workers were harassed by the "Patriotic Leagues," paramilitary street activists recruited from right-wing middle- and upper-class families. Systematic government repression followed. Virtually all anarcho-syndicalist and other leaders who did not choose exile or go underground were arrested and imprisoned. The leaderless workers were further demoralized by a wave of lockouts, as employers revoked many of the concessions made between 1917 and 1920.

There was hope that the antilabor policy would change when Alessandri assumed office, and so it did. For the first half of 1921 the Alessandri government intervened (under authority of the Yáñez decree) in a number of strikes, favoring workers in their mediation. But conflict mounted and Alessandri drew attacks from all sides—from the right for being too soft on labor and from the left for winking at aggressive employer tactics. In July 1921 Alessandri finally opted for the employers. He intervened in a bitter tram strike in Santiago, helping the company to break the strike. A wave of successful lockouts followed. By the end of 1921, the government had gone back to systematically repressing labor.

While organized labor was struggling against adverse economic and political conditions, President Alessandri pushed his proposals for a labor code and social welfare package, introduced in Congress in 1921. Conservatives balked at these ideas, since they preferred the status quo—where unions had no legal status and therefore faced continuous assaults from both employers and government. Some Conservatives also feared that Liberals might pick up new voters among urban workers. The impasse between Liberal president and Conservative Congress continued until 1924. Then the military intervened.

From Instability to Popular Front

A military junta took partial control of the government in early September 1924, and three days later issued a manifesto listing legislative

demands, with Congress dutifully approving every one. Included was a package of labor measures. Most important was an elaborate labor code that subjected unions to close government supervision.

Chile was following Bismarck's system of social welfare benefits created in the German empire of the 1880s. As in the German case, this social advance was *not* the product of a political process where workers played a direct role. Rather, it was a preemptive move by the government, under military pressure, to head off further mobilization by workers' organizations. In Chile this apparently progressive step resulted from the pressure of a government cadre that had much to fear from worker mobilization—the military officer corps.

Alessandri, in the meantime, was losing ground in his struggle with the army and took a leave to Italy. He was recalled after a second military coup in January 1925. At this point, ironically, the officers now in control felt they needed both Alessandri and urban labor support to bolster their legitimacy. The new military government intervened frequently in strikes, usually on the side of the workers. It seemed that organized labor might be on the verge of gaining power; some even thought the revolution was at hand. Fear spread among the elite, which could see its power slipping away.

The revolution was not at hand. Alessandri returned from his leave in March 1925 and soon taught organized labor another lesson about its vulnerability. In a clash with nitrate workers in June 1925, the government cracked down. For the next two years labor battled not only government hostility but also economic recession and unemployment. In January 1927 labor leadership ill-advisedly attempted a general strike. Its divisions became all too obvious and the strike failed.

Colonel Carlos Ibáñez emerged as the strongman from the political instability following Alessandri's resignation in 1925. In May 1927 he was formally elected president by the Congress, and he proceeded to consolidate a dictatorship that lasted to 1931. Chileans, having prided themselves on their relative democracy and free political competition, were shocked, The general-president jailed opponents, especially labor leaders, and suspended civil liberties.

Ibáñez had activist economic ideas to go with his authoritarian politics. The government now greatly enlarged its role in the economy. That meant speeding the construction of railways, roads, and power facilities. Not surprisingly, Ibáñez also stepped up spending for the military. Much of the financing came from abroad—loans and especially U.S. investment in mining. The world economic expansion of the 1920s made it all possible.

The Wall Street crash of 1929 abruptly ended that era, in Chile as elsewhere. Mineral exports fell disastrously and foreign financing dried up. Desperate attempts to create a national cartel for nitrate sales abroad failed. Protests against the government grew. An ever-widening spectrum of the society, now including professional persons as well as

workers, joined the attack on the dictator. Ibáñez finally gave in. In July 1931 he resigned, joining the ranks of the other South American heads of government who had the misfortune to be governing when the Great Depression hit.

For the next year Chile lacked a stable government. The interim regimes included a thirteen-day interlude of a "Socialist Republic" in which Colonel Marmaduke Grove became the best known figure. Although an ineffectual government, this regime dramatized the rise of a new institution, the Socialist Party, formally founded a year later. Another presidential election was finally held. The winner was a familiar face: Arturo Alessandri.

The once-fiery Alessandri was now more interested in order than change. His government cracked down on the opposition, especially the left. In 1936, when a wave of strikes broke out, Alessandri took tough measures. He proclaimed a state of siege, closed Congress, and sent labor leaders into exile. The army took over the railways, always a focus of labor conflict.

In economic policy the Alessandri presidency was quite successful. An ultra-orthodox finance minister, Gustavo Ross, drastically reduced public sector spending and dismantled some of the key government agencies that Ibáñez had created. Thanks to recovery of world demand for Chilean exports, especially minerals, the foreign trade balance improved dramatically. The official unemployment figure, at 262,000 in 1932, dropped to less than 16,000 by 1937. Inflation remained a problem, however, as wage increases seldom kept pace with price increases.

Chile approached the presidential election of 1939 with apprehension. In 1935 the world communist movement, dominated by the Soviet-directed Comintern, had called for a coalition strategy in battling fascism, in effect encouraging communist parties to seek alliances with parties of the left and center (as quickly happened in Spain and France). In 1936 the idea bore fruit in Chile, as the Communists and Radicals joined forces in a "Popular Front," which by 1938 came to include a broad spectrum of parties: Radicals, Socialists, Communists, Democrats, plus a new Confederation of Chilean Workers. After the Socialist Marmaduke Grove withdrew his candidacy the nomination went to Pedro Aguirre Cerda, a wealthy Radical known for his reformist ideas on the agrarian question.

The incumbent political alliance nominated Alessandri's finance minister, Gustavo Ross. It was a choice bound to anger middle-class opinion. Ross presented an inflexible and backward-looking image, despite the relative prosperity brought by his policies. The campaign was bitterly fought, and Aguirre Cerda won by the narrowest of margins—a mere 4000 out of 241,000 votes cast.

Despite its narrowness, or perhaps because of it, this election set the political context for years to come. Centrist voters had tipped the balance by opting for the left. At the same time, however, they were voting

for a reformist, so the outcome seemed ambiguous. What sort of a mandate would the resulting government have?

The Popular Front government soon suffered the strains inherent in such a heterogeneous coalition. The Radicals were the dominant element, and the least radical in ideology. They focused on economic development, not social welfare, and they were accused by some of exploiting power for the sake of old-style patronage.

The other elements in the Front were far from united themselves. The Communists and Socialists were natural antagonists, because many Socialists were ex-Communists who had refused to knuckle under to Party discipline. Both the Communists and Socialists tried to mobilize rural workers, thereby alarming the powerful landowners—and placing themselves in competition with each other.

The Congress was controlled by the right-wing opposition, but popular support for the left was on the rise. In the congressional elections of 1941 the Communists received 12 percent of the vote, up from only 4 percent in 1937. The Socialists (and kindred groups) got 20 percent. Although the rightist parties, when combined with the Radicals, won a majority, conservatives saw in the elections a growing threat from the left.

The policies of the Popular Front were anything but menacing. Economic policy concentrated on an expanded economic role for the national government. In 1939 a new state corporation was created: CORFO *(Corporación de Fomento)*, which was to stimulate economic development by strategic investments in both the public and private sector.

The Popular Front lost even its titular leadership when poor health forced President Aguirre Cerda to resign in 1941. The following president was another Radical, Juan Antonio Ríos (1942–46), who had to face the uncertainties of wartime. Ríos struggled to keep Chile neutral in the spreading world conflict. Under U.S. pressure to join the Allies he feared a reaction from the German colony in the south of Chile. He also feared possible Japanese attack on Chile's long, undefended coastline. In January 1943 Chile finally broke relations with the Axis. Chile had faced a dilemma much like that of Argentina, which delayed the break with Germany and Italy until early 1945.

The succeeding president, Gabriel González Videla (1946–52), was once again a Radical. He accepted Communist support in the campaign, and rewarded them with three seats in his cabinet. This modest throwback to the Popular Front did not last long. 1946 brought a series of violent strikes. Protest centered in the mining fields of the north, but soon spread throughout the country. A call for a general strike provoked strong police measures, and riots ensued. A full-scale social conflict loomed. The government declared a state of siege and suspended civil liberties. Strikes continued into 1947.

González had by now purged the Communists from his cabinet. The

strikes gave the right its chance to mount an offensive. The rightists had been alarmed by the steady rise in the Communist vote, which came to 18 percent in the municipal elections of 1947 (up from 12 percent in the congressional elections of 1941). The Chilean government now decided to move against the left, and in this it had plenty of support from abroad. The U.S. government was launching a major campaign in Latin America to isolate the left, especially the communist parties, and the U.S. embassy strongly encouraged the Chilean conservatives. The left fought back by attacking the González government and the United States. The climax came in 1948: by an act of Congress, where the left was greatly outnumbered, the Communist Party was outlawed and its members banned from running for office or holding official posts. A witch hunt followed. The Radicals showed their true colors. Together with the rightists, the centrist Radicals had again shown how they were prepared to use "legal" means to eliminate their most dangerous adversaries from the political game. For the left, the Popular Front became an object lesson and they vented their anger on González Videla.

The Era of Party Politics

The final demise of the Popular Front ushered in a period of intense political competition based on party organizations. During this period the Chilean political system displayed several identifying characteristics.

First, elections were extremely competitive. There were many different parties, so it was rare for any one of them to receive more than one-quarter of the total vote. This fact accounted for a second feature: in search of governing majorities, the parties had to take part in coalitions. Alliances were fragile, however, and political leaders were constantly in quest of new arrangements and intent on mending fences. Underlying this was an increased tendency toward ideological polarization. In one opinion survey 31percent of the Chilean population classified themselves as rightists, 24 percent described themselves as leftists, and the rest were centrists or undecided. Because of this fragmentation, the parties of the center could, through bargaining and skillful maneuver, have great leverage on coalition formation and electoral results.

Third, the system was highly democratic. In contrast to Argentina, where trade unions had uneasy relationships with political parties, the Chilean labor movement was closely identified with various parties, mainly on the left, so it did not form a separate power center. Measured as a percentage of registered voters, electoral participation was high (around 80 percent, compared to 50–60 percent in the United States), and registration grew rapidly in the early 1960s. And election results were accepted by almost all Chileans as binding.

The 1952 presidential election brought back another figure from the

past: General Carlos Ibáñez. Now in his midseventies, the former dictator proclaimed himself the only answer to Chile's many problems. This
caudillo put himself forward as a true nationalist, but his appeal was
really aimed at the right and center, who were once again worried
about the left. Socialists and Communists formed another electoral alliance, although the latter were hobbled by their illegality. The election
results were indicative of the path Chile was to follow for decades: a
deeply divided vote with no candidate or party getting a clear majority.
Ibáñez took office with a plurality of 47 percent.

Ibáñez had claimed to be the apolitical man able to solve all the political problems, but not surprisingly he failed to deliver on his promises.
His prime economic problem was inflation, which had hit Chile earlier
and harder than most of Latin America. Because he faced a major
deficit in the balance of payments, Ibáñez had to look abroad for help.
The logical source was the International Monetary Fund (IMF), created
to help member countries with temporary balance of payments problems. Unfortunately for the Chileans, it was not simply a matter of
arranging a foreign loan. The IMF was obligated by its charter to require evidence that any country getting help had a convincing plan for
correcting the causes of the payments deficit. As applied in the mid-
1950s, that policy meant the IMF must oversee the borrowing country's
economic policies. As a result, the IMF came to be seen by most Chileans (and by most other Latin Americans) as an extension of U.S. economic and political power.

Ibáñez was thus caught in the typical policy dilemma produced by
inflation. His government had to act because it was running out of foreign exchange and could not import much-needed goods from abroad.
Yet the sources of foreign financing offered their help only on the
condition that they have a veto over basic policymaking. Financial pressure was driving Chile to compromise its national autonomy. Ibáñez
knew that the left would be in full cry against him if he acceded to the
IMF terms. He decided to take the gamble.

His government soon paid the price. The initial measures were for
austerity. An early target was the public utilities, which invariably
charged very low rates in times of rapid inflation, since managers hesitated to pass rising costs on to customers and thus spark popular protests. An increase in bus fares, for instance, provoked a furious response. Riots began in Santiago and spread to other cities. Given the
strength of labor and the leftist parties, Chile was a difficult place for
anti-inflation policies. In the end Ibáñez was not able to live up to his
grand claims. He had proved to be a tired old general who had little
political base and even fewer political ideas.

The 1958 election produced a new president with a familiar family
name: Alessandri. He was Jorge, the son of Arturo Alessandri. Jorge,
although considering himself an independent, had run as the leader of
the right, on a combined Conservative and Liberal ticket. His oppo-

nents were Salvador Allende, a medical doctor and long-time politician who represented the Socialist-Communist alliance (FRAP), and Eduardo Frei, an ambitious young idealist who headed the Christian Democrats (PDC), a relatively new party on the national scene. Alessandri won a plurality of the vote (31.6 percent), as against 28.9 percent for Allende and 20.7 percent for Frei, with the remaining 18.8 percent split between the Radical candidate and a maverick priest. The Congress readily confirmed Alessandri's election, as required by the constitution when no candidate won an absolute majority. The election had once again shown the Chilean electorate to be deeply divided.

Alessandri was hardly the leader to bridge those divisions, even though he enjoyed personal popularity. An austere figure, he was the very opposite of the ebullient personality types who created the "populist" political style in Latin America.

The new president was an authentic representative of conservative political and economic thought in Chile. He strongly believed in free enterprise economics, including monetary orthodoxy and an open door to foreign investment. His government attacked the serious inflation with an orthodox IMF-style stabilization policy: budget cutting, devaluation (to a fixed exchange rate), and an appeal for new foreign investment.

Alessandri's stabilization efforts were undercut by a bitter battle over copper policies. The government tried to convince the U.S. copper companies to increase their investment. The idea was to get more of the processing of the copper to be done in Chile. This would increase the economic returns to Chile, as well as make Chile more self-sufficient in marketing the final product. But Chilean nationalists were incensed: they wanted to expropriate the companies, not just encourage their investment. Government policy carried the day, but copper company investments did not increase and Chile did no better at exploiting its only major asset in international trade.

Other orthodox economic policies showed some success in the short run. In 1957 and 1958 inflation had hovered between 25 and 30 percent. In 1959 it went up to 39 percent, but then came down to 12 percent in 1960 and only 8 percent in 1961. Export earnings failed to increase significantly, however, and a freeing of import controls led to huge trade deficits. A fixed exchange rate was supposed to restore confidence, but as the trade deficits mounted it only encouraged the speculators to abandon Chilean currency while there was still time.

Alessandri had also hoped his orthodox policies could make progress against the mounting social problems being created by Chile's slow and uneven economic growth. Large-scale public works projects were launched, financed mainly by foreign funds. A principal source was the United States, where worry over the Cuban threat had led to hurried formulation of the Alliance for Progress. Alessandri even dared to tackle the agrarian question, long a forbidden subject in his political

ranks. Although the law passed in 1962 was thought by all on the left to be ludicrously inadequate, in fact it did furnish the basis for an aggressive expropriation program.

Not surprisingly, none of the Alessandri policies did much toward solving the grave socioeconomic problems facing Chile. The steady exodus of the rural poor to the cities, especially Santiago, continued. There they were ill-housed, ill-fed, and ill-educated. Furthermore, there was little work. These "marginals" were the tragic underside of capitalist urbanization in a Third World country. By the 1960s about 60 percent of the Chilean population lived in urban areas.

Alessandri would have liked to govern a tranquil land. Events soon ruled out that dream. In the early 1960s the Chilean political scene began to change significantly. There was first the great growth of the electorate. Just over 500,000 in 1938, by 1963 it had reached 2,500,000—a fivefold expansion in twenty-five years. Second, a realignment of political forces had occurred. There were now four main groupings: (1) the right, including the Conservative and Liberal parties; (2) the centrist Radicals, long the masters of opportunism; (3) the Marxist left, comprised primarily of the Communists and Socialists; and (4) the Christian Democrats, located in the center, a reform-oriented party now building its electoral following. In the 1963 municipal elections each of these four won roughly equal percentages of the vote. The biggest net gainers were the Christian Democrats, who were attracting votes from both left and right.

As the presidential election of 1964 neared, the polarization sharpened. A widely discussed indicator was a March 1964 special election to fill a congressional seat in the province of Curicó. Although previously a Conservative Party stronghold, its overwhelmingly rural voters gave the FRAP candidate 39 percent, while the center-right candidate got only 33 percent and the Christian Democrats drew 28 percent. Assuming these results were typical for the nation as a whole, Liberals and Conservatives quickly decided their only salvation lay in an alliance with the Christian Democrats. They dissolved their Democratic Front and began courting the PDC. The political isolation of the Chilean countryside appeared to have ended. No longer could the landowners and their agents take the votes of the rural poor for granted.

The 1964 presidential election loomed as a crucial one for both Chile and Latin America. The left once again ran Salvador Allende. FRAP's strident criticisms of capitalism seemed all the more relevant, now that a classically conservative government had so recently failed.

The 1964 election was to be very different from 1958. A relatively new party, the Christian Democrats, had burst onto the scene. When the rightist parties decided to endorse the PDC candidate, Eduardo Frei, the Christian Democrats gained an enormous boost. It was a pragmatic decision, made out of the fear that the FRAP might win a plurality victory, as almost happened in 1958. The rightists decided this de-

FIGURE 4-2   A Genealogy of Major Political Parties in Chile

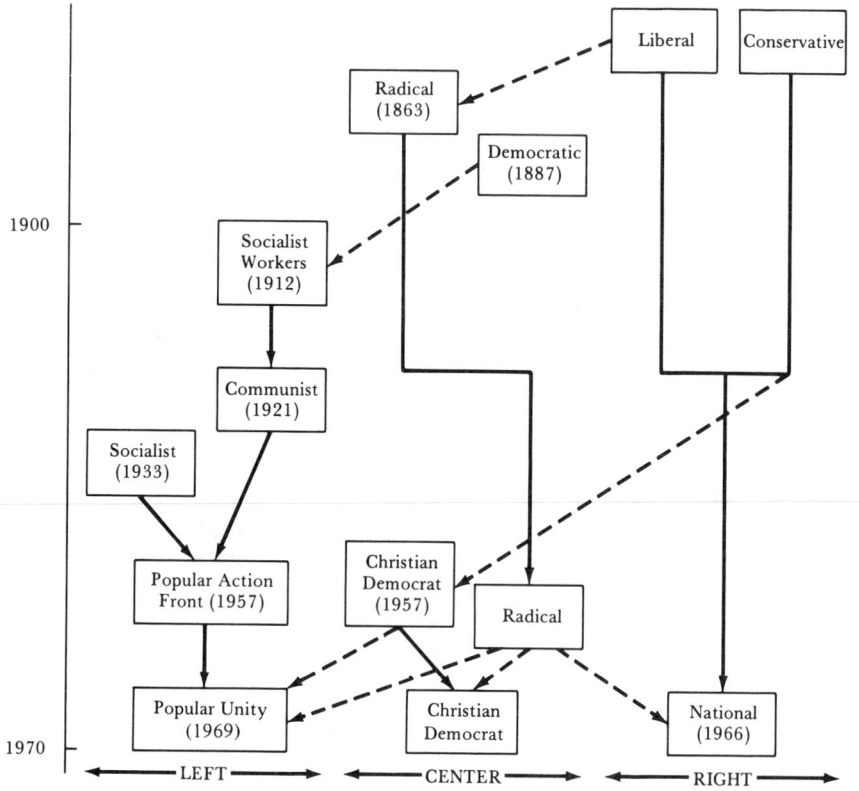

Note: Broken arrows indicate splinter groups or temporary coalitions.

spite their misgivings about the PDC's reformist ideas, which many conservatives saw as dangerously close to the formulas of the left.

The campaign was hard-fought and aroused strong interest throughout the Americas. The FRAP called for a clear-cut repudiation of capitalism and imperialism. Chilean landlords and U.S. copper companies were depicted as the arch-villains. Allende called for nothing short of a sweeping transformation of Chilean society, in order to lead Chile to socialism.

The PDC campaign was a high-powered operation designed to convince the electorate that Frei could bring significant change without violating Chile's traditional freedoms. It was to be a "Revolution in Liberty." In fact, however, the Christian Democrats were promising reforms, not revolution. The reforms added up to a more efficient capitalist economy, to be achieved by limited government intervention to carry out such measures as agrarian reform (through the expropriation of underused land), increased public housing, and greater control over

the U.S. copper companies (through Chilean acquisition of part ownership).

Frei and the PDC wasted no time in branding FRAP as an extension of Moscow. Cleverly written campaign cartoons and radio jingles played on fears of "another Cuba" in Chile, fears known to be highly exploitable. The U.S. government, as well as West European Christian Democrats, also took a strong interest in this contest between reformism and Marxism. The Central Intelligence Agency would later admit to contributing more than 50 percent of Frei's campaign expenses. In this and other ways, probably including money and support from U.S. business firms, the Chilean voter felt the effects of strong North American interest.

It may have been a case of overkill. Frei won the election more handily than anyone had expected, with 56 percent of the vote. Allende got 39 percent, well over his share in 1958. The difference, of course, was that this time it was a two-way race. A third candidate, Julio Durán of the Radicals, was deserted by many from his own party and finished with only 5 percent. The triumph belonged to Frei, but the Revolution in Liberty owed its birth far more to the absence of a rightist candidate than to any sudden change of mind by Chilean voters.

Frei's government began with high expectations. The left had been decisively defeated. The voters had given a mandate for change. Now the Christian Democrats would have to move quickly and decisively.

Their plate was certainly full. First priority was given to economic policy. One of the hottest issues was copper: how to improve Chile's share in the exploitation of its most valuable export. Here, as elsewhere, the Frei strategists sought a middle way. Outright nationalization (with compensation) would be too expensive, they reasoned, since Chile would have to come up with enormous dollar payments. Simply encouraging the U.S. companies to increase their investment under the old terms was equally unacceptable, since it would be a backward step on the path toward greater national control. Their solution was centrist: the Chilean government would buy into part ownership of the companies, with the proceeds to be reinvested by the companies in expanded facilities, especially for processing. The goal was to double copper production by 1970. If successful, the plan would increase both national control and export earnings.

The proposal was savagely attacked by the left, which branded it a sellout. The Christian Democrats decided to make it a major issue in the congressional elections of March 1965, where they won an overwhelming victory. They promptly used their new congressmen to vote through the copper plan ("Chileanization") in November 1965. The opposition on the left and from many mine workers remained intransigent. Frei got his agreements with Anaconda (which became 25 percent government owned) and Kennecott (now 51 percent government owned), the two leading companies, but over the next five years, cop-

per production increased by only 10 percent. Export earnings doubled, but that was because of a rise in the world price of copper, not because of output. Furthermore, a big share of those higher earnings went to the companies, due to the terms of the new contracts. The company lawyers had thought ahead more skillfully than Frei's technocrats, a point made much of by the left.

The agrarian sector was another key policy area. Chile had long suffered under one of Latin America's most archaic rural structures, with the marginalized rural masses daily becoming more desperate. The Christian Democrats pushed through a land reform act in 1967 which was another compromise. Elaborate provisions were made for identifying land to be expropriated, compensation to be given, and land distributed to 100,000 peasants by 1970. Frei's planners hoped that cooperatives—always a mainstay in Christian Democratic thinking—could furnish the facilities necessary to get new farms running profitably. The program went more slowly than hoped, and by the end of Frei's term there were only 28,000 new farm ownerships, a number whose significance was obscured by the high expectations the Christian Democrats had aroused.

The United States continued to take strong interest in the fortunes of the Frei government. It had all the marks of the kind of reformist regime that the Alliance for Progress was designed to support. The United States, as well as the multilateral agencies such as the Inter-American Development Bank and the World Bank, gave Chile extensive financing. In the short run this helped Chile's annual balance of payments. In the long run it added to Chile's foreign debt.

In the political sphere the Christian Democrats attempted to deliver on their promise of a new kind of popular participation. Rejecting the massive state role that the solutions of the left would inevitably bring, they pushed for *promoción popular,* which was supposed to be a new kind of grass-roots political activity. In practice it meant a mixture of communitarianism, self-help, and cooperatives. Above all, it meant heading off the left, which had through its unions and party structures (both communist and socialist) gotten a head start in organizing at the mass level. The net effect was a dogfight to win elections throughout the society: in unions, student associations, cooperatives, bar associations, and every kind of professional group. Politics was penetrating more and more deeply into Chilean society.

The 1965 congressional election victory proved to be the high point of the Christian Democrats' political fortunes. In the 1967 and 1969 municipal elections they lost ground. In 1969 they lost their majority in the Chamber of Deputies. Efforts at reformist socioeconomic change proved difficult, and the odds on their Revolution in Liberty lengthened.

The left, so close to victory in the 1958 presidential election, was struggling to create another coalition for the 1970 campaign. The

right, which had backed Frei out of sheer opportunism, now indulged its longstanding preferences. Conservative voices suggested that they could win with their own candidate, especially if he were the ever-popular Jorge Alessandri.

Time had run out on the Revolution in Liberty. The reformist gains had been substantial, if measured by past Chilean standards, but they were no longer enough. Because the constitution forbade consecutive reelection of the president, the Christian Democrats had to find a new candidate. Frei had been an overshadowing figure, but not without detractors within the party. In fact, the Christian Democratic left wing had veered sharply in the direction of radical change. As the presidential election of 1970 approached, it looked as if the electorate would return to the sharply fragmented voting pattern common before 1964.

The 1970 presidential election in some ways resembled 1964. But this time the right decided to run its own candidate, Jorge Alessandri, the magic name in twentieth-century Chilean politics. The divided Christian Democrats nominated Radomiro Tomic, whose leftist stand precluded any possible electoral alliance with the right. The Communists and Socialists, now united under the rubric of *Unidad Popular* (UP or Popular Unity), once again chose Allende. The UP relentlessly attacked the Christian Democrats' record under Frei, charging a sellout to imperialism and domestic oligarchs. Alessandri offered an old-fashioned conservative recipe, almost oblivious to the bitter ideological controversy swirling about him. Tomic sounded remarkably like Allende. He favored radical change, including complete nationalization of the copper companies.

When the votes were counted, the tally looked a bit like 1958. Allende had won a plurality, but it was far from a decisive result. His vote total was 36.3 percent. Alessandri, whom the U.S. ambassador had confidently predicted would win, got 34.9 percent, and Tomic only 27.8 percent. The left was jubilant, but their more sober leaders were aware of the fragility of Allende's mandate.

Allende's first problem was the requisite confirmation from the Congress. The right saw Allende's impending presidency as a grave threat, and some militant members, especially within the army, began conspiring to block him. One military plot, initially supported by the U.S. government, aborted when General René Schneider, the army commander-in-chief, was murdered in a bungled kidnap attempt. Allende was eventually confirmed by Congress three days later, and Chilean democracy seemed to have survived its first challenge after the election of a Marxist.

Socialism via Democracy?

Allende's three-year presidency was rich in significance, for both the history of Chile and that of Latin America, although the nature of that

significance is still bitterly argued. The president and his advisers decided that despite the narrowness of their election victory they would seek radical change—but by legal means. Was such a course of action possible? How, asked the more radical UP militants, could socialism be introduced in a genteel fashion?

Allende's initial economic strategy was similar to that of Perón in 1946 and Castro in 1959: freeze prices and raise wages. The result was an immediate boom in consumer buying. This caused significant short-term redistribution of income. Merchants' inventories were quickly depleted, while producers put a hold on all production orders until they could see how long price controls would prevent them from recovering the inflationary costs they faced in producing new goods for the retail market. Here Allende had pursued an essentially populist strategy in order to expand his political support.

Allende's other economic policies flowed from campaign promises. Top priority went to the complete nationalization of the copper companies. Most significant, the congressional vote on this issue was unanimous. That spoke volumes about the growth of nationalist sentiment in Chile and the perceived failure of Frei's Chileanization policy. Later, the Allende government argued that no compensation was due the companies because of what the president charged were their previous illegally high profits. That aggressive stand provoked Christian Democratic opposition and furnished U.S. government hard-liners with "proof" that Chile had declared war on private property in the hemisphere.

The UP government also extended state control into many other economic sectors. Coal and steel were nationalized, along with 60 percent of the private banks. As the "transition to socialism" continued, more and more firms and businesses were nationalized, with Allende's hand often forced by workers who occupied management offices and refused to leave until expropriation was announced. Foreign firms were a favorite target, with such well-known names as ITT and Ford falling victim.

This attack on foreign capital was bound to aggravate tensions with the United States. The Chilean government could never hope to come up with the foreign exchange needed to compensate the owners. The refusal (and/or inability) to compensate gave the Nixon administration legal grounds to mount an "invisible blockade" against Chile in the international economy, although the United States took steps before the nationalization even began. This included a hold (with minor exceptions) on any loans from the World Bank or the Inter-American Development Bank, not to speak of the U.S. Export-Import Bank. Private foreign investment also came to a halt, so Allende faced a severe shortage of foreign financing. In his last year (1972–73) West European and socialist bloc countries began opening lines of credit to Chile, but their effect had yet to be felt.

In the rural sector the Allende government moved with speed. The

expropriations came faster than the government's ability to ensure the services (credit, access to supplies, equipment) needed by the new small owners or by the state-controlled cooperatives. Furthermore, the government was increasingly outflanked by peasants, often organized by leftist radicals, who seized land on their own. By 1973 the Agrarian Reform Institute had lost control of the situation in many areas. Landowners hired armed guards, tried to fight back legally, or simply fled the country. The longstanding agrarian problem was being met by radical means, as the state rapidly displaced the *hacendados.*

In its overall management of the economy the Allende government took an early chance. Facing a majority opposition in the Congress, Allende's political strategists decided to go for a constitutional amendment that would create an assembly of the people in place of the Congress. The populist policies of 1970–71 (freezing prices, raising wages) were intended partly to build support for the management. It was a risk because the populist measures were bound to be inflationary. A great deal was riding on the gamble to win increased constitutional power, because the Congress could block so much of the program Allende sought to carry out.

Not surprisingly, the Congress rejected the amendment in 1972. At that point Allende and his advisers decided to pause in order to consolidate their political gains. They planned eventually to submit the amendment to a popular plebiscite—thereby bypassing the opposition-dominated Congress. The proper moment never came, however; at least they could never identify it. As 1972 continued, the government was preoccupied by the enormous dislocations hitting the economy. First, there were the distortions resulting from the attempt to enforce price controls. More and more products disappeared from the legal markets and made their way into black markets. Because it was a legal government in a system of limited constitutional powers, the Allende regime was ineffectual in combating the burgeoning underground economy. Second, there was extensive sabotage or deliberate diversion by producers, landowners, and merchants who wanted either the UP experiment to fail or wanted to make quick profits, or both. Finally, there was the inefficiency of an inexperienced government trying to take over and run huge sectors of the economy. New and often untrained bureaucrats, frequently appointed for political rather than technical qualifications, could hardly master tasks that still bedevil their counterparts in tightly controlled socialist societies.

The result was that by early 1973 Chile was in the grip of runaway inflation. The annual rate was exceeding 150 percent (it would climb even higher than that) and the government looked helpless. Monetary authorities were simply pumping out money to cover huge budget deficits. An overvalued exchange rate was encouraging imports, while low world copper prices depressed export earnings. The balance-of-payments deficits in 1971 and 1972 could be covered out of the foreign

exchange reserves left by the Frei government, but that recourse was gone by 1973. The new credits from the socialist bloc and West Europe were yet to materialize, so the foreign accounts had become a major economic worry.

The widespread nationalizations had demoralized the private sector, while disorganization prevented rapid investment and expansion in the hugely augmented public sector. The economy was in a shambles.

But how could the transition to a socialist economy be smooth? In Cuba (see Chapter Eight) there certainly was dislocation in the early years, and Chile faced much greater obstacles. Allende did not have the power Fidel enjoyed in Cuba. Chile was still a pluralist democracy. The opposition still controlled the Congress. The economy was still open to international blackmail.

Added to these inherent difficulties was the intransigence of the opposition. At no time, one should remember, did the Allende government get over half the vote. He became president in 1970 with a smaller percentage of the vote than he had received when losing in 1964 (36.3 percent, compared with 38.9 percent). In the local elections of April 1971 the UP parties received 49.7 percent, their high point. Subsequent months saw furious battles in every political arena—elections in unions, students' groups, and professional associations.

The UP itself was weakened by splits within its own ranks. The far left, led by the MIR (the *Movimiento de la Izquierda Revolucionaria*, or Movement of the Revolutionary Left), pressed for more radical action. They wanted faster nationalizations, tough police action against the opposition, and rule by decree. The moderates within the UP, including the Communists, urged caution, arguing that precipitous action would play into the hands of the rightists who could manipulate the military and the middle class.

By mid-1972 the political climate had become superheated. Massive street mobilizations by both pro- and anti-Allende forces became routine. In August shopkeepers staged a one-day boycott to protest government economic policies. In October a series of protests began to sweep the country. They began with the truckowners' suspension of shipments. Small businessmen, private farmers, and pilots joined in strikes. Although often orchestrated by opposition politicians, these eruptions showed that widespread sectors of the Chilean public would confront the government in the streets. Virtually all of them stood to lose if a socialist society were achieved; they were determined not to go down without a fight.

The government had its own mass support. Especially in the major cities, UP could on command turn out several hundred thousand disciplined marchers. Their ranks included the many Chileans who had already begun to experience significant changes—higher real wages, subsidized fresh milk, a role in administering their community or workplace. They also responded to the new nationalism—the takeover

of the copper companies, the tough line toward all foreign firms, the highly publicized welcome to Fidel Castro when he came to Chile in 1971.

March 1973 brought another political test. At stake was the composition of the Congress. It was a crucial moment, because the opposition hoped to gain a two-thirds majority and thereby be able to impeach Allende, or at least hold commanding legal leverage over him. The elections brought a huge turnout. When the votes were counted, the government had done better than even it had dared to predict: UP got 43 percent, reducing the opposition's majority from 32 to 30 (out of 50) in the Senate and from 93 to 87 (out of 150) in the lower house. UP leaders jubilantly pointed to the increase in the leftist vote over 1970, noting that no previous Chilean president had ever been able to increase his support in a midterm congressional election. The opposition pointed to its 55.7 percent vote as equivalent to Frei's landslide victory in 1964; they also reminded UP that its 43 percent was down from the 49.7 percent it had won in the 1971 municipal elections.

The election returns could be used to buttress almost any political position. One thing was certain: the opposition had not gotten the big electoral boost it had sought. And since it lacked a two-thirds congressional majority, any attempt at impeachment was foreclosed. Allende may have lacked an absolute majority, but he had rock-hard support among the workers and an increasing number of rural laborers.

There had never been a shortage of plots to overthrow the elected Marxist government. The rightist *Patria y Libertad* (Fatherland and Liberty) had already been engaged in terrorist attacks against government officials and vital economic installations. By 1973, however, more and more of the middle class had come to think there was no democratic solution to the crisis. They saw Allende (or the more radical left, which he couldn't or wouldn't control) as threatening the very basis of private property in Chile.

In April a strike of copper workers began, giving the opposition ideal grounds to claim multiclass resistance to Allende. In July the truck-owners struck again, triggering a wave of strikes by middle-class associations such as lawyers, physicians, and architects. The pro-Allende mass labor organizations staged huge counterprotests, with marches including hundreds of thousands. Chile was in the grip of a feverish political battle. Terrorist incidents became frequent. Few thought it could remain peaceful until 1976, when the next president would be elected.

Allende knew this. He had long since rejected his far left's advice to resort to extralegal means, and he knew the Christian Democrats were the only political force that was strong enough and possibly willing to reach an agreement to reduce levels of conflict and thereby keep the country's democratic system intact. Allende negotiated with Frei and his fellow leaders, but after extended deliberation they refused. They did not want to be drawn into joint responsibility for a collapsing gov-

ernment unless Allende promised them more than he was prepared to give; equally important, they suspected they had much to gain from further discrediting of UP. They may even have suspected that a military coup would restore them to power.

Allende believed he had no other choice than to increase military participation in his government. Although it might give short-term stability, since the armed forces were obliged to carry out commands and maintain order, it might also open the way to military intrigue and to opposition charges of politicizing the armed forces. Allende sensed the danger and in August attempted to shuffle his army commanders. But it was too late.

By early September the military conspiracy to depose the UP government was in high gear. Santiago seethed with rumors about Brazilian money and "destabilization" experts pouring into Chile. Strikes and counterdemonstrations had further slowed an economy already hit by hyperinflation and capital flight. Allende now knew that the fate of Chile's socialist experiment was in the hands of the military. General Carlos Prats, army commander-in-chief and minister of defense, was a key figure. Unfortunately for Allende and the UP, Prats' military prestige was slipping badly. Late in August he was submitted to the indignity (from the military viewpoint) of a noisy demonstration in front of his residence by army officers' wives demanding he resign. Outvoted at a generals' council the next day, he resigned both his army and cabinet posts. His successor as army chief was General Augusto Pinochet, widely believed to be as constitutionalist as Prats.

The military now lost little time. On September 11, 1973 a well-coordinated coup began. Early that morning the Special Services *carabineros,* long thought to best embody Chile's tradition of nonpolitical police, were still guarding the presidential palace against possible attack. Ominously, they pulled out when informed that their commander had joined the unfolding coup. At 6 A.M. Allende was told that the navy had seized Valparaíso, and he decided to go immediately from his residence to La Moneda, the landmark presidential palace in the heart of Santiago.

The rest of the morning saw frenzied activity at the palace, as the defenses were prepared. Allende began receiving offers of safe exit to exile. To one air force general's offer, he reportedly replied, "Tell General von Schouwen that the president of Chile does not flee in a plane. As he knows how a soldier should act, I will know how to fulfill my duty as president of the republic."

Allende had not lived the life of a revolutionary. He had spent three decades as a parliamentary politician, ceaselessly negotiating to create and maintain coalitions. Not a few cynics (on both the left and right) had suggested he was too fond of the good life to make a revolutionary. But Allende now chose to stand and fight. Just before noon air

The presidential palace in central Santiago burst into flames under rocket attacks by the Chilean armed forces during the coup of September 11, 1973. (United Press International.)

force Hawker Hunters attacked the palace with rocket fire, spreading flames through the building that had seen so many peaceful transfers of power. As army troops prepared to storm the palace, Allende committed suicide. In an early presidential speech Allende had noted, "Our coat of arms says 'By reason or force,' but it puts reason first." That order was now reversed.

The military commanders expected resistance, especially from the industrial worker housing areas, but government supporters had few arms. Resistance was scattered but the repression was rapid and brutal. We shall never know how many died—at least 2000. It was the most violent military coup in twentieth-century South American history and it had happened in a country that prided itself on its deeply grounded democratic traditions. The "transition to socialism" that so many on the left thought to be "irreversible" was about to be reversed.

In retrospect, it appears that Allende's downfall resulted largely from the interplay between social classes and political parties that characterized contemporary Chile. The left drew its support mainly from the urban working class. It met opposition from a cohesive upper class whose agrarian and industrial components were united by family ties

and objective interest, and this unified elite was able to gain the allegiance of middle-sector groups (and some traditional peasants) and, most important of all, militant lower-middle-class activists, such as storekeepers and truckers, in a common stand against a socialist order. Between 1970 and 1973 the worker-based Allende movement was unable to form an enduring coalition with the other strata in Chilean society. That explains its inability to win a clear majority at the polls, hence its ultimate vulnerability.

This is not to discount the effects of opposition from the United States, which worked steadily at "destabilizing" (that is, overthrowing) the Allende regime by passing dollars (which were highly valuable on the black market) to conservative groups and subsidizing anti-Allende strikes. But the U.S. intervention was not the deciding factor in the government's downfall, since the Allende administration had a mountain of troubles of its own. Nonetheless, the United States once again placed itself squarely on the side of counterrevolutionaries.

The Pinochet Regime

The new military government promptly set out to impose a bureaucratic-authoritarian regime on Chile. Proclaiming its goal as "national reconstruction," the junta set about to destroy—not merely reform— the country's political system. Congress was dissolved, the constitution suspended, and parties declared illegal or placed "in recess": there was to be no more political bickering, no more *politiquería*. The junta further imposed a state of siege, called a 9 o'clock curfew, and set strict limits on the media. In October 1973, one month after the coup, the military took over the proud universities as well.

The armed forces sought to revamp longstanding relationships between state and society in modern Chile. One critical component of this plan was the unity of the military, led by army general Augusto Pinochet. Another was the disarray of civilian society, which made it possible for the regime to dismantle (or at least repress) such intermediate institutions as political parties and labor unions and to establish direct authority. Political activity in its traditional sense came to a halt. In January 1974 General Pinochet announced that the military would remain in power for no less than five years.

As the generals consolidated their political power, a group of civilian technocrats introduced far-reaching changes in economic policy. Known as "the Chicago boys" because so many had been trained at the University of Chicago, these economists believed strongly in the efficiency and fairness of market competition. What had restricted Chile's growth, they reasoned, was government intervention in the economy— which reduced competition, artificially increased wages, and led to inflation. To put the laws of supply and demand back to work, they set

out to reduce the role of the state and also to cut back inflation. The ultimate goal, Pinochet once said, was "to make Chile not a nation of proletarians, but a nation of entrepreneurs."

The regime's programs had a clear effect on inflation, which was running at an annual rate of around 500 percent at the time of the coup. By 1976 it was down to 180 percent, by 1978 it was around 30–35 percent, and by 1982 it was down to 10 percent. From 1983 to 1987 inflation fluctuated between 20 and 31 percent. This was a far better performance than that of Argentina, Brazil, or Mexico, and here the junta could justifiably claim success. They could make a similar claim for export diversification (copper now accounting for less than half export value) and for growth, which averaged over 7 percent from 1976 through 1981. But it was achieved at the cost of lower real wages and declining social services.

The Chicago-trained technocrats' goal was to open Chile to the world economy, drastically reducing tariff protection, government subsidies, and the size of the public sector. In late 1973 the state owned nearly 500 firms. The junta returned about half to their original owners and opened bids for many of the rest. Lack of true competition made for low sale prices, benefiting local business conglomerates and multinational corporations such as ITT.

Economic policymakers also reduced barriers to imports on the ground that quotas and tariffs protected inefficient industries and kept prices artificially high. The result was that many local firms lost out to multinational corporations. The Chilean business community, which strongly supported the coup in 1973, was badly affected. Emphasis was put on export promotion and attraction of foreign loans, both public and private. Ironically, Chile was attempting to create a free-market economy with assistance largely drawn from international organizations and other governments, not private banks and companies.

The financial crash of 1982, triggered by Mexico's de facto default on its foreign debt and the contraction of the world economy caused by the U.S. recession, hit Chile even harder than the rest of Latin America. The gross domestic product plunged 14 percent that year, as unemployment (including those on government make-work programs) rose to include a third of the labor force by mid-1983. Pinochet installed a new set of conservative technocrats, who launched an even more radical economic restructuring. They stimulated investment, greatly increased exports, and sharply reduced unemployment. They also significantly lowered the external debt through innovative schemes such as a partial conversion into investment capital at home. But wages remained chronically low and the systematic privatization of social services left many poor Chileans without the essentials of life.

On the political front the Pinochet regime never hesitated to use repression, especially at any hint of labor unrest or popular protest. Its

brutal tactics earned widespread condemnation, as critics denounced repeated and persistent violations of human rights. In September 1976 a car bomb in Washington, D.C., killed Orlando Letelier, a former Allende ambassador to the United States and at the time of the bombing an effective lobbyist against U.S. government aid to the Pinochet government. The assassins' link with Chilean intelligence was clear, but Chile contemptuously rebuffed the Carter government's attempt to extradite the accused members of the Chilean military. The election of Ronald Reagan came as a great relief to the Pinochet government, which soon found Washington seeking closer relations. Starting in 1985 even the Reagan government pressured Pinochet to liberalize, but to no avail, at least for the moment.

Through clever political maneuvering Pinochet achieved supreme authority, and what had at first been a thoroughly institutionalized military regime became highly personalized—as Pinochet alone commanded power. A plebiscite in 1978 produced the widespread appearance of support for Pinochet's "defense of the dignity of Chile." Another plebiscite in 1980 approved a constitution that confirmed Pinochet's hold on his office until 1990.

The non-Marxist opposition politicians, who had once benefited from South America's most stable democracy, were deeply divided. Their attempts to include the Communist and Socialist Parties (or at least not repudiate them) in a united opposition played into Pinochet's hands as he exploited middle- and lower-middle-class fears of a return to the chaos of 1973.

The Communist Party organized an armed wing which attempted to assassinate Pinochet in 1986. The president barely escaped. Subsequent government raids uncovered huge caches of weapons, while the police and military blitzed working-class barrios, arresting opposition activists of every hue. The assassination attempt dramatized Pinochet's claim that Chile must choose between him and the revolutionary left.

But 1988 brought a dramatic gamble which the dictatorship lost. Pinochet, reacting to international pressure to liberalize and feeling confident in a recovering economy, risked another plebiscite on his one-man rule. The opposition now united to mount a highly effective television campaign (aided briefly by U.S. media consultants), for the "no," which triumphed by a decisive 55 to 43 percent. The die was now cast for a return to elected government.

After a tense interval, Pinochet accepted the result, knowing that the constitution ensured his continuation as army commander-in-chief until 1998. The next step was the 1989 presidential election, won by the longtime Christian Democratic leader (and implacable Allende foe) Patricio Aylwin, supported by a coalition of seventeen center and center-left parties. The extreme left failed to win a single congressional seat, as the once-powerful Communist Party dissolved into a bitter fight between reformists and hard-line Marxist-Leninists.

Redemocratization

Aylwin assumed power in 1990, committed to the restoration of Chile's democratic institutions, investigation of past human rights abuses, and rapid improvement in the living conditions of the poor. His heavily technocratic cabinet (conspicuously not apportioned by party strength) was also committed to maintaining the essentials of Latin America's leading economic success story (at least by orthodox standards): relative price stability, booming exports (buoyed by high copper prices), record foreign investment, impressive foreign debt reduction, and significant progress in privatizing much of a once highly inefficient public sector.

Chile's newly restored democracy also faced formidable obstacles: an ever-alert army still headed by an unrepentant Pinochet (although he was now tarnished by family financial scandals), a pro-military judiciary, a rightist-dominated Senate, sporadic terrorism from left and right, and the explosive issue of what to do about past human rights abuses—with its potential to ignite civilian-military conflict.

The governing coalition (the *"Concertación"*) held together for another presidential election in 1993. Once again the Christian Democrats furnished the president. He was Eduardo Frei, the son of Chile's president from 1964 to 1970. His winning margin was 58 percent, 3 points higher than that of Aylwin in 1989. The campaign was notable both for its lack of passion and for its high participation. The uncharismatic Frei, whose name was his greatest asset, promised "growth with equity." The once powerful Communist Party continued to be thoroughly marginalized, while most of the left remained loyal to the coalition. Most important, there was general acceptance of the rules of the democratic game, although the latter suffered from some constraints left over from the Pinochet era.

Chile's notable accomplishment continued to be its noninflationary rapid growth. Between 1986 and 1993 growth had averaged 6.3 percent, by far the highest in Latin America. The foreign debt was significantly reduced and new foreign capital was readily attracted. Privatization had gone virtually to the maximum, as Chilean investors now turned to Argentina for new privatized utilities to buy. Most impressive were the high savings and high investment rates. Chile now resembled the East Asian "tigers" in its ability to sacrifice consumption for future productivity. The Chilean government confidently predicted it would be the next new member of the North American Free Trade Area (NAFTA) that encompassed Mexico, Canada, and the United States.

How well were the benefits of this growth distributed? The income distribution data showed that Chile still differed little from the extreme inequality typical of Latin America. Real wage rates, although rising, had not attained their 1970 level by 1993, and many of the new jobs were part-time, or low-paying, or both. On the other hand, the standards in health care and primary education were acknowledged to be

high. Obviously Chile would continue to be closely watched as an important test of capitalist development in Latin America. In the words of one economist and veteran Chile watcher: "The Chilean economy is a tiger . . . however, it is a young, inexperienced, and timid tiger, still trying to find her feet. Much remains to be done."

FIVE

BRAZIL
Development for Whom?

With an area of more than 3 million square miles. Brazil occupies nearly half of South America. The 3900-mile Amazon River reaches from the steamy jungles of the interior to the Atlantic coast. Land ranges from the semiarid northeast, plagued by recurrent droughts, to the rich forests and fertile plateaus of the center and the south. The country abounds in natural resources, including iron and other industrial minerals, and it has long been regarded as a potential world power—the "sleeping giant" of Latin America. Perhaps because of this anticipation, and perhaps for other reasons, Brazilians tend to have an optimistic, ebullient outlook on life. One saying sums it up: "God is a Brazilian."

Brazil's relatively nonviolent acquisition of independence from Portugal in 1822 left the country with an auspicious start. The lack of large-scale conflict meant that physical and economic destruction was minimal, especially in comparison to the devastation wrought in the Río de la Plata region, in Venezuela, and in central Mexico. Nor did Brazil have to cope with the problems of demobilizing a massive military apparatus in the postwar period. And most important, the transition of the Portuguese monarchy to Brazil provided a coherent political structure endowed with the authority of time-tested tradition. There were struggles, to be sure, and Brazil would meet a crisis of political legitimacy before the end of the century. But it did not face the same kind of instability that other Latin Americans faced at the outset of independence.

The economy was mainly agricultural, and sugar was by far the largest commercial crop. By 1822 the population included about 4 million inhabitants, roughly half of whom were slaves of African birth or descent. The social order consisted principally of two tiers, the landowning aristocrats and the labor-producing slaves, a dichotomy that would come to be aptly and sympathetically described by Gilberto Freyre in his classic book, *The Masters and the Slaves*. There were some merchants and lawyers and other professionals, mainly in the cities and

especially in Rio de Janeiro, but society was dominated by the forces of the countryside.

Dom Pedro I (1822–1831)

Too often we define historical eras by the rise and fall of heads of state or government. Yet at times it makes sense. Few would contest the label "Victorian" for the era of Queen Victoria's reign from 1837 to 1901. And in nineteenth-century Brazil, many basic social issues were bound up with the fate of the crown. Most obvious was the consolidation of Brazil's independence. Related issues involved the centralization or decentralization of authority and executive versus legislative power. These questions had to be faced immediately after independence because both the elite and the emperor wanted to write a Brazilian constitution.

Dom Pedro I had become the first emperor of a newly independent Brazil in 1822, when the Brazilian aristocracy forced a break with Portugal. A year earlier Pedro's father, Dom João VI, had left Brazil to resume the throne in Portugal, but only after advising his son to remain in Brazil (to which the royal family had become very attached), even if it meant creating a separate monarchy. Dom Pedro I called for a Constituent Assembly and the resulting elections in 1823 revealed several political divisions. Most basic was the split between the Brazilian Party and the Portuguese Party, the latter consisting of those who had opposed Brazilian independence and wanted to resubordinate Rio de Janeiro to Lisbon. Its leaders were primarily Portuguese-born, mostly military officers, bureaucrats, and merchants. The Brazilian Party was led by José Bonifácio Andrada e Silva, a São Paulo landowner who was the leading spokesman for Brazilian liberalism and the leading minister of Dom Pedro's government.

Despite majority support in the Assembly, José Bonifácio's cabinet had to resign after three months because the emperor continually endorsed the Portuguese Party's protest over the government's anti-Portuguese measures. Heated polemics continued and street fights broke out, as an extremist faction of the Brazilian Party called for decentralized rule and piled abuse on the crown. Amid the furious debate the emperor simply dissolved the Assembly in November 1824. Shortly thereafter he unilaterally decreed a constitution for Brazil. It included many features from a draft prepared by Antonio Carlos Andrada e Silva, José Bonifácio's brother, but reserved greater powers for the *Poder Moderador* (the "Moderating Power"), which was to be the monarch himself. Most important was the power to dissolve the Chamber of Deputies and to appoint and dismiss ministers. Citizen voting was tied to a high minimum-property test, thereby severely limiting public participation in an imperial government that was to be highly central-

ized. Ironically, this unilaterally decreed constitution included passages from France's 1789 Declaration of Human Rights.

The story of this constitution demonstrated several things about the new nation of Brazil: (1) the monarch had seemingly preserved his absolutist initiative by dissolving the elected Assembly and imposing his own constitution; but (2) the constitution, while favoring the crown in the division of powers, was more liberal than absolutist, more akin to the contemporary English parliamentary system than to the French; and (3) the commitment to human rights, however qualified by the real intentions of Dom Pedro and his loyalist advisers, thenceforth became a lodestar in Brazilian history, an ideal to which libertarians and reformers would continuously repair. The struggle over the new country's political structure had ended ambiguously: a liberal charter imposed by an emperor who was thereby establishing limits on all future governments.

The absolutist aspects of events in Rio stirred concern in the Northeast, the region that had proved most receptive to the liberal ideas of abolition, federalism, and republicanism. Back in 1817 republican conspirators in Pernambuco province had stubbornly resisted the discipline of Rio. Dom Pedro's imposition of the constitution in 1824 provoked a new rebellion, which dramatized the key issues at the heart of Brazilian politics for the rest of the empire.

The Pernambucans declared their independence anew in a manifesto poetically proclaiming the "Confederation of the Equator." This climaxed a violent reaction to Rio's attempt to depose a locally popular junta and impose one less radical. After gaining the support of other northeastern provinces, the rebels called for their own Constituent Assembly. The movement split apart on the slavery issue, however, as one leader shocked his colleagues by calling for an end to the slave trade through the northeastern port of Recife. Most of the rebel organizers feared a mobilization of the lower orders, and not without reason. Discontent of marginal free persons, many of color, was threatening to turn the anti-Portuguese, anticentralist agitation into a social revolution.

The rebels' internal divisions in Pernambuco came as the military pressure from outside was growing. The emperor had hired English and French ships and mercenaries, and they taught the insurgents a bloody political lesson. Most of the rebel leaders were executed. There were limits to the range of permissible social protest in Brazil.

Rio's domination came only with British help, and that aid had its price. Having secured a favored foothold in the Brazilian economy since 1810, Britain now found itself underwriting the transition to Brazilian independence.

Britain could help consolidate the newly independent Rio government by facilitating diplomatic recognition from the world's principal

powers. That goal was achieved by a series of 1825 agreements Britain negotiated with Portugal and Brazil. They provided that the Portuguese king, now Dom João VI, was to recognize Brazil as a separate kingdom; that British exports to Brazil would continue to receive a preferential tariff rate; and, not least important, that Brazil would pay Portugal an indemnification of two million pounds sterling for damages suffered in the struggle for independence. (This was exactly the debt that Portugal owed to Britain; the negotiators kept this provision secret.)

The following year, 1826, Britain got from Brazil a treaty commitment to end the slave trade by 1830. The British wanted this commitment for several reasons. One, usually stressed by modern-day economic historians, is that Britain feared that slave-produced sugar from Brazil would prove cheaper in the world market than sugar from the British West Indies, where slavery had recently been abolished. Another reason, also important, was the pressure on the British government generated by British abolitionists. The new Brazilian government, with little enthusiasm and less genuine commitment, gave the British the clause they demanded. Further concessions to the British were made in an 1827 trade treaty which put Brazilian exports to England at a disadvantage with exports from British colonies. Much of the Brazilian elite saw the concessions as excessive and only explicable by Dom Pedro's apparent desire to retain British good will toward Portugal, which desperately needed continued British economic help. Criticism would have been even more strident if the two-million-pound payment had been made public.

In the end, Dom Pedro's loyalty to Portugal proved his undoing in Brazil. His imposition of the new constitution had by no means ended the struggle over the division of governmental powers. In 1826 the emperor became the target of a new wave of attacks, whose authors ranged from the "moderates" wanting more power for the legislature and revisions in the treaties with Britain, to the "extremists" demanding a decentralization of power and autonomy for the provinces. The emperor's critics dominated the expanding press with their drum-fire of invective.

In this same period Dom Pedro suffered a serious reverse in foreign policy. What is modern-day Uruguay had been annexed to Portuguese America in 1821 as the "Cisplatine Province." But in 1825 local guerillas seized power and proclaimed union with the United Provinces of the Río de la Plata (present-day Argentina). The resulting war between Brazil and the United Provinces ended in 1828 with a treaty that created an independent state, Uruguay. The British, again intermediaries in arranging the treaty, hoped for a buffer state between Argentina and Brazil. This setback to Brazilian ambitions in the Río de la Plata basin was financially and politically expensive for the emperor. But it soon faded in significance when compared to the quagmire of the Por-

tuguese royal succession into which Dom Pedro had been drawn since 1826.

When Dom João VI died, in 1826, Dom Pedro, his legal successor, knew that the Brazilians would never agree to a reunion with Portugal. But the emperor had become increasingly absorbed in trying to protect his daughter's succession rights in Portugal. That drain of energy made him less able to deal with the aggressively antiabsolutist political forces in Brazil. He found his position increasingly untenable, as his opponents mobilized street crowds to protest his preference for an absolutist ministry. On April 7, 1831, Dom Pedro I abdicated, departing the land whose independence he had helped to secure less than a decade earlier.

Dom Pedro's abdication was a victory for the Brazilian Party and a defeat for the beleaguered absolutists. It also created a power vacuum, because the emperor's son, later to become Dom Pedro II, was only five years old. His father had deliberately left him behind in the care of other royal family members in order to maintain the Braganza family's claim to the Brazilian throne. Who would exercise power in his name? Would the huge and thinly settled lands of Portuguese America hold together? Or would it follow the example of Spanish America, which immediately fissured into the patchwork of nations we see today?

For nine years after Dom Pedro I's abdication, a regency, of varying structure and membership, exercised executive power. In 1834 the constitution was amended (by the "Additional Act") to give increased powers to the provinces, partly in response to separatist sentiments that were rocking key areas of the empire. The most violent separatist movement was in the province of Pará in the Amazon valley; the most dangerous, because of its location in a province bordering Argentina, was the *Guerra dos Farrapos* in Rio Grande do Sul.

Dom Pedro II (1840–1889)

Dom Pedro II's accession to the throne in 1840 unified the divided elite. Brazil had survived the separatist challenges and halted the drift toward social revolution. Decentralization was discarded as the emperor assumed the wide powers (the "Moderating Power") his father had spelled out in the 1824 constitution. The young emperor and the politicians now settled into an era of relatively harmonious parliamentary politics.

The two decades after midcentury were the golden years of the empire. Executive power was exercised by the emperor and his ministry, the latter dependent upon retaining the confidence of the lower house. Yet the legislature's ultimate power was more apparent than real, because the emperor could dissolve the Chamber at will, thereby necessitating new elections. Until the late 1860s, however, Dom Pedro II exercised his power discreetly and the system seemed to function well.

By 1850 two distinctive political parties had emerged—both owing their origin to the Brazilian Party of the 1820s. The parties were Conservative and Liberal, although historians have long cautioned against taking these labels too seriously. In 1853 the two parties collaborated to form a "conciliation cabinet," which held power, except for the 1858–62 interval, until 1868.

The empire's most important test in foreign policy came in the Río de la Plata basin, the site of a long-time rivalry among Paraguay, Uruguay, Argentina, and Brazil. The Brazilian government became alarmed over the strength and intentions of Juan Manuel de Rosas, the autocratic ruler of Argentina, who was claiming the right to control all traffic on the Río de la Plata. This was a grave threat to Brazil, since the economics of its southern provinces relied heavily on access to the Plata basin river system.

At the same time Brazil was being drawn into a dangerous political battle in Uruguay, where Brazilians had gained a financial and commercial foothold. Brazilian troops were sent into domestic Uruguayan battles on the side of the "Colorado" faction, which prevailed. The Brazilians then turned to face Rosas. They were encouraged by the French and British, who chafed over the tough terms Rosas had imposed for economic access to Argentina. The anti-Argentine coalition prevailed. Foreign troops, assisted by Argentine rebels (the latter representing the soon-to-be dominant liberals), defeated the Rosas forces in 1852, sending Rosas to a permanent exile in England.

But that did not bring all combat to an end. Even with Brazilian support the Colorados lost control in Uruguay. Since the victorious Blancos could no longer look to Rosas for help, they turned to Francisco Solano López, the dictator of Paraguay. Argentina, now controlled by the liberals, joined Brazil in support of the Colorados in Uruguay. The deepening foreign involvement there soon led to direct military confrontation between the Brazilians and the Paraguayans. Solano López wanted to expand his rule by allying with the Uruguayan Blancos to conquer the Brazilian province of Rio Grande do Sul. He invaded both Argentina and Brazil in 1865, pushing them and the Colorado government of Uruguay into a military alliance.

The ensuing war lasted five years. The Paraguayan army proved to be well trained, superbly disciplined, and extraordinarily brave. The Brazilians bore the brunt of the fighting on the other side. At first they suffered humiliating reverses, but then triumphed after greatly expanding their army.

The Paraguayan War had important consequences: (1) access to the Río de la Plata river network was guaranteed, thus pleasing European traders, Argentine liberals, and the Brazilian military; (2) the two major powers, Argentina and Brazil, cemented close relations; (3) Brazil consolidated its position—both political and financial—in Uruguay; and (4)

Paraguay was left with half (it is thought) of its population dead and the country in ruins.

The war also had a profound effect on politics within Brazil. The long combat forced Brazil to enlarge its army, whose officers soon became important actors in Brazilian politics. It had also provoked the emperor into unprecedented steps in asserting his authority. Pedro II demanded Paraguay's unconditional surrender, while the Liberals, who held a majority in Chamber, wanted by 1868 to negotiate. He dismissed the Liberal cabinet, which had strong majority support in the chamber, and called for new elections. Some radical Liberals reacted angrily by forming a splinter group that in 1870 became the Republican Party. And the war threw a new light on that most national of institutions, slavery. The slaves recruited for the Brazilian army performed well in battle and were given their freedom in compensation. Their combat effectiveness must have given pause to more than a few of the white officers who were later called on to hunt down fugitive Brazilian slaves.

The End of the Empire

The final two decades of the empire were dominated by debate over the legitimacy of two institutions: slavery and the monarchy. Both came under scrutiny during the Paraguayan War.

Although the slave trade effectively ended in 1850, slavery was by no means dead twenty years later. The rapidly growing coffee plantations demanded labor, and the planters turned to an obvious source: slaves from the economically decadent Northeast. Slave owners there became alarmed at this loss of "capital," and succeeded in getting provincial laws that forbade slave exports to other provinces. Such measures came late, however, and even if every slave in the Northeast had moved south they could not have furnished the labor needed in the coffee economy of the late 1880s.

The only solution, according to the coffee planters, was increased immigration. In 1886 the province of São Paulo launched a major effort to attract European immigrants to Brazil, but the *paulistas* found themselves unable to attract the amount of cheap labor they needed. Why? Partly because of the persistence of slavery. This conclusion led some of the elite to become pro-abolitionist on the pragmatic grounds that immigrants could never be recruited quickly enough unless Brazil's retrograde image in Europe was transformed. Abolition would be the most obvious step.

The manner in which Brazil carried out abolition was unique in the Americas. Brazilian slavery was a nationwide institution, thus preventing the kind of sectional conflict that occurred in the United States. Furthermore, Brazilian slaves had worked in virtually every job category, including many "skilled" ones. No less important, a large number

of free persons of color had already established themselves economically, providing examples to the newly freed. Brazil had also escaped the extreme racist view that dismissed all persons of color as irremediably inferior. The large mixed-blood population, a few of whose members had reached prominent national positions by 1889 (such as the novelist Machado de Assis and the engineer-abolitionist André Rebouças), showed that some mobility was possible.

Abolition in Brazil was a seventeen-year process marked by three laws. The first came in 1871, when the Congress passed the "law of the free womb," which provided freedom for all children thenceforth born of slave mothers. But the masters were given the option of retaining labor rights over these children until the age of twenty-one.

It was not until the 1880s that the abolitionist movement was again able to force slavery to the center of the political arena. The abolitionists were led by urban professionals, especially lawyers. Prominent among them was Joaquim Nabuco, a Pernambucan deputy of impeccable social origins. Led by such orators as Nabuco, the abolitionists became the empire's first nationwide political movement, organizing clubs, newspapers, and public rallies in the major cities. They raised significant sums to finance their propaganda and to buy the freedom of local slaves.

This mobilization had its impact on the Parliament, which in 1885 passed the second abolitionist law. This one granted freedom to all slaves sixty or older, without compensation to the master. Cynics ridiculed the measure, pointing out that few slaves ever survived to such an age, and if they did their masters would be delighted to be freed from caring for them. The new law did little to defuse the agitation of the abolitionists, some of whom began inciting slaves to flee from or rebel against the masters. By 1887 slavery was visibly disintegrating. Soldiers in the army, charged with catching and returning fugitive slaves, found their job more and more repugnant. In 1887 army officers formally refused to carry out this mission any longer.

By 1888 slave owners had had ample time to prepare for the transition to free labor. The final step was the "golden law," passed in May of that year, which freed all remaining slaves without compensation. The law was approved by an overwhelming vote in both the Chamber of Deputies and the Senate. The political elite had managed to preserve a consensus while dealing with a volatile socioeconomic issue. This success at incremental reform helped to perpetuate the Brazilian elite's self-image as conciliatory. Remarkably enough, this image has come to be shared by many of the nonelites, showing how their modern-day rulers have manipulated Brazilians' understanding of their own past.

The other major drama of the late empire was the rise of republicanism. It had erupted earlier in the century, usually linked to regional demands for autonomy. The Republican Party, founded in 1871, also

had a strong regionalist cast, especially in São Paulo. The birth of this party could be traced to Liberal deputies' reaction to Dom Pedro II's imposing, in 1868, a Conservative ministry in the face of a Liberal majority in the Chamber. In 1870 a group of indignant ex-Liberals founded the Republican Party.

At first the Republicans appeared harmless. Up to 1889 they never controlled the Chamber of Deputies, and they had a very uneven following. It was strongest in São Paulo, Rio Grande do Sul, and Minas Gerais; weakest in the Northeast. The Republicans wanted to replace the empire with a republic headed by a directly elected president, governed by a bicameral legislature (with direct elections for both), and organized on federalist principles. In effect the Republicans wanted to trade Brazil's English-style constitutional monarchy for a U.S.-style federal republic.

During the 1880s republicanism made great inroads among the younger generation—the university-educated sons of planters, merchants, and professional men. Often they combined republicanism with abolitionism. Both sentiments were reinforced by the teachings of the Brazilian Positivists, a dedicated group that had penetrated faculties of higher education, especially in military colleges. Thus the 1880s saw a convergence of intellectual and political movements that were simultaneously eroding support for the monarchy and for slavery.

However, it was not high-minded debate that sealed the empire's fate. It was the discontent of an institution the emperor had too long neglected: the army. In the late 1880s recurring friction mounted between army officers and civilian politicians—often over the officers' rights to express publicly their political views. Because of the Paraguayan War, Brazil had created a much larger military than was wanted by the politicians of Dom Pedro II, who provided meager financing for modernization of the army. By the 1880s there was a disproportionately high ratio of officers to troops. That led to frustration over delayed promotions among junior and middle-level officers, who became especially receptive to the abolitionist and republican sentiments that were so influential among their civilian counterparts.

The final agony of the empire came in 1889. The emperor had insisted on trying to rule with a Conservative ministry, despite its minority position in the Chamber. In June the emperor invited the Viscount of Ouro Preto to form a cabinet. He succeeded and formulated an ambitious reformist program including administrative decentralization. But it was too late. A military plot developed in November. Led by Marshall Deodoro da Fonseca, the conspirators demanded Ouro Preto's resignation. An ultimatum was given the emperor: the monarch was finished and he must leave. Dom Pedro II and his family calmly left for exile in Portugal. The republic was proclaimed the next day, November 16, 1889.

The empire had fallen with little upheaval. Although the planters

had long feared that abolition would doom agricultural exports, they soon came to their senses. They now realized they could preserve their economic (and therefore political) dominance in a world without monarchs or slaves. Neither the abolition of slavery nor the overthrow of the empire in themselves brought structural change in Brazil.

Overview: Economic Growth and Social Change

In the mid-nineteenth century the Brazilian economy began a fundamental transition, not tied to any legal or constitutional changes, that has continued well into the twentieth century. It has also had a profound impact on Brazilian society and on relations between social classes.

Like most of Latin America, Brazil has exported a few primary products to the North Atlantic economies at the center of the world-system. But in contrast to many other countries of the continent, Brazil has passed through a chronological sequence of dependence on the exportation of different products at different points in time. The repeated pattern of boom-and-bust has made it difficult to achieve sustained growth. And since the various products have come from differing areas of the country, these cycles have created regional pockets of prosperity and decline.

After independence was achieved, sugar continued to be the most lucrative export, as it had been during the eighteenth century. Produced mainly on large plantations in the Northeast, where labor came from slaves, sugar accounted for 30 percent of Brazilian exports in 1821–30. It then began a long decline, as competition mounted in the Caribbean, and by 1900 it contributed only 5 percent of the overall export amount (though domestic consumption was and has remained substantial).

Rubber production started in the early nineteenth century, principally in the Amazon, and steadily increased. By 1853 the port of Belém was exporting more than 2500 tons of natural rubber. Demand in the industrial world grew enormously in the wake of Charles Goodyear's discovery of the vulcanization process—which prevented rubber from getting too sticky in hot weather and brittle in the cold. A spectacular boom arrived in 1900–1913, when rubber amounted to about one-third of all the country's exports. Then the British capitalized on the more efficient rubber plantations they had developed in the East Indies, and the world price went into sharp decline. The Brazilians were unable to withstand the competition, and the rubber boom came to a sudden and permanent end.

It was coffee that provided the most durable stimulus to economic change in the postindependence era. Coffee production began to develop in the Caribbean in the early nineteenth century and then took hold in Brazil, where it enjoyed excellent natural conditions. The vol-

In the late nineteenth century sacks of coffee left São Paulo's plantations on mule trains and eventually reached overseas destinations. (Courtesy of the Library of Congress.)

ume of Brazilian exports held fairly steady until the 1890s, then entered a period of spectacular growth. In 1901 Brazil exported nearly 15 million sacks of coffee (at sixty kilograms each) and produced nearly three-quarters of the total world supply. Early in the century coffee yielded about one-half of the country's foreign exchange.

Coffee thus became a central feature of Brazilian life. The state of the economy was directly related to the international market: when coffee prices were high, the prospects for Brazil were positive; if they were down, so was the national outlook. And the domestic consumption of coffee has long been an essential aspect of social life, as Brazilians conduct meetings and discussions over cup after cup of steaming coffee, usually taken with large quantities of sugar. A proverb gives the recipe for coffee: "As strong as the devil, as black as ink, as hot as hell, and as sweet as love."

Coffee production flourished in southern-central Brazil, particularly in the state of São Paulo. It requires ample land, a fair investment, and much labor. Coffee trees yield full production only after six years, and they need steady care. Berries need to be gathered, washed, and shelled. The beans inside the berries must be dried, sifted, sorted, sacked, and stored. This requires labor.

Like Argentina, Brazil turned its eyes to Europe. First the state of São Paulo and then the national government initiated programs that attracted millions of European immigrants, especially in the last quarter of the nineteenth century. The largest portion, perhaps a third of the total, came from Italy. But the relative size of the immigrant population never reached the same level as in Argentina. The peak for Brazil was 6.4 percent in 1900, and it declined after that.

In retrospect it is ironic to realize that this wave of European new-comers probably helped distort, at least regionally, Brazil's long-term economic development. Although ample labor was available in the Brazilian center and Northeast, where the number of jobs had fallen disastrously behind the increase in workers, the prophets of immigration opted for Europeans, who would presumably be better workers and more reliable future citizens. So the Brazilian government paid the ocean passage of millions of Europeans, while millions of Brazilians in Minas Gerais, Rio de Janeiro, and the Northeast could not afford to move south. Great contributions were made by the transplanted Europeans and Japanese; but each of those jobs might have been held by a Brazilian who would have been rescued from the economically moribund regions.

Technology was harder to obtain. There was and is no quick substitute for technology—one of the most essential features of modern economies. In the capitalist world it has proved extremely difficult to buy technology by itself. The Brazilians, like other populations outside the dynamic North Atlantic industrial complex, found themselves having to accept direct investment by foreign firms in order to get technology. The telegraph system, for example, arrived with British and American firms, which installed and operated their own equipment. The same held true for railroads, electric utilities, and shipping—most of the infrastructure needed to sustain the growing agro-export economy. These were highly visible investments and later became convenient targets for nationalist attacks.

Capital was also sought abroad. Much came with the technology described above. It also came in the form of loans to Brazil on the state and national level. In 1907, for example, the states of São Paulo, Minas Gerais, and Rio de Janeiro signed a coffee marketing agreement to be financed by foreign creditors. The state governments planned to repay the loans with the receipts from export taxes on coffee. Such commitments obligated Brazil not only to repay the loans but also to finance the remission of profits (and eventually capital) on direct investments by foreigners. The crucial question was the *terms* on which all these transactions took place. Available data suggest that the profit rate on foreign-owned railways, to take an obvious example, did not exceed rates for comparable investments in Britain. But this question has yet to be systematically researched.

Throughout the years between 1889 and 1930 the center of the Brazilian economy moved south and southwest. The primary push came from the "march" of coffee, as planters found it cheaper to break new ground than to recycle the plantation soils whose yields were dropping. The result was a path of abandoned plantations, stretching from Rio de Janeiro and Minas Gerais down into São Paulo and its vast interior.

The reliance on coffee entailed large-scale risks. One was overpro-

FIGURE 5-1   Coffee Exports from Brazil, 1860–1985

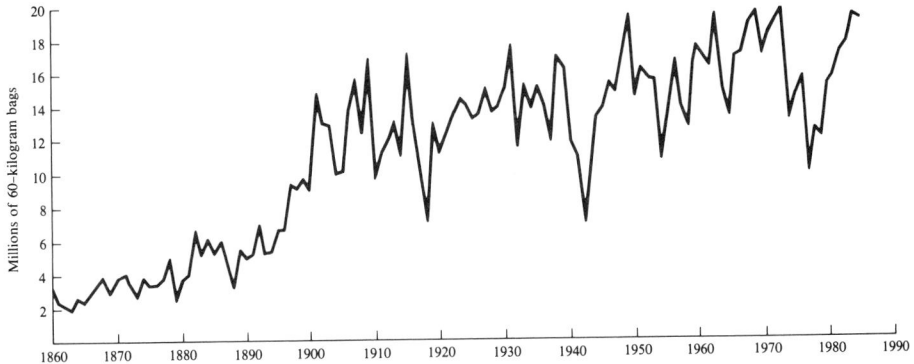

Sources: Werner Baer, *Industrialization and Economic Development in Brazil* (Homewood, Ill.: Richard D. Irwin, 1965), pp. 266–267; James W. Wilkie, Enrique C. Ochoa, and David E. Lorey, eds., *Statistical Abstract of Latin America*, 28 (Los Angeles: UCLA Latin American Center, 1990), Table 2426.

duction. It was difficult to anticipate demand six years in advance, and therefore to plan when trees should be planted. Furthermore, the yield would depend on weather and other unforeseeable factors. In 1906, for example, Brazil produced 20 million sacks of coffee for a world market that could absorb only 12 or 13 million. A political question promptly arose: What should be done with the surplus?

A related uncertainty came from the rise of foreign competition, especially in Africa and other parts of Latin America. Brazil's share of the world market declined from 75 percent in 1900 to 67 percent in 1930 to only 32 percent in 1970 and 18 percent in 1978. With time the country gradually lost its near-monopoly on supply.

A third source of vulnerability came from wide fluctuations in the world price. This reflected not only the effects of competition but also changes in demand. Coffee is essentially a luxury, though habit-forming, and its consumption can be reduced in times of hardship. Between 1929 and 1931, after the Great Depression struck, the price of coffee plummeted from 22.5 cents a pound to merely 8 cents. Frequent oscillations led to wide variation in Brazil's foreign-exchange earnings from year to year—and in government revenues, which came primarily from export duties.

To illustrate both the growth and the uncertainty of the Brazilian coffee sector, Figure 5-1 displays the volume of the country's coffee exports during the period from 1860 to 1985. The rise in output and commerce is clearly visible. So are the fluctuations, which mainly reflect the instability of world demand.

A final hazard derived from the small number of purchasers. In the late nineteenth and early twentieth centuries Brazil sold between three-fifths and three-quarters of its exports to only three countries: the

United States, Britain, and Germany. The United States was the largest single buyer and became as well the largest source of foreign investment (accounting for roughly half the total by 1960). The reliance on two or three customers created not only close but highly unpredictable ties to outside economies, as Brazil discovered after the crash of 1929. It could lead to political problems as well.

Prominent politicians and economists regarded this vulnerability as an inevitable result of Brazil's "agrarian vocation." Brazil, they argued, had no choice but to buy needed foreign finished goods with the funds earned by export and augmented by direct foreign investments or loans. Any significant attempt to industrialize, they reasoned, would produce inferior goods and jeopardize relations with foreign buyers and creditors. Furthermore, Brazil could not hope to take the United States as a model for industrialization "because we don't have the superior aptitudes of their race," in the words of a Brazilian cabinet minister of the 1890s. Brazil must live, and would be forced to live, with what God gave it: a comparative advantage in a few agricultural exports.

Since the late empire, however, a handful of intellectuals and men of business began disputing this logic. They argued that Brazil should stimulate home industry and monitor carefully economic ties with foreign countries, especially Britain and, later, the United States. These critics had little influence on key policy areas such as tariffs or exchange rates. Yet they did succeed in creating a "nationalist" critique that was to prove important after 1930.

Official encouragement of industrialization came forth in 1890, when a tariff revision provided mild protection for local manufacturing from foreign competition (and also lowered the duties on capital goods required for production). Engineering schools sprouted in Recife, São Paulo, Pôrto Alegre, and Bahia. By 1907 the country had about 3000 industrial establishments, most of them small, textiles and foodstuffs being the principal products. By 1920 the number of firms had grown to more than 13,000.

Brazil's industrial sector underwent large-scale expansion in the 1930s and 1940s, as the Great Depression and World War II reduced the available supply of manufactured goods from abroad (as happened elsewhere in Latin America, too). As with coffee, the center of industrial growth was in the state of São Paulo—where 15 percent of the nation's population was producing about 50 percent of the country's manufactured goods by 1940.

The upsurge continued thereafter, and Brazil moved into such heavy industries as steel and automobile production. Between 1947 and 1961 manufacturing output increased at an annual rate of 9.6 percent, compared to 4.6 percent for the agricultural sector. By 1960 industrial production amounted to more than 25 percent of the gross domestic prod-

uct, and by 1975 it was up to nearly 30 percent. This diversification of the economy helped reduce Brazilian dependence on the outside world, and lent credibility to claims that the country would someday join the ranks of superpowers.

These economic transformations brought about far-reaching changes in the Brazilian social structure. One result was urbanization. In 1920 about 25 percent of the population lived in urban areas, and by 1992 about three-quarters lived in cities. But there are two unusual features in this trend. First, the tendency toward urbanization in Brazil has been *later* and *slower* than in many other Latin American countries. Second, Brazil does not have a single predominant city (like Buenos Aires or Montevideo, for instance). São Paulo and Rio de Janeiro have both become megalopalises, with millions of inhabitants and the amenities and complications of urban life, but between them they contain only about 10 percent of the national population of 154 million. Urbanization has taken place in Brazil, but the cities coexist with a large and populous countryside.

Consequently Brazil has developed an intricate social system. The upper-class elite includes landowners, frequently divided among themselves, as when *paulista* coffee planters rose up in the nineteenth century to challenge the sugar barons of the Northeast. In the course of the twentieth century there appeared as well an industrial elite, a new and vigorous group that would struggle for status and wealth, sometimes using the power of the state in pursuit of its prosperity.

The popular masses were varied, too. There has been, and remains, a large-scale peasantry, consisting of those who eke out meager livings from the soil. There is a rural proletariat, in the coffee fields and elsewhere, a stratum that performs wage labor in the countryside. And in the interior there are indigenous and other groups that have little contact with national society and who are for the most part marginal to it.

An organized working class of substantial size, at least 4 million by 1970 and 6 million by 1980, emerged within Brazilian cities. Its struggles with employers and its constant manipulation by the state provide one of the central themes in twentieth-century Brazilian life. There is also a large stratum of chronically unemployed city dwellers, often migrants from the backlands who find shelter in the slums of the large cities.

In between the upper and lower classes, middle sectors gradually appeared. Their size is difficult to estimate: they may now include as much as 30 percent of the people in some cities, though their share of the national population is less than that (perhaps 10 to 15 percent). They play important roles in commerce and the professions, and they have a particularly intimate relationship with one major institution: the military.

Social status in Brazil is not just a function of occupation or wealth.

It is also a matter of race. The massive importation of slave labor from Africa brought an additional ethnic dimension into Brazilian society and this in turn has affected customs and attitudes.

There tends to be a strong correlation between race and social standing in Brazil: most on top are white, most blacks are on the bottom, and mixed-bloods are largely in between. Some institutions, such as naval officers and the diplomatic corps, long remained white. But in Brazil race is not defined purely on biological grounds. It is a social concept, open to interpretation. To be "black" one has to be totally black (in contrast to the United States, where partly black in ethnic origin means black). Mulattoes in Brazil have considerable opportunity for upward mobility, and for this reason miscegenation has been viewed by one scholar as a kind of social "escape hatch."

This is not to say that Brazil constitutes a racial paradise. Prejudice and bias have existed. For the last century most of the Brazilian elite has placed its faith in *branqueamento* ("bleaching"), presumed to be the inevitable result of miscegenation, with the unequivocally racist intention of purging Brazil of the presumably "negative" influence of black blood. The overall correlation between status and race continues to exist, despite the denial of well-to-do Brazilians. Several recent studies by Brazilian demographers have shown significant differences in income by race (controlling for all other factors), based on official 1976 and 1980 data. The conclusion to be drawn is that race is a separate and significant variable in the Brazilian socioeconomic system. But mobility exists, marriage across color lines is common, and attitudes are more open than has been true in North American history.

Racial differentiation in Brazil has posed one obstacle, however modest, to the formation of durable coalitions between social strata and groups. Another obstacle is the sheer geographical size of Brazil. Distance (and poor communications) made it for a long time implausible to imagine a long-lasting alliance between urban workers of São Paulo, for example, and the landless peasants in the Northeast. Such natural divisions enabled Brazil to attempt political solutions that would have been immediately impossible in more densely populated and integrated countries such as Cuba.

The First Republic (1889–1930)

Although the military overthrew the empire, civilian politicians shaped the new republic (see Figure 5-2). A Constituent Assembly was elected and produced a new constitution in 1891. It was a virtual copy of the U.S. constitution, as its principal author, Bahian delegate Rui Barbosa, had intended. Brazil became a federation of twenty states and the Brazilian president was to be elected directly and empowered to intervene in the states in case of threatened separation, foreign invasion, or conflict with other states. Suffrage was restricted to literate adult male citi-

FIGURE 5-2 *Café com Leite:* Brazilian Presidents by State, 1889–1930

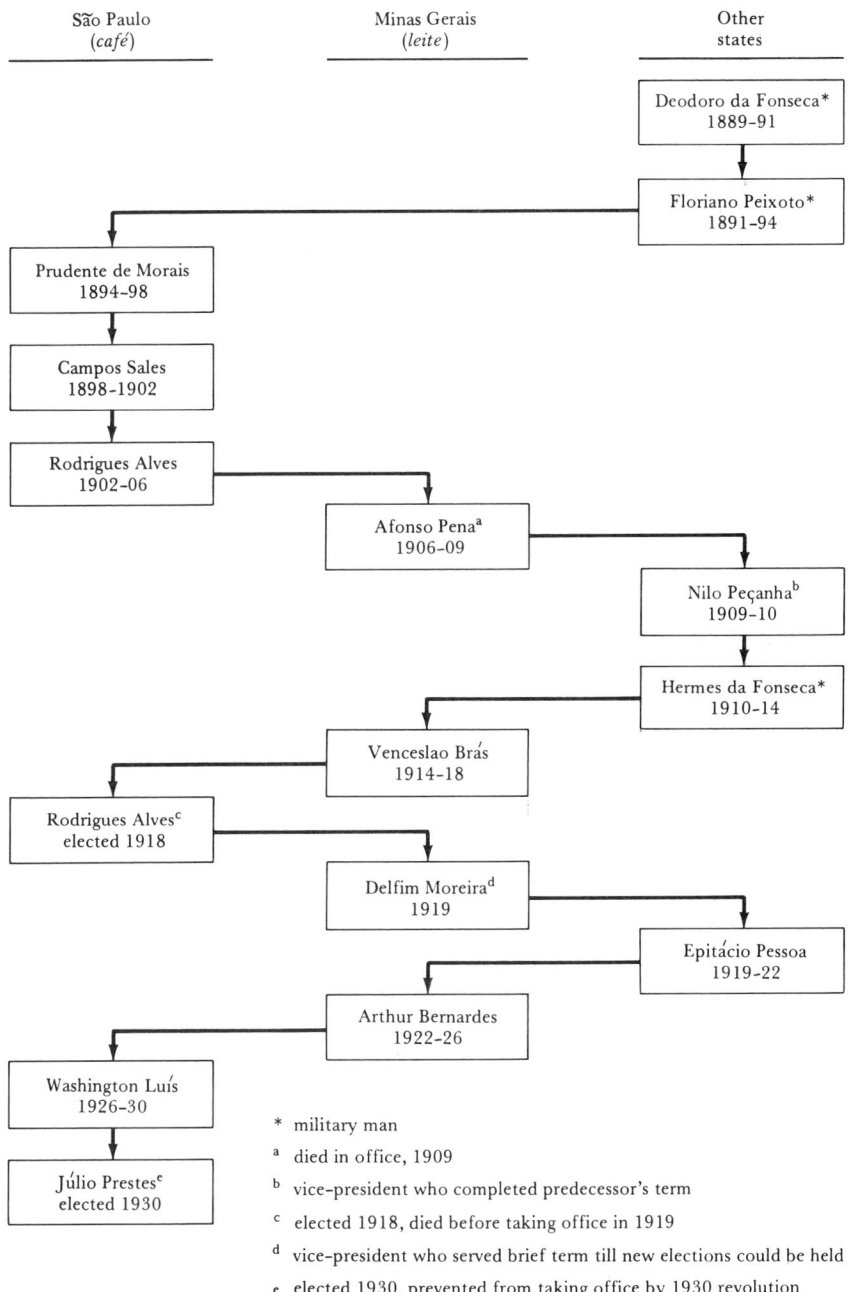

| São Paulo (*café*) | Minas Gerais (*leite*) | Other states |

* military man
a died in office, 1909
b vice-president who completed predecessor's term
c elected 1918, died before taking office in 1919
d vice-president who served brief term till new elections could be held
e elected 1930, prevented from taking office by 1930 revolution

Explanation: The power of two preeminent states—São Paulo, a coffee-producing area, and Minas Gerais, a dairy state—led to a series of presidential bargains during the Old Republic (1889–1930) which Brazilians have nicknamed "the politics of coffee and milk" (*café com leite*). Rio Grande do Sul was another significant state, and the military played an important intermediary and supervisory role.

163

zens. This resulted in fewer than 3.5 percent of the population voting in any presidential election before 1930, and only 5.7 percent in 1930.

After electing Deodoro da Fonseca president, and another officer, Floriano Peixoto, vice president, the Assembly rapidly collided with Deodoro over his financial policy and his interventions in the new state governments. In November 1891, plagued by ill health, Deodoro resigned, passing power to Floriano Peixoto, the so-called Iron Marshal. Floriano soon encountered a rash of revolts erupting throughout Brazil. In Rio Grande do Sul, revolt was part of the deadly conflict between local factions; in Rio de Janeiro, it was a naval revolt led by monarchist officers. Both rebellions were crushed, as the new republic used censorship, martial law, and executions. When Prudente de Morais of São Paulo was elected in 1894 as the first civilian president, the new regime had gained stability. But it came at the price of having to recognize the legitimacy of the entrenched oligarchical regime in each state.

Who were those oligarchs and how did they hold power? In every state a tightly organized political machine emerged, dedicated to monopolizing power. In states such as São Paulo and Minas Gerais, where the Republican Party had been strong before 1889, the "historic Republicans" controlled state government. In Bahia and the Northeast, which had few Republicans before 1889, power went to those politicans who most quickly established credentials as new-born Republicans. The resulting power structure was a "politics of the governors" at the national level, and the "rule of the colonels" *(coronelismo)* at the local level. The colonels were actually small-town and rural bosses who could produce bloc votes in any election. In return, they obtained control over state and national funds spent in their area of influence. At the state level the political leaders used their deals with the colonels to bargain on the national level with leaders from other states.

The chief prize in such bargaining was the presidency. As might be expected, the states enjoyed very unequal influence in this process. São Paulo and Minas Gerais were the most important, with Rio Grande do Sul able to tip the balance when the two larger states were at odds. Bahia, Rio de Janeiro, and Pernambuco were second-level states, often serving as power bases for dissident presidential nominees, although none was ever elected during the Old Republic (1889–1930).

The constitutional decentralization created by the republic allowed several states with dynamic economies, especially São Paulo, to gain virtual autonomy over their own development. Between 1890 and 1920 the state of São Paulo more than trebled its population. It had contracted a foreign debt larger than the national government and was accounting for 30 to 40 percent of Brazil's national output. Able to impose its own taxes on interstate commerce, it had, from the standpoint of the federal government, achieved a remarkable self-sufficiency. Only a loose federal structure could have allowed São Paulo's extraordinary burst of economic development ("the locomotive

pulling the twenty empty boxcars," said *paulista* chauvinists), later to propel Brazil's rise to world prominence in the mid-twentieth century.

Brazil's relatively smooth-running political machine ran into trouble soon after the First World War. The political system created by the Republicans in the 1890s had not survived long in its original form. The first major crisis grew out of preparations for the 1910 elections. The "official" choice for president was Governor João Pinheiro of Minas Gerais, who died unexpectedly in 1908. The crisis deepened when the incumbent president, Afonso Pena, died in 1909, eighteen months before the end of his term. A bitter struggle ensued, with Marshal Hermes da Fonseca, son of the republic's first chief executive, becoming the "official" candidate. He won, but for the first time there was a significant opposition movement. It supported Rui Barbosa, the liberal crusader from Bahia.

During Hermes da Fonseca's presidency (1910–14) many of the smaller states experienced bitter battles within the political elite—usually between the incumbent machine and dissenters. The lines of conflict often followed prerepublican patterns of family and class struggles. These battles made it impossible to return to the smoothly functioning "politics of the governors," not least of all because President Hermes usually sent federal troops to the dissidents' aid. Formally, at least, the system continued functioning until October 1930. The "official" presidential nominees were invariably elected, and the federal Congress remained under the control of the state machines.

Yet the constitutional structure and political culture of the Old Republic had become targets for criticism and ridicule from every quarter. Prominent among the critics was a new generation of the elite, born with the republic. Most were educated as lawyers. They denounced as corrupt the way the politicians were running the republic. Most traced this to the republic's founders, whom they accused of having imposed on Brazil a liberalism for which it was utterly unprepared. Necessary changes could be found only after a careful analysis of where Brazil stood—economically, socially, politically, and intellectually. In a word, Brazilian problems need Brazilian diagnoses and Brazilian solutions. They described themselves as "Brazilians who think like Brazilians: American, Latin, and tropical." Leaders of this group included Oliveira Vianna, sociologist and lawyer; Alceu Amoroso Lima, literary critic and essayist; and Gilberto Amado, essayist and politician. Their mentor was Alberto Tôrres, a restless Republican of the older generation who had become disillusioned during his years as governor of the state of Rio de Janeiro and subsequently as a justice of the federal Supreme Court.

Criticism from intellectuals was paralleled by a mutinous mood among younger army officers. There was a series of barracks revolts in 1922 and 1924, led by lieutenants (*tenentes*). The 1924 revolts, which began in São Paulo and Pôrto Alegre, were the most serious. But the

rebel officers fled and held out for two and a half years as guerrillas on a 25,000 kilometer march through the interior of Brazil. It was dubbed the "Prestes Column," after Luíz Carlos Prestes, a rebel lieutenant who was later to lead Brazil's Communist Party for more than thirty years.

The rebels' formal manifestos were vague, emphasizing the need for fair elections and honest government, along with attention to the nation's social needs. A more immediate complaint focused on professional concerns—anachronistic training, obsolete weapons, unresponsiveness from civilian governments, and poor prospects for promotion. This frustration was reminiscent of the late empire, when army officers had both professional and intellectual reasons for supporting a coup against the crown.

Another powerful political current of the 1920s was the Democratic Party, founded in São Paulo in 1926. Its leaders, typified by coffee baron Antonio Prado, agreed that the Old Republic was a fraud, not because of too much democratic liberalism, but because of too little. Many of the party's votes came from the urban professionals, disgusted at seeing their votes canceled out by rural voters mobilized by the federal government's machine. They wanted what the European middle classes had won in the nineteenth century: political power through an electoral system that gave them a major role in government. It was no accident that this current of "liberal constitutionalism" made its strongest showing in São Paulo, the center of the fastest economic growth and urbanization. It was the voice of "modern" Brazil speaking out against the disproportionate influence of their country's "backward" regions.

Economic development in the late nineteenth century had created a working class in three or four major cities. Workers' first organization came in "mutual-aid societies," which did very little. They were superseded in the early 1900s by anarchist and anarcho-syndicalist organizers who were far more militant. In the decade after 1910 the anarchist and anarcho-syndicalist unions staged a variety of strikes, including several attempted general strikes. Sooner or later, they met heavy repression. The Spanish- or Italian-born leaders were deported, while Brazilian leaders were jailed, beaten, and harassed. By 1921 the organized urban movement was a shambles.

In subsequent years social welfare laws were passed, as a tardy carrot to accompany the omnipresent stick. But Brazilian workers had many fewer organizing rights and welfare provisions than, for example, Chilean workers in the same era. One reason was the continuous labor surplus in Brazil, especially for unskilled or semiskilled jobs. In the face of such numbers, Brazilian workers found it hard to organize themselves.

One result of the repression was the decline of anarchist and anarcho-syndicalist leadership and their replacement, in many cases, by communists, whose Brazilian party was founded in 1922. The commu-

nist presence furnished a new target for the authoritarians among civilians and military. By 1930 urban labor, although growing steadily in economic importance, was a political orphan. Meanwhile, employers saw no reason to change the autocratic manner in which they had long dealt with their workers.

Getúlio Vargas and the Estado Nôvo

The world economic crash of 1929 hit Brazil, like the rest of the Americas, very hard. As world trade contracted, the coffee exporters suffered a huge drop in foreign exchange earnings. Despite the crisis President Washington Luís clung to a hard-money policy. In effect that meant guaranteeing convertibility of the Brazilian currency *(mil reis)* into gold or British sterling. The Brazilian government should have frozen all transactions immediately. The gold and sterling reserves were quickly exhausted, forcing the government to suspend convertibility of the *mil reis.* The government was left in a deepening balance-of-payments crisis, and the coffee growers were stuck with an unsellable harvest.

Given coffee's great importance to the Brazilian economy, one might have expected the government to rush in with help. But it did not. Instead, it tried to please foreign creditors by maintaining convertibility. Such were the money principles preached by the foreign bankers and economists who set the terms for Brazil's relations with the world economy. At a critical moment, the Brazilian government decided to stick with an economic policy which had no support from even one major sector in Brazilian society.

Not surprisingly, Washington Luís did not last out his term. As in 1889 it was the military that did the job. An opposition movement had coalesced around Getúlio Vargas, a Riograndense politician who had run for president earlier in 1930 and been defeated by Júlio Prestes, the "official" candidate who had been endorsed by Washington Luís. In his campaign Vargas had *not* challenged the political system. He ran strictly from within the elite. His supporters were dissenting factions in several states, outsiders anxious for their chance at power. It was only after the election, the loss of which Vargas had expected, that a successful conspiracy arose among the disgruntled politicians and officers.

The coup of October 1930 did not constitute a "revolution." The top military commanders deposed Washington and then passed power to Vargas as president of a provisional government. The cabinet invoked revolutionary power in order to take the arbitrary and ad hoc steps they thought necessary. Yet 1930 is a watershed in modern Brazilian history, even if it was not comparable to the Mexican Revolution of 1910–20 or Cuba's transformation after 1959.

When Getúlio Vargas moved into the presidential palace in November 1930, few guessed how important a leader he would become. He was there only because a conflict within the national political elite was

turning into armed warfare. It never reached a climax only because the military intervened. After the senior commanders had deposed Washington Luís, some officers wanted to retain power themselves, but the pressure whipped up by the Liberal Alliance's mobilization was too great: after only four days in power the three commanders transferred power to Vargas, who became provisional president. He held that post for four years. Since there was no legislature, the president governed by decree. Meanwhile, important shifts were occurring among the nation's political forces.

First, Vargas moved swiftly to replace the governors in all the states except one, Minas Gerais. The replacements, or "interventors," reported directly to the president. This was a prelude to later battles over state autonomy, such as the president's drive to reduce the size of the state militia. Such activism from the central government often threw the state machines off balance and gave benefit to the dissenting factions, many of which had supported Vargas in the 1930 election. As in the Hermes da Fonseca presidency, political rivalries within states were being settled by decisions in Rio.

A second major development was a realignment of political forces in São Paulo. Vargas' interventor (João Alberto) had proved inept and tactless in handling the touchy *paulistas*. Their heightened sense of state loyalty and their fury at João Alberto united São Paulo against Vargas. Its leaders demanded that Vargas fulfill his promise to call a Constituent Assembly that would write a new constitution. In 1932 the *paulista* frustration finally erupted into an armed rebellion. The state militia, aided by enthusiastic volunteers, fought federal forces to a standstill for four months in the Constitutionalist Revolution. The rebels had to surrender because they were trapped by the federal forces' encirclement of São Paulo City. By their armed attempt at separatism, the *paulistas* further discredited the cause of decentralized government and strengthened the hand of the centralizers in Rio de Janeiro.

A third significant political development was the disintegration of the *tenente* movement. These young military officers had never achieved a cohesive organization. Some accompanied Vargas into power in 1930. Others founded the October 3rd Club to focus effort on achieving more radical social changes than those propounded in the 1920's, but their movement was isolated and vulnerable. Before long police raided the club premises and the group disintegrated.

Meanwhile, Vargas was strengthening his own network of political allies and collaborators. His success became obvious during the Constituent Assembly of 1933–34. The new constitution, the second for republican Brazil, was not startlingly different from the first (1891). State autonomy was reduced: states could no longer tax goods shipped interstate. Yet it continued the bicameral legislature, which was to be directly elected, as was to be the president (except the first). Some nation-

alist measures appeared for the first time, placing restrictions on foreign ownership of land and on alien participation in professional occupations (such as law and medicine). The modest nature of these constitutional changes simply confirmed that the revolution of 1930 had grown out of elite infighting and the continuous tension between centralization and state autonomy. The Constituent Assembly's most important act was to elect Vargas as the first president with a four-year term under the new constitution.

In 1934 Brazil entered one of the most agitated periods in its political history. Attention focused on two nationally based and highly ideological movements, both committed to mass mobilization. One was Integralism, a fast-growing rightist movement with affinities to European fascist parties. Founded in late 1932 and led by Plínio Salgado, an ambitious if minor literary figure, the Integralists claimed a rapidly growing membership throughout Brazil by 1935. Their dogma was Christian, nationalist, and traditionalist. Their style was paramilitary: uniformed ranks, highly disciplined street demonstrations, colorful green shirts, and aggressive rhetoric. They were essentially middle class and drew support from military officers, especially in the navy. Unknown to the public, the Integralists' ambitious activities were financed in part by the Italian embassy.

At the other end of the spectrum was a popular front movement, the National Liberation Alliance (*Aliança Libertadora Nacional,* or ALN), launched in 1935. Ostensibly a coalition of socialists, communists, and miscellaneous radicals, it was in fact run by the Brazilian Communist Party, which was carrying out a Latin American strategy formulated at the 1935 Comintern meeting in Moscow. The first stage of the strategy in Brazil would be an above-ground mobilization on conventional lines: rallies, local offices, and fund-raising efforts to forge a broad coalition on the left in opposition to the new Vargas government, the Integralists, and the Liberal Constitutionalists.

By mid-1935 Brazilian politics had reached a fevered pitch. The Integralists and the ALN were feeding off each other, as street brawls and terrorism increased. Brazil's major cities began to resemble the Nazi-Communist battles in Berlin of 1932–33. But the ALN was far more vulnerable than the Integralists. There could be little doubt where the government preference lay. In July 1935 the government moved against the ALN, with troops raiding offices, confiscating propaganda, seizing records, and jailing leaders.

The communists now moved to the second stage of their strategy: a revolutionary uprising. It was to be triggered by a barracks revolt, led by party members or sympathizers among officers in the army. The insurrection began in November 1935 in the northeastern state capital of Natal, spreading within days to Recife and Rio. From the rebel standpoint, it was a disaster. Although the Natal rebels controlled the

city for several days, their comrades in Recife and Rio, who lacked the advantage of surprise, were contained in their garrisons and quickly forced to surrender.

Vargas and the military now had a perfect opportunity to request a "state of siege," which would revoke normal constitutional guarantees. The Congress rapidly voted it. Vested with its new emergency powers, the federal government imposed a crackdown on the entire left—with arrests, torture, and summary trials. The Integralists were elated. With their chief rival eliminated, they began to smell power. What could be more logical than for Vargas to turn to the only cohesive nationwide movement on the right?

It took two years for that illusion to be destroyed. Plínio Salgado and his collaborators were becoming more and more convinced that they would reach power by the 1938 presidential election, if not by other means. But Vargas had other ideas. On November 10, 1937, he took to the radio and read the text of yet another constitution to a nation that had just witnessed yet another military intervention. That morning the Congress had been dissolved, its premises occupied by soldiers. Brazil thus entered the *Estado Nôvo*, a legal hybrid combining elements of Salazar's Portugal and Mussolini's Italy. All the democratic hopes and promises that had grown since 1930 were gone. Brazil had suc-cumbed to its own brand of authoritarianism.

Brazil's lurch into dictatorship in 1937 certainly fit the era. But was there more than a superficial similarity between Brazil's *Estado Nôvo* and European fascism? Where, for example, was the mass mobilization so typical of Hitler's Germany and Mussolini's Italy? Were the Inte-gralists to play that role? Many—both inside and outside Integralist ranks—certainly thought so.

Like the Nazis in 1932, the Integralists in 1937 debated not *whether* they should enter government, but *on what terms*. Salgado, their leader, rejected Vargas's tentative offer of a cabinet post. Salgado thought he could hold out for more. In fact, Vargas and the military were playing their own game.

By early 1938 the greenshirts had become very frustrated. Soon after the coup the government had banned all paramilitary organizations. The obvious target was the Integralists, some of whom decided to take matters into their own hands. In February they organized an armed assault on the presidential residence. It was a *tour de force* of military incompetence, but the president's security forces seemed equally inept. There was a shoot-out and a standoff during the early morning hours at the palace gates. The farce ended at dawn, when army units arrested the remaining Integralist besiegers. The government cracked down and the Integralist movement in effect disappeared, as Salgado fled into exile.

Vargas could now survey a political scene that no longer offered any organized opposition. In the coup Vargas had appointed himself to

another presidential term, to last until the elections, scheduled for 1943. Few took that commitment seriously, given the ease with which Vargas had aborted the election that was to have been held in 1938. That skepticism was well founded. When 1943 arrived, Vargas announced that the wartime emergency precluded elections. He remained president until October 1945.

What was the significance of Vargas's authoritarian rule from 1937 to 1945? First, Vargas and his political and technocratic collaborators got a free hand in maneuvering to maximize Brazil's advantage in a capitalist world-system moving toward war. At stake were two central and related questions about Brazil's international role. Who could best help the Brazilians to modernize and equip their armed forces? And who could offer the most favorable conditions in foreign trade?

Before the coup of 1937 Nazi Germany had offered attractive terms in both areas. Brazilian generals, such as Pedro Góes Monteiro and Eurico Dutra, admired the German war machine and found the German arms deal tempting. By the same token, Brazilian economic policymakers were intrigued by trade terms under the German "blocked" currency scheme, which was a form of barter. When Brazil sold to Germany it had to agree to accept only German goods in payment.

Strategy and ideology were also at stake in these negotiations. The pro-German faction within Brazil, strongest in the military, was countered by a pro-U.S. faction. The latter argued that Brazil had opted for the Allies in the First World War and had most to gain by sticking with the United States. Many of the Brazilian elite therefore saw the flirtation with Nazi Germany as dangerous and short-sighted.

Meanwhile, the U.S. military and State Department were sparing no effort to pull Brazil back into the U.S.-dominated hemispheric orbit. They succeeded, but only after strenuous U.S. effort and German failure to offer the armaments Brazil wanted. From then onward Brazil became a vital cog in the Allied war machine, furnishing essential raw materials (like quartz and natural rubber) and air and naval bases that became critical in the "Battle of the Atlantic." Brazil even sent a combat division to Italy in 1944, where it fought alongside the U.S. Fifth Army.

Vargas had dealt shrewdly with the United States. In return for its raw materials and bases, Brazil got the construction of a network of air and naval installations along the northern and northeastern Atlantic coast. The United States also promised to help finance construction of Brazil's first large-scale steel mill, at Volta Redonda. It was the first time an American government committed public funds to industrialization in the "developing world."

The *Estado Nôvo* furnished a centralized apparatus through which Vargas and his aides could pursue economic development and organizational change. The federal government assumed an aggressive role in the economy, organizing and strengthening marketing cartels (in cocoa, coffee, sugar, and tea), and creating new state enterprises, such as

the National Motor Factory (to produce trucks and airplane engines). Vargas also overhauled the federal bureaucracy, creating a merit-oriented system to replace a patronage-ridden structure. Finally, one of the most important measures was a new labor code (1943), which spelled out rules of industrial relations that were to last until the 1980s. The law permitted unions to organize by plant and industry, although only on a local basis. Statewide and nationwide unions were forbidden. Only one union was permitted in each plant—under the scrutiny of the labor ministry, which controlled union finances and elections. Unions were in effect tied to the government, but the union leaders who "co-operated" could profit personally. This semicorporatist labor union structure was paralleled by a semicorporatist structure among the employers. These arrangements gave the federal executive a mechanism for controlling the economy. But Brazil of the early 1940s was *not* a modern, industrialized, urbanized society. Outside of a few key cities the corporatist structure left untouched most of the country, which was a vast, disconnected rural expanse.

The *Estado Nôvo* also had its darker side. The security forces had a virtual free hand. Torture was routine, against not only suspected "subversives" but also foreign agents (German businessmen were especially vulnerable). Censorship covered all the media, with a government news agency (*Departamento de Imprensa e Propaganda,* or DIP) furnishing the "official" version of the news. There were resemblances to Germany and Italy, but the Brazilians stopped well short of those extremes.

Brazil's economic history from 1930 to 1945 is not easy to capsulize. Coffee continued to be the primary foreign exchange earner, although helped during wartime by the boom in other raw materials shipped to the United States. Industrial growth continued in São Paulo and, to a lesser extent, in Rio. The war cut off trade with Europe, with most shifting to the United States. Brazil's wartime alliance with the United States also resulted in a U.S. technical mission to Brazil, which recommended programs for the country's economic development. Here again the U.S. government endorsed basic development in Brazil as the payment for the Vargas government's wartime cooperation.

Vargas had in 1943 promised elections, for which he would be ineligible. As the war continued, Vargas knew that a wave of democratic opinion was building and he anticipated events by adopting a new, populist stance after 1943. The urban working class was now the object of government attention through such media as the nightly nationalist radio broadcast ("The Hour of Brazil"), and moves were made toward creating a Labor Party. Vargas was trying to create a new electoral image—something he had been able to neglect earlier in the *Estado Nôvo*.

Events moved rapidly in 1945. Vargas hoped to play down the contrast between the defeat of fascism in Europe and continued authori-

tarianism at home. He quickly demobilized the returning Brazilian Expeditionary Force, fresh from its sobering encounter with the Wehrmacht in Italy. In May 1945, with victory over the Axis a foregone conclusion, Vargas' government issued a tough antimonopoly decree aimed at restricting the role of foreign firms in the Brazilian economy. It was part of the turn toward populism begun in 1943. The U.S. government was disturbed and put Vargas, and later Juan Perón in Argentina, on its list of Latin American presidents that had to go. There were plenty of Brazilians who shared the U.S. view. The Liberal Constitutionalists believed that foreign capital should be welcomed into Brazil. And they saw this issue as one that might help them gain the power they thought had been within their grasp in 1937.

There were other signs of Vargas' shift to the left. In early 1945 he decided to release leftist political prisoners. Most prominent was Luís Carlos Prestes, the leader of the Brazilian Communist Party who had been jailed since 1938. The relaxation of police control greatly helped the Communist Party, the best organized force on the left.

The polarization accelerated as the year went on. The anti-Vargas forces included the liberal constitutionalists, many military officers, and most state political bosses. On the other side were assorted populists, some labor union leaders, and the ideological left, which included socialists and Trotskyites, although the communists were strongest. The confrontation climaxed in October 1945, when the army gave Vargas an ultimatum: resign or be deposed. He refused to step down, so the military declared him deposed. Vargas then acceded and flew off to a self-imposed exile on his ranch in Rio Grande do Sul.

The Second Republic (1946–1964)

Three principal political parties emerged in 1945: the UDN *(União Democrática Nacional)*, the PSD *(Partido Social Democrático)*, and the PTB *(Partido Trabalhista Brasileiro)*. The UDN was a coalition of anti-Vargas forces dominated by the liberal constitutionalists. The PSD was more heterogeneous; it included many political bosses and bureaucrats and some prominent industrialists. The PTB, smallest of the three, was created by Vargas in 1945, when he was still trying to shape the upcoming elections. The PTB was aimed at urban labor with a political approach supposedly modeled on the British Labor Party. These three remained Brazil's principal parties until 1965. They were often described as non-ideological, personalistic, and opportunistic—in short, not to be regarded as modern political parties. As we know from U.S. history, however, parties need not be ideologically precise in order to seek and gain office according to consistent patterns.

Elections for a Constituent Assembly had been called before Vargas' fall, and, when held in December 1945, they proved to be among the

freest in Brazil's history. The newly elected president, with 55 percent of the vote, was General Eurico Dutra, a close Vargas collaborator in the *Estado Nôvo*. The chief opposition candidate was Air Force Brigadier Eduardo Gomes, a throwback to liberal constitutionalism. He won 35 percent of the vote. The Communist candidate received 10 percent of the vote, which greatly encouraged the left. President Dutra and his advisers began watching closely the growth of the left and its links to urban labor.

In 1946 the Constituent Assembly produced another constitution, one that resembled the constitution of 1934. There was decentralization and a return to the classic guarantees of individual liberty. The elections that produced the Constituent Assembly had highlighted some other trends. They showed that the traditional political machines could still predominate in a national vote. That was hardly surprising, since Brazil was still a mainly rural society, and electoral manipulation was easiest in the countryside. Nonetheless, the extensive Communist vote showed that new forces were at work on the urban scene.

Soon after the war Brazil began struggling with the issue of how to finance its economic development. Total reliance on domestic capital was never seriously considered by Brazilian decision makers. One result of Brazil's involvement in the war had been to push the government toward economic planning. In wartime the objective was to maximize mobilization, but the same approach could be applied to peacetime economic development. Instead, the Dutra government (1946–51) avoided planning and returned to reliance on coffee exports, dropping most of the measures taken by Vargas to stimulate industrialization. This policy made Brazil once again highly vulnerable to changes in the world demand for coffee.

On the political front, the Dutra regime soon decided to repress the left. The Communist Party, legalized in 1946, had shown surprising strength in São Paulo and Rio de Janeiro. Labor unions, despite the corporatist legal structure, were gaining de facto autonomy, to the worry of employers and conservative politicians. Washington was closely monitoring these events, which had parallels elsewhere in Latin America. As would happen one year later in Chile, the Brazilian Congress in early 1947 voted to revoke the Communist Party's legality. Police raided its offices and seized its publications. The ministry of labor intervened in hundreds of labor unions and arrested or dismissed their officers, appointing government stooges in their place. The years 1945–47 had proven to be a rerun of 1930–35: a political opening, then a burst of activism on the left, climaxed by government repression. Henceforth the left was outlawed, and Communist Party candidates had to resort to electoral guises.

Vargas had not accepted his exit in October 1945 as the end of his career. Only two months later he was elected senator from two states, and chose to represent Rio Grande do Sul. During the Dutra presi-

dency, Vargas worked steadily to retain national visibility and maintain his political contacts. Soon his friends and allies were urging him to run for president. He did not need much convincing.

In the presidential campaign of 1950 Vargas was supported by most of the PSD and PTB. His principal opponent was former *tenente* Juarez Távora, running under the UDN banner. There was also a splinter PSD candidate. Vargas conducted a shrewd campaign, attacking the Dutra regime for neglecting economic growth and for favoring the rich. Yet his position was moderate enough to appeal to the landowners who controlled votes in states such as Minas Gerais. Vargas won by a plurality (48.7 percent) and began his third presidency—the only one he gained by popular election.

In returning to power by popular vote, Vargas reversed the victory that his opponents, especially the liberal constitutionalists, had won in 1945. He had beaten them at the game they thought was theirs. They exploded, some even calling for the army to block the return of the ex-dictator. But it was to no avail.

Vargas made economic policy his top priority and he promptly assembled a team of young technocrats—engineers, economists, and planners. They formulated an eclectic strategy designed to maximize the inflow of capital and technology from both public and private sources abroad. The prospects looked favorable. In 1949 the U.S. and Brazilian governments had launched a joint study of the Brazilian economy. The commission's report in 1953 spotlighted inadequate energy and poor transportation as the prime obstacles to rapid economic development. The U.S. government indicated interest in channeling public funds for investment in these areas, and the Brazilian government created new federal agencies to handle the investment projects now in prospect.

Vargas' economic strategy also had its nationalist side. Profit remissions by foreign-owned firms were a frequent target for nationalist attack. By 1951 the remissions were high, encouraged by the growing overvaluation of the Brazilian currency (which progressively increased the value of the companies' earnings in U.S. dollars). In 1952 Vargas denounced the foreign firms and threatened new controls. The outflow slowed, however, and he held back from attempting to impose comprehensive controls.

Another target for the nationalists was oil. Since the late 1930s Brazil had been working on a national oil policy. Argentina and Mexico had already opted for state monopolies. In both cases, nationalist sentiment was a strong political force. Throughout Latin America international oil companies were regarded with strong suspicion. Brazil was no different. In 1951 Vargas proposed a mixed public-private corporation (to be called PETROBRAS) that would monopolize the exploration and production of oil.

The proposal touched off the most heated political debate since

1945. Nationalism proved very strong, especially among army officers. Bitter controversies arose, with state monopoly advocates questioning the patriotism of free enterprise supporters, and vice versa. In 1953 the Congress created an even stronger monopoly than proposed by Vargas. The debate had sharply polarized opinion, reducing the room for political maneuver.

Vargas had been elected in 1950 on a moderate platform, and the party line-up in Congress required him to maintain that course. But economic pressures were forcing hard choices on the government. First, Brazil's rate of inflation turned up from 11 percent in 1951 to 20 percent in 1952. Second, the foreign trade balance went into the red. Third, the U.S. president elected in 1952, Dwight Eisenhower, threw into doubt the loan commitments the Brazilians thought the United States had made for the Joint Commission-endorsed infrastructural investments.

These reverses gave ammunition to Vargas' enemies on both the left and right. The left charged Vargas with selling out to the imperialists, who wanted to keep Brazil in the role of a primary-product exporter. The right, on the other hand, charged that Vargas was alienating the trading partners and foreign creditors on whom Brazil had to depend. Most politically conscious Brazilians fell between these extremes. Yet economic and political pressures were making moderation more difficult, spelling danger for Vargas and his government.

In 1953 Vargas reorganized his cabinet to face the economic crisis. Inflation and the balance-of-payments deficit were related problems because Brazil had clung to an overvalued exchange rate which, combined with Brazilian inflation, had made imports cheaper and exports more expensive. An economic stabilization program was urgently needed. In the short run that would mean falling real wages and strict controls on business credit and government spending. Such a policy was bound to be unpopular.

To lead the effort Vargas recruited Oswaldo de Aranha, his longtime political lieutenant, as minister of finance. Aranha pursued classic stabilization measures with apparent success in 1953. As 1954 approached, however, a bitter fight loomed over wage policy. Under the *Estado Nôvo* the ministry of labor fixed the minimum wage. It had not been increased for many years, despite the accelerating inflation. Aranha's objective was to prevent an increase so large as to wreck the anti-inflation program. For this Aranha would have to deal with the minister of labor: João Goulart, a young PTB politician and Vargas protégé from Rio Grande do Sul. Goulart was closely identified with the PTB left and the militant labor leadership.

By 1954 Aranha and Goulart were pulling in opposite directions: Aranha toward austerity and Goulart toward a populist, redistributionist path. Vargas had to decide the issue. In February, apparently opting for Aranha's austerity, he dismissed Goulart. The left, strength-

ened by its success in the fight over oil policy, now attacked Vargas for pandering to the imperialists with his stabilization program. Vargas cleared the air on May 1, 1954, when he announced a 100 percent increase in the minimum wage—higher even than Goulart had recommended.

This battle now merged into a wider political crisis. Vargas' bitterest enemies had found an issue on which they thought they could beat him: corruption. Evidence of financial scandals began to surface. The anti-Vargas propagandists closed in on the weary president. Unbeknownst to him, the palace security chief had arranged an assassination attempt on Carlos Lacerda, a sensationalizing journalist who was leading the attack on Vargas. The bullet meant for Lacerda killed an air force officer who was a volunteer bodyguard for Lacerda, who suffered only a minor wound. The officer's death brought the military into the crisis. When their investigation pointed to the presidential palace, the senior officers demanded Vargas' resignation. Realizing he was trapped and isolated, Vargas put a bullet through his heart on August 24. He left behind an inflammatory suicide letter, blaming his demise on sinister forces, domestic and foreign, and proclaiming a highly nationalist position. By his sensational exit, Vargas exacted revenge on his tormentors. Lacerda had to flee Brazil, and the anti-Vargas factions, especially among the UDN and the military, found themselves on the defensive.

Caretaker regimes governed Brazil until the 1956 inauguration of Juscelino Kubitschek, elected to a full presidential term in 1955. Much to the surprise of both supporters and opponents, Kubitschek served out his full term. He was an ebullient PSD politician and former governor of Minas Gerais with a reputation as a skillful campaigner. Although he won the presidency with only 36 percent of the vote, he quickly moved to gain broader support.

Mindful of how often the military had intervened in politics, Kubitschek mollified them with large weapons purchases. He was also fortunate to have a war minister, General Henrique Teixeira Lott, who enjoyed great prestige among the officers and who proved staunchly loyal to the president. Kubitschek also had an effective PSD-PTB coalition in the Congress. Finally, the "Target Program" of economic development, plus the audacious idea of building a new capital, Brasília, in the interior, combined to generate enthusiasm which muffled the bitter political conflicts from the mid-1950s.

(The futuristic city of Brasília, built from scratch in four years on a completely undeveloped plateau site 600 miles from the old capital of Rio de Janeiro, also captured the imagination of the outside world. André Malraux called it "the capital of hope." President Eisenhower was one of many heads of state to attend its inauguration in 1960. Brazil was now on the world map for the daring, if controversial, urban planning and architecture developed in Brasília.)

No small part of Kubitschek's political success was due to his own

talents. It took great political skill to orchestrate the coalition that sustained his policies. Kubitschek's motto had been "fifty years of progress in five," and the economic leap forward was impressive.

Yet it would have been too much to expect Kubitschek's political strategy to endure forever. The PSD-PTB alliance in Congress was coming apart, growing discord among military officers precluded any repeat of General Lott's role, and the economy had once again run into inflation and balance-of-payments deficits. Kubitschek briefly tried economic stabilization in 1958–59, but scuttled it when the IMF demanded austerity measures that would have prevented Brazil from reaching the president's economic "targets." Kubitschek pressed on with his economic program, and that created mammoth problems for his successor. When he left office in January 1961 no one doubted that a reckoning with foreign creditors was at hand.

The president who inherited this challenge was Jânio Quadros, one of Brazil's most talented and most flawed politicians. A whirlwind success as governor of São Paulo, Quadros won big in the 1960 presidential election, running with UDN endorsement. His campaign featured a broom as the symbol of his fight against corruption. That talk buoyed the liberal constitutionalists, who believed that at last power was near.

Quadros began by embracing a tough stabilization program, while at the same time launching an "independent foreign policy" especially aimed at new overtures toward the socialist camp. After seven months of idiosyncratic rule, however, Quadros suddenly resigned in August 1961. His reasons have never been entirely explained—apparently he expected the Congress to reject his resignation and offer him increased powers. He was wrong; the Congress promptly accepted his resignation. Quadros, the most charismatic populist politician of modern Brazil, faded into a retirement punctuated by occasional oracular pronouncements.

Quadros' self-engineered demise was demoralizing for the anti-Vargas factions Getulistas and other Brazilians who believed that his moralistic promises and his administrative success in São Paulo boded well for the new federal government. Worst of all from the UDN viewpoint, Quadros' departure meant power would now pass to the elected vice president—Vargas's former labor minister, João Goulart, the epitome of populism and anathema to the conservative military.

The military was in no mood to agree to Goulart's succession to the presidency. But the "legalists" among the officers argued in favor of observing the constitution. A compromise was reached. The Congress created a parliamentary system in which Goulart was president but was obliged to govern through a cabinet accountable to the Congress. It was an unworkable hybrid, designed solely to reduce Goulart's power. The new president assumed his diminished powers in September 1961 and promptly started a campaign to get the parliamentary innovation repealed. January 1963 brought success when a plebiscite restored the

The glittering capital of Brasília boasts extraordinary modernistic architecture. *Above*, headquarters of the governor of the federal district; *below*, the legislative palace, whose twin towers and buildings contain the separate houses of the national congress. (Courtesy of the Consulate General of Brazil, New York.)

full presidential system. By then Goulart had precious little time left from the 1961–66 presidential term.

Goulart's presidency proved ill-starred from the beginning. The odds were against him on almost every count, but he made them worse with his inexperience, weakness, and indecision. By 1963 inflation and the balance-of-payments deficit had grown even more difficult to deal with. Goulart chose his own stabilization team, headed by the brilliant intellectual-politician Santiago Dantas and the noted economist Celso Furtado. Dantas worked out a detailed plan, duly negotiated with the U.S. government and the International Monetary Fund. It called for the usual: reduction of government deficit, tough controls on wages, and painful reductions in credit. It was the same medicine that had been served up in the stabilization efforts of 1953–54, 1955–56, 1958–59, and 1961.

For Goulart, stabilization presented special problems. A tough wage policy, which always meant falling real wages, would strike at the social group to which Goulart was most committed. Furthermore, meeting the harsh terms of foreign lenders would invite attack from the nationalists, another area of his prime support. To what end should Goulart sacrifice his slender political capital? Even if he could bring off stabilization, his term would probably end before Brazil could resume rapid growth.

Notwithstanding the gloomy prospects, Goulart endorsed the Dantas-Furtado plan. But he did not stay with it for long. In a few months the political costs had become all too evident. Dantas quietly resigned, Furtado had already left Brasília, and any further serious stabilization effort was thereafter out of the question.

Stabilization was not Goulart's only worry. Since 1961 the Brazilian political scene had been heated up, stoked by passionate opinion on both left and right. The military, as always, was a key factor. Some of the officers who had fought Goulart's accession to power in 1961 were still fighting. They had begun an ongoing conspiracy to overthrow Goulart. Many of the ideas and personnel of the conspiracy could be traced to the 1954 military cabal against Vargas. What steadily increased the strength of the conspirators was the increasingly radical tone of the political combatants.

The left of the political spectrum had become very crowded. A rising sense of confidence had gripped the radical nationalists, who included Catholic literacy teachers, labor union militants, Trotskyist student organizers, and artistic idealists, all spreading a revolutionary message through popular culture as well as "high" art forms. By early 1964 the radical left had gained government blessing, sometimes even government financing and logistical support.

Conservatives were incensed over nationalist inroads among two groups. One was the military. Brazilian enlisted men had traditionally enjoyed only restricted political rights. Voting was not among them.

The radicals began to organize among these "neglected" ranks to the point of advocating unionization of enlisted men. This scandalized the officers, who were hardly about to learn collective bargaining. Even politically centrist officers could understand that threat.

The other new area of mobilization was the countryside. In 1963 rural unionization was legalized and competing groups, including several on the left, vied to win sponsorship of local syndicates. Yet the rural sector was an unpromising arena for the Brazilian left to test its power. There was always excess labor, and landowners traditionally ruled with an iron hand. This rural unionization campaign, combined with a few land invasions, provoked landowners to take decisive action. They pressured the pro-landowner politicians, who were numerous in a federal Congress which overrepresented rural districts.

Although Goulart's opponents had made an impact in the Congress by early 1964, they did not have the votes to impeach him. The old PSD-PTB alliance still operated. It might not back a stabilization program, but it was also not ready to serve the anti-Goulart conspirators. The plotters saw only one way out: a military coup.

The president's military advisers had warned him about the conspiracy, but they assured him that only a minority of officers were involved. That may have been true but it was rapidly changing. Now even centrist officers were leaning toward a coup. The principal factor pushing them was the radical move to the left already under way, either by the president or by those who controlled him.

The U.S. government was taking a strong interest in Brazil's emerging political confrontation. Both the U.S. ambassador, Lincoln Gordon, and the U.S. military attaché, General Vernon Walters, were in close touch with the conspirators, both military and civilian. The United States had a contingency plan to support the anti-Goulart rebels with fuel and weapons, if needed. As it happened, they were not. On March 31 speculation ended as a military revolt, surfacing first in Minas Gerais, spread across the country. Within twenty-four hours João Goulart had fled into exile in Uruguay.

On April 1 the leader of the Congress, in Goulart's absence, declared the presidency vacant. Although his action lacked any legal foundation, the Congress acceded. Into the power vacuum moved the military conspirators and their civilian allies. Brazil once again opted for the authoritarian path to development.

In retrospect, the breakdown of Brazilian democracy (such as it was) bore a close connection to the interplay of social-class relations. The populist policies of Getúlio Vargas constructed a hierarchical order through which the state created and controlled institutions for organizing urban workers. This posed a significant but ultimately acceptable challenge to the upper and middle classes, the latter represented largely by the military. But in 1964 Goulart presented, or appeared to present, a much more fundamental threat. By mobilizing peasants as

well as workers, and by using radical rhetoric, he seemed to be creating the conditions for a *classwide* worker-peasant alliance against the socio-economic establishment. Both the suddenness and the simultaneity of these movements startled and alarmed elites and onlookers, who recoiled at the prospect of such far-reaching and communistic alterations in the country's political structure. A classwide coalition was simply not acceptable. The military exercised its longstanding veto power and went on to create a bureaucratic-authoritarian regime.

Military Rule

The conspirators of 1964 were surprised at the speed with which the Goulart government collapsed. The radical nationalists, so confident of their mobilization, found themselves without followers. The justice minister called for a general strike on April 1, but the response was minimal. Goulart's zig-zagging and the divisiveness within the left had undercut any effective mass support. The rebels encountered little or no resistance as their troops seized command of government.

From 1964 to 1985 Brazil was governed by a succession of authoritarian regimes, each headed by a four-star general. Despite variations in structure and personnel, all were coalitions of military officers, technocratic administrators, and old-line politicians.

The most important group was the military. Army officers have had a long history of intervention in politics since the empire was brought down. There was the military regime of Floriano Peixoto (1891–94), and later the military interventions in state politics in 1910–14, followed by the revolts of junior officers in 1922 and 1924. In 1930 the military ended the Old Republic by delivering power to Vargas, whom they kept in power by the coup of 1937, only to depose him in 1945. It was a military manifesto that led to Vargas' suicide in 1954 and it was a "preventive" coup in 1955 that ensured Kubitschek's succession to the presidency. Finally, the military led the fight against Goulart's succession to the presidency in 1961 and then conspired to bring him down in 1964. Army officers were seen by all to be vital actors in Brazilian politics.

In the years after 1945 the army officer corps has been buffeted by conflicting political currents. The 1950s brought a polarization between nationalist and anticommunist positions. The former favored statist solutions in key economic sectors (such as oil) and a relative independence in foreign policy (strongly against, for example, sending troops to fight in Korea alongside the United States). As their label would indicate, the anticommunists identified with the United States in the deepening Cold War and saw the nationalist left as a stalking-horse for pro-Castroites or communists.

Officer opinion turned decisively against the populists, of whom Goulart was a principal example. The Goulart government's inability

to get control of the economy (Brazil was near default to foreign creditors in March 1964); the mobilization of the lower sectors; and the direct threat to military hierarchy all pushed centrist military officers toward supporting a coup. By early 1964 the conspiracy was headed by General Humberto Castello Branco, the staunchly legalist army chief of staff who had supported Goulart's succession to the presidency in 1961.

Once the military had deposed Goulart, a new question faced the conspirators: the form and direction of the new government. The hard-liners argued that Brazilian democracy had been corrupted by self-seeking and subversive politicians. The country needed a long recuperation, which would require such measures as purging legislators, suppressing direct elections, and firing civil servants. The hard-liners' economic views were less easy to discern. They obviously detested the radical nationalists and the populists.

The moderate military composed another group. They believed a relatively brief interval of administrative and economic reorganization could return Brazil to the electoral democracy recently endangered by irresponsible politicians. The violation of democratic and constitutional principles bothered the moderates, who hoped for a rapid return to legal normalcy.

General Castello Branco was quickly chosen by the (purged) Congress as the new president and served until 1967. He considered economic stabilization to be his government's top priority. The immediate need was to bring inflation under control and to improve the balance of payments. Roberto Campos, a well-known economist-diplomat, was made planning minister and became the dominant figure in economic policy making. Inflation was reduced, but by much less than had been hoped, and a surplus was achieved in the foreign accounts, mainly because a government-induced recession greatly reduced demand for imports. Campos' team also attempted to reorganize and update Brazil's principal economic institutions. The banking system was overhauled (a proper central bank was finally created), a stock market and a government securities market were institutionalized for the first time, labor regulations were revised to make easier the discharge of employees, and export regulations were simplified. Campos had long argued that capitalism had not failed in Brazil because it hadn't yet been tried. This was his chance. The short-term results were disappointing, but Castello Branco and Campos did not despair; they saw their efforts as unpopular in the short run, but indispensable for sound growth in the future.

The hoped-for economic upturn did not occur in 1965–66 and Castello Branco was persuaded to extend his presidential term a year in the hope that the economy would improve. In fact, the country's economic problems could not be resolved even in his two and a half years.

The second military government, that of President Artur da Costa e Silva (1967–69), brought an ugly turn in politics. The president had

hoped to preside over a liberalization, but events proved otherwise. Until 1967 the authoritarian government had shown considerable tolerance for the opposition, at least in comparison to Spanish American military governments of the 1960s and 1970s. But tolerance invited mobilization. In 1967 and 1968 the opposition mounted a series of protests, climaxing in mass demonstrations in Rio.

The hard-line military, now opposing any compromise between democracy and a "tough" government, argued for a crackdown. In November 1968 a series of industrial strikes spread from Minas Gerais into the industrial heartland of São Paulo. The Costa e Silva government hesitated, then reacted by strongly repressing the strikers. A pattern was set: an authoritarian government resorting to dictatorial measures to carry out its version of rapid economic development. It was a growth strategy based on repression of labor unions, avid recruiting of foreign investment, and high rewards for economic managers.

In 1969 Brazil was hit with new levels of political violence. The militant opposition had produced a guerrilla network, mainly in the cities. In September 1969 President Costa e Silva suffered a debilitating stroke and the guerrillas made it the occasion to kidnap the U.S. ambassador, whom they subsequently released, in return for the government's releasing from prison fifteen political prisoners and the publishing of a revolutionary manifesto in all the media. For the next four years Brazil experienced guerrilla warfare. A small cadre of revolutionary activists kidnapped foreign diplomats, holding them hostage to ransom other revolutionaries in prison. The guerrillas also wanted to draw the government into a more systematically repressive role, thereby hoping that more of the populace would be alienated.

By 1973 the guerrilla movements were vanquished. They had exhausted their human resources to achieve meager results. They rescued a number of their revolutionary comrades from prison and torture, but in so doing they reinforced the repressive apparatus and made credible the hard-liners' argument that any political opening meant civil war.

When General Ernesto Geisel assumed the presidency in 1974 he repeated the earlier moderates' hopes of a return to democracy and the rule of law. A major obstacle was the security apparatus, including both army and civilian units, which had gained great influence within the government. Their unsavory methods, including torture, had facilitated the liquidation of the revolutionary opposition, but had given them a powerful veto over liberalization.

President Geisel's commitment to redemocratization came from his close personal link to the legalist tradition of Castello Branco. Geisel saw this process not as a response to pressure, but as the working out of a democratic commitment inherent in the military intervention in 1964.

The fundamental problem for Geisel, as for all the preceding military governments, was the inability to win a free popular election. This

would not have mattered if the military had not taken the democratic rules so seriously. But they did, and the result was an endless series of improvisations to make the voting results fit their preferences. The depth of the problem was shown in October 1974 when the new government, in contrast to its predecessor, allowed relatively free congressional elections. The result was a landslide for the opposition party. The lesson was clear: if given a choice, the public, especially in the urban industrialized centers, would vote against the government.

After 1967 the Brazilian economy returned to a growth path, duplicating the record of the 1950s. From 1968 to 1974 the growth rate averaged 10 percent, and exports more than quadrupled. As though to mark the end of an era, manufactured goods replaced coffee as the country's leading export product. Outside observers soon talked of the "Brazilian miracle." It was achieved by low wages and easy credit to purchasers of consumer durable goods.

But the "miracle" began to fade by the end of the decade. In 1980 inflation was more than 100 percent, and Brazil faced a serious and growing deficit in its international balance of payments. The foreign debt mounted and industrial production sagged. Furthermore, industrial labor had bestirred itself in São Paulo, staging a series of strikes in 1978, 1979, and 1980. The church, in the person of Cardinal Arns, supported the strikers and helped dramatize the disproportionate share of the sacrifice they had borne during the "miracle." To deal with the balance of payments crisis the government was forced to run a recession in 1981. Naturally this was unpopular, as unemployment spread in the industrial heartland of São Paulo. In 1982 Planning Minister Delfim Neto and his fellow policymakers hoped to engineer an economic recovery, all the more since it was to be an election year.

These hopes were soon dashed by the world recession, which depressed the value of Brazilian exports, while high interest rates kept the cost of servicing the foreign debt at a crippling level. By the end of 1982 Brazil gained the dubious honor of having the largest foreign debt in the world ($87 billion—although many observers thought its short-term obligations ran the total up to $100 billion), and, like Argentina and Mexico, had to suspend payments on principal. To get the essential "bridging loans" to meet immediate obligations, Brazil agreed to an IMF-architected economic plan that involved a brutal reduction of imports in order to earn a trade surplus.

From Liberalization to Redemocratization

These bleak economic prospects spelled trouble for the "redemocratization" process begun during the years (1974–79) of President Geisel. After several false starts, President João Figueiredo (1979–85) worked hard to deliver on the promise of direct elections in 1982. For the first time since 1965, Brazil directly elected all its state governors in Novem-

ber 1982. Also elected were all the federal deputies, a third of the federal senators, and virtually all the mayors, state representatives, and local council members. The opposition Party of the Brazilian Democratic Movement (PMDB) won a smashing victory in the most developed states, winning the governorships of São Paulo, Rio de Janeiro, and Minas Gerais. The government party, the PSD, lost control of the Chamber of Deputies, but retained control of the electoral college, which would elect the new president in 1985.

That election proceeded in a very Brazilian way. The opposition party (PMDB) candidate was Tancredo Neves, a skillful, old-style politician from Minas Gerais. He shrewdly began by reassuring the military of his moderation. Meanwhile, Paulo Maluf, the government party (PSD) candidate and former São Paulo governor, alienated his party by his heavy-handed campaign. Enough PSD electoral college delegates defected to elect Tancredo.

Tancredo did not live to fulfill the great hopes the public had in him. On the eve of his inauguration he underwent emergency intestinal surgery from which he never recovered. Former Senator José Sarney, the vice president elect, became president. Ironically, Brazil's first civilian president in twenty-one years was a previous PSD leader and former pillar of the military regime. He, too, had defected to run with Tancredo.

Tancredo had promised, if elected, to attack Brazil's myriad social and economic ills. One step in that direction would be an end to urban labor's manipulation via the semicorporatist structure Vargas had architected. Government efforts to reformulate the structure bogged down in fierce cross fire between employers and the new, more militant union leaders. Meanwhile, strikers continued to face repressive police, who at least were less brutal than under the military rule.

Another key socioeconomic problem was the gross inequality in rural land ownership. The Sarney government announced an unrealistically ambitious goal of redistributing millions of acres to the rural landless. The landless were often led by Roman Catholic clergy who sometimes organized land invasions. The landowners responded with an angry campaign (and hired gunmen in some cases), and the program virtually died.

The best that could be said of the Sarney presidency was that the military remained on the sidelines and the president was committed to redemocratization. The economy took Brazilians on a roller coaster. Inflation reached 227 percent in 1985 but came down dramatically with a 1986 wage-price freeze. The spurt of economic growth in 1985–87 was followed by stagnation. The Sarney government ended in an orgy of inflation. The scene was now set for some new leader, capable of bringing new solutions to Brazil's pressing problems.

The new face was Fernando Collor de Mello, a young and previously unknown former governor of the poor northeastern state of Alagoas.

He mounted a lavishly financed television-based campaign aimed at the more than three-quarters of Brazilian homes with TV. His chief opponent in the 1989 campaign was the former labor union leader Luis Ignácio Lula da Silva ("Lula"). Collor won in a runoff, although Lula's percentage of the vote (47 percent) had reached a level unprecedented for the left.

By mid-1991, after fifteen months in office, Collor proved a bitter disappointment. He had begun, à la Jânio Quadros, with a highly autocratic style and a personal arrogance ill-suited to Brazilian politics. Admittedly, his task was made more difficult by the extremely fragmented Congress he faced.

Collor chose to bet on economic stabilization. Unfortunately, his program relied on such short-term gimmicks as the freezing of financial assets and the immediate abolition of indexation. Both proved ineffective after only a few months. Collor also ordered the massive firing of civil servants, hoping to reduce the swollen government payroll and thus gain fiscal relief. The dismissals were rapidly reversed on the grounds that job tenure was constitutionally guaranteed. By early 1991 the stabilization plan had come apart. Inflation hit an annual rate of 1585 percent, fiscal control was lost, and indexation was back. The Brazilian economy returned to its pattern of drift, discouraging foreign and domestic investors alike.

Collor had also begun an ambitious program of neo-liberal reforms. It included privatization, deregulation, and opening of the economy through lower tariffs. Many of these proposals aroused strong opposition from industrialists and from nationalists in the Congress. The government's single victory in this sphere was the sale of a major state-owned steel mill, which greatly increased its profits and productivity once in private hands.

Collor failed to see any of his programs through. In little more than two yeas he lost his mandate. His nemesis proved to be the specter he had campaigned against in 1989: corruption. Investigative reporters, a disgruntled presidential brother, and a congressional inquiry furnished proof that Collor was enmeshed in a vast web of bribery orchestrated by an influence peddler called P. C. Farias. Collor turned to television for his defense weapon, but his telegenic skills had worn thin. Public indigntion led to a civil campaign for the president's impeachment and removal. In September 1992 the Chamber of Deputies overwhelmingly voted his impeachment, and Collor resigned only hours before the Senate approved his conviction on grounds of official malfeasance.

The vice president who succeeded him was Itamar Franco, a former Senator and political nonentity whose personal honesty was his greatest recommendation. But his government, which lacked any party base, also lacked political direction. Inflation soared to an annual rate of 2490 percent in 1993. By hemispheric consensus, Brazil was regarded as the sick man of South America.

Itamar's government finally found an anchor when Fernando Henrique Cardoso became finance minister in late 1993. Cardoso gathered a group of talented technocrats well experienced in the politics of stabilization. In July 1994 they launched yet another anti-inflation program. But this one, far better conceived than its predecessors, showed immediate results.

Cardoso capitalized on this success and the resulting mood of confidence to run for president in October 1994. Overcoming his past reputation as a leftist intellectual, Cardoso, a former Senator from the Brazilian Social Democratic Party (PSDB), won the endorsement of the conservative party. Without a significant right wing candidate in the fray, Cardoso won 54 percent of the vote, easily defeating Lula, again the runner-up. As the Cardoso government assumed power in early 1995, its major challenge was to continue the stabilization program. That would require, above all, winning congressional cooperation in fiscal policy. The government sought also to resume the privatization program, notably stalled under Itamar. The public appeared ready to accept privatization, but tightening congressional spending would be much more difficult.

Despite the new president's rhetoric, his government was unlikely to be able to attack the social problem for some years. Stabilization and economic restructuring would preoccupy Brasília, and government budgets would remain tight.

Brazil had spent the early 1990s on an emotional roller coaster. The decade opened with the dazzling promise of Collor. Then came his tawdry fall amid carnival-like celebrations. Next was the bizarre behavior of Itamar, a widower who changed policies almost as often as his fiancées. If 1994 brought an unprecedented fourth world championship in soccer, it also brought the violent death of Ayrton Senna, the world's leading race car driver, whose funeral paralyzed the country for two days. By 1995 the shell-shocked Brazilians were ready for a strong hand at the helm. Their choice, Cardoso, was a former sociology professor who was once a high priest of the *dependencia* school of social analysis. His job now was to help render that term obsolete for Brazil.

SIX

PERU
Soldiers, Oligarchs, and Indians

Any understanding of modern Peru must begin with its geography. Located on the Pacific coast of South America, and approximately twice the size of Texas, the country has three geographical regions: coast, sierra, and *montaña*. The coastal area, arid and dry, has for centuries been dominated by the city of Lima, whose population in 1993 came to about 4.2 million (nearly one-fifth the national total of 23 million). The sierra is the Andean mountain range, a world of snowcapped peaks and chilly valleys that contain such ancient provincial cities as Cuzco and Ayacucho. And the *montaña,* on the eastern slope of the Andes, is a jungle region whose tropical forests extend to the upper reaches of the Amazon River, where the town of Iquitos is the major headwater port. Because of its inaccessibility the *montaña* has until recently seen little settlement, although today colonization has made it a dynamic and growing region.

These geographical features have helped to create disparate regional economies. The coast has given rise to commercial agriculture and fishing industries. The sierra has been an area for mining, livestock, and subsistence agriculture. Notwithstanding its natural resources the *montaña* has not enjoyed sustained growth; there has been a rubber boom and, in recent years, cultivation of coca leaves (for the international narcotics trade), but no sustained prosperity. More recently, however, the successful cultivation of coffee, sugar, and fruit (for the Lima market) has energized the regional economy.

Ethnic variations compound these differences. Once the center of the Inca empire, Peru continues to have a large Indian population. The 1940 census, the last to use racial categories, classified 46 percent of the population as Indian. This was probably an undercount in ethnic terms, since the definition of "Indian" depended more on physical appearance than ethnic heritage. The current figure is in all likelihood about 35 percent. It has been estimated that there are now about 3 million speakers of Quechua in Peru, and a sizable number of Aymará speakers as well. Indians live (and have lived) mainly in the sierra, of-

ten in tightly knit traditional communities, perpetuating folkways that go back to Inca days.

Peru has a relatively small layer of whites, a stratum that originated with the Spanish conquerors and now makes up about one-tenth of the national population. Aside from blacks and Asians, most of the rest consists of *mestizos*—perhaps one-third or more of the total. As in other Latin American countries, the *mestizos* occupy an ambiguous position in Peruvian society, representing both the burdens of an oppressive past and, occasionally, the prospects for a radically different future. *Mestizos* are often known in Peru as *cholos,* a term sometimes used as a pejorative for Indians trying to pass as mixed-bloods.

Not surprisingly, this mixed ethnic background had led to widely divergent interpretations of Peruvian history and society. Some writers have violently denounced the country's Spanish legacy. Alberto Hidalgo, for instance, proclaimed in the early twentieth century: "I hate Spain because it has never done anything worthwhile for humanity. Nothing, absolutely nothing. . . . Spaniards are brutes by nature." But Spain has also had its apologists. Bartolomé Herrera, a nineteenth-century priest, had only praise:

> The work which the Spaniards accomplished . . . was the greatest work which the Almighty has accomplished through the hands of man. To conquer nature, to master inward fears, to dominate far-off places through the formidable power of the intrepid heart, to accomplish all this and to take as the trophy of victory a new section of the world with an immense population which for thousands of years had been lost to civilization, and then to infuse this world with Christianity, to introduce the fire of life into millions of moribund souls, to broaden by millions of leagues the sphere of human intelligence, was an accomplishment of unparalleled splendor.

The implication was clear. Peru should place its power, and its hopes, in its citizens of European descent.

Such commentators also usually denounced the Indians. They were depicted as lazy, shiftless, hopelessly addicted to the chewing of coca leaves. The Indians stood beyond redemption; they posted an unyielding obstacle to national progress. If Peru had problems, they could be traced to the stubborn survival of the heirs of the Incas.

The *mestizo* has also come in for criticism. Just as José Vasconcelos was propounding his pro-*mestizo* ideas about "the cosmic race" in Mexico, Alejandro O. Deústua offered in 1931 a biting critique of Peru. "Among us," he said,

> the problem of the *mestizo* is much more grave than in other countries. The product of the Indian in his period of moral dissolution and the Spaniard in his era of decadence, the *mestizo* has inherited all the defects of each without being able to conserve the remains of the gentlemanly life of the conqueror. . . . The mixture has been disastrous for the national culture.

Miscegenation was for Deústua not a sign of social progress but a symptom of backwardness. Peru was condemned by the racial composition it had inherited.

During the twentieth century, Peruvian thinkers have sought to resolve the haunting problem of national identity. Some have found inspiration in the country's Indian heritage. A notable case was José María Arguedas, the only novelist able to penetrate both the indigenous and creole worlds of Peru. He was a *mestizo* who had lived in Indian communities as a child and was fully bilingual in Spanish and Quechua. In *Canto Kechwa* (1938), he argued that

> The indigenous is not inferior. And the day on which the people of the highlands who still feel ashamed of the Indian discover of their own accord the great creative possibilities of their Indian spirit, on that day, confident of their own values, the *mestizo* and Indian peoples will definitely prove the equality of their own creative ability with that of the European art which now displaces and puts it to shame.

Arguedas found the burden of his own ethnic heritage too much to bear and committed suicide in great despair.

Others have searched for a fusion of Spanish and Indian components, as did Víctor Andrés Belaunde in a book entitled *Peruanidad* (1957). Yet the basic question persists: Can Peru become a unified nation?

The Independence Period

During the colonial era Peru was a major source of income for Spain. The silver mines of Potosí in upper Peru (now Bolivia) produced vast amounts of wealth, particularly in the sixteenth and seventeenth centuries, and rich veins were subsequently found in the Peruvian sierra. The trappings of Spanish civilization soon appeared. The University of San Marcos was founded in 1551. The inquisition set up court in 1569, as the church became a powerful institution. Lima, an impressive metropolis for its place and time, was worthy of its name: the City of Kings.

Peru suffered a prolonged economic crisis during the late eighteenth century. Silver production slumped, although there was a short-lived boom in the 1790s. The Bourbon free-trade policies reduced Peru's share of trade with Upper Peru and Chile, as goods now came in overland from Buenos Aires. The creation of a viceroyalty in the Río de La Plata region curtailed the activity and prominence of Lima's royal bureaucracy. Túpac Amaru II (the *mestizo* José Gabriel Condorcanqui) led an unsuccessful, bloody Indian revolt in 1780–81. Although the uprising was unrelated to the Spanish crown's administrative changes, many of the elite thought the revolt's message was clear: Peru was in decline.

Paradoxically, the economic trend did not produce a widespread in-

dependence movement. Lima's intellectuals learned of the Enlightenment and launched a liberal journal, the *Mercurio Peruano,* in 1791. But they did not clamor for independence from Spain. Rather, they argued for concessions within the colonial framework, for policies that would bring back the privileges and prosperity of the pre-Bourbon era. There was a brief separatist movement—in the sierra, under a *mestizo* (socially known as an Indian) named Mateo García Pumacahua—but it was quickly snuffed out. While the cause of independence was sweeping through the rest of Spanish America, Peru remained a loyalist stronghold of the crown.

When liberation came, it was from outside. Late in 1820, having led his troops over the Andes from Argentina to Chile, José de San Martín reached the southern coast of Peru. Several months later the Spaniards evacuated Lima and on July 28, 1821, San Martín proclaimed the independence of Peru. Recognized as the "protector" by the local populace, he began making plans to establish a monarchy—and commissioned an agent to search for a suitable European prince. This brought opposition from liberals, who wanted a republican form of government, and the project disappeared after San Martín's fateful meeting with Bolívar in late 1822 and subsequent withdrawal from the scene.

A special assembly passed a constitution in 1823 and José de la Riva Agüero became the country's chief executive, but independence still had to be won. The following year Bolívar resoundingly defeated the Spaniards at the battle of Junín, and Andrés Santa Cruz delivered the coup de grâce at Ayacucho. For all practical purposes Peru thus became free, though Spain refused to recognize the independence of its colony.

Bolívar then proposed a confederation of Peru with Upper Peru and Gran Colombia (Ecuador, Colombia, and Venezuela)—under his leadership, of course—and secured acceptance from a timid Lima assembly in 1826. Opposition mounted and Bolívar went northward to quell the criticism, having reluctantly approved a plan for a separate Bolivia.

Caudillismo and geopolitics took hold in Peru. A series of military chieftains battled for the presidency, with coups and countercoups the order of the day between 1828 and the early 1840s. General Agustín Gamarra, having captured the presidency in 1839, attempted to subdue and annex Bolivia, but met his death on a battlefield in 1841.

Peruvian politics in the postindependence era presented a paradoxical scene. Having defeated Spain through the help of outsiders, Peru found it difficult to assert autonomy from neighboring states. Having refrained from open conflict until the 1820s, Peru fell under the sway of military dictators. And having cast off the burdens of colonial rule, Peru found many saying its own society was in steady decline.

The economy was exceedingly weak. Fighting in the early 1820s had left Callao, the principal port (near Lima), in a state of ruins. Landed estates along the coast and in the sierra had been ravaged. Commerce

remained in depression. The mines were in disrepair. The national treasury was nearly empty, and from the 1820s onward the government began to accumulate a series of foreign debts (mainly to British lenders) that would later prove to be nearly ruinous.

Nor had conditions improved for the Indians, who at this time composed about 70 percent of the total population (between 1 and 1.5 million). The traditional tribute, formally abolished with the expulsion of Spain, was replaced by a head tax, the so-called *contribución de indígenas*. And under the "liberal" doctrines of the era Indians were seen as individuals, not communities, so they no longer enjoyed their previous special protection. Some sought refuge as peons on estates or as workers in mines. Others tried to pass as *mestizos* and find employment in the cities and towns.

The Guano Age

Nature came to Peru's rescue. For centuries the coldness of the country's offshore waters, due to the Humboldt current, had attracted large numbers of fish. The fish in turn attracted birds, who left their droppings on islands near the coast. Atmospheric dryness aided the preservation and calcination of these deposits, known as guano, which had a high concentration of nitrogen. Guano, as the Incas had known, was a first-class fertilizer.

In 1841 the first shipment of Peruvian guano reached the port of Liverpool. Thus began a half-century of export-led growth in Peru, and a period of apparent prosperity.

It was a special bonanza for the hard-pressed government. The islands with the guano deposits were on public property, not private land. This posed both a practical and a theoretical question for liberal policymakers: How should a government committed to laissez-faire principles take advantage of this virtual monopoly?

The answer was found in the "consignment" system, through which the government would lease out (usually exclusive) exploitation rights to a merchant house or partnership. Under such contracts the government would obtain a fixed share of the total value of sales, perhaps two-thirds or so, rather than a tax on profits. The merchants, usually foreign, would receive reimbursement for costs plus a percentage of the sale. The strategy seemed sensible enough, since it involved the liberal state in a partnership with private enterprise to the advantage of both.

But the consignment system had some important ramifications. One was to place the state in constant conflict with the merchant houses. With complete control of supply, the government wanted to sell the guano for as high a price as possible. This could mean holding back on shipments in order to keep prices up. Because of the commission arrangement, however, the merchant was more interested in the total

volume of sales than in the *price* of any individual shipment. After all, the merchant had the contract for a specified period of time—during which the point was to sell as much guano as possible. High prices would help, of course, but it would be more profitable from the merchants' viewpoint to sell large amounts of guano at moderate prices than to sell very small amounts at high prices. As a result, the Peruvian government and its consignees constantly bickered throughout the guano age.

A second factor derived from the size of required investment. Though the guano itself was readily accessible, piled in open-air mounds, its exploitation required large-scale capital investment: in ships, warehouses, transportation, lodging, and wages. Moreover the government continually demanded cash advances from its leaseholders. This restricted the pool of possible consignees to merchants with strong capital reserves, and given the plight of the local economy, few Peruvians could take advantage of the opportunity.

A third factor, related to the second, derived from the foreign debt. As early as 1822 the Peruvian government negotiated a loan of 1.8 million pounds sterling from a British banking house. Subsequent loans and accumulated interest led to a massive debt, and Peru fell behind on its payments. British bondholders became impatient and quickly saw the guano bonanza as their salvation.

They were right. The guano trade eventually passed largely into British hands. In 1849, just as the Peruvian legislature was approving a resolution granting Peruvian nationals preference in the consignments, the government negotiated a contract with the London firm of Anthony Gibbs. The agreement authorized Gibbs to buy up debt securities from British bondholders at the going market price (about 40 percent less than face value) and then allowed Gibbs to use these bonds at face value to pay for guano shipments. The government thus took a sizable loss, while Gibbs could expect to make a handsome profit on sales of guano in the European market.

Peru became highly dependent on guano, thereby creating a single-product export economy. By the early 1860s the government was earning about 80 percent of its revenues from guano: this testifies to both the state's reliance on its consignments and the relative paucity of other exports. At the same time, about *half* the government's receipts on guano were destined for English bondholders. The guano boom therefore provided little stimulus for long-run economic growth: as Fredrick Pike has observed, "the greater the windfall gains became, the less self-sustaining the economy grew."

In 1845, as the guano trade was expanding, Peru came under the rule of its strongest nineteenth-century leader: Ramón Castilla, forty-six years old, the intelligent, dark-eyed son of a part-Indian mother and a Spanish-Italian father. A military officer who had distinguished himself at Junín and Ayacucho, Castilla sought to modernize the Peru-

vian armed services. Upon reaching office he submitted to Congress a national budget, the first in the nation's history. He promoted public works, including the construction of one of Latin America's first railroads (extending from Lima to Callao). In foreign affairs he revived Bolívar's dream of Spanish American unity, partly because he feared the southward drive of the United States—which was bringing Mexico to its knees in 1846–48.

Castilla was succeeded in 1851 by José Rufino Echenique, whose main contribution involved consolidation of the domestic national debt. Under this policy the Echenique government recognized as valid all manner of claims from citizens, usually his upper-class cronies, who testified to losses sustained during the Wars of Independence and subsequent conflicts. The Echenique administration paid out about 19 million pesos in this fashion, and according to one estimate at least 12 million pesos went for unfounded claims. The source of the payments needed no explanation: the governmental share of guano sales. Notwithstanding the corruption, the financial reorganization led to some important internal capital accumulation.

In 1854 a disgusted Castilla ousted Echenique and resumed control. Castilla promptly issued two far-reaching decrees: one abolishing the *contribución de indígenas,* the other emancipating black slaves. Social legislation, too, was facilitated by the guano trade, since the government could now afford to forgo the head tax on Indians and to indemnify plantation owners for their slaves. Peru promptly began to tap another source of labor—Chinese coolies, about 100,000 of whom came as indentured workers from the mid-1850s to the mid-1870s and worked primarily in the coastal economy (guano, sugar, cotton).

During this second term, which lasted to 1862, Castilla continued to support military professionalization and public education. He presided over assemblies that produced two constitutions, a federalist document in 1856 and a centralized charter in 1860. And he continued to press for continental unity. As Castilla explained his rationale: "The relative weakness of the South American republics, divided and isolated among themselves, is in the judgment of this government the deplorable cause of the fact that on many occasions we have been treated with grave lack of respect, as if for the great international potentates there did not exist a common law of nations."

After Castilla's exit, troubles developed with Spain, which had never officially recognized Peruvian independence. Spain protested the alleged mistreatment of Spanish immigrants in Peru and in retaliation seized some guano-rich islands about 100 miles south of Callao. A provisional settlement sparked protests and a revolt; Mariano Ignacio Prado came to power, and in 1866 he declared war on Spain. The conflict was short-lived, and in 1869 Spain at last extended official recognition of Peru's independence.

Governments continued to face a spiraling debt. President José Balta

(1868–72) turned the problem over to Nicolás de Piérola, a thirty-year-old *wunderkind* and something of a dandy, and Piérola negotiated an agreement with the Parisian firm of Adolph Dreyfus. Under the plan Dreyfus assumed complete responsibility for the government's foreign debt and forwarded additional loans—in exchange for a monopoly over the purchase and sale of guano. In time the agreement unraveled, after Dreyfus put his own shares on the open market, but in the short run it provided some stability to an ever-changing and chaotic situation. Piérola also made arrangements for Henry Meiggs, the adventurous and unscrupulous North American entrepreneur, to expand railroads throughout the country. The Dreyfus and Meiggs contracts were immediately attacked by nationalist critics as sellouts, but they were approved anyway.

The foreign debt continued to grow. Manuel Pardo, candidate of the country's first political party, the *Civilistas,* held office from 1872 to 1876. Thoughtful and statesman-like, a born aristocrat and self-made millionaire, he wrestled with the debt, promoted public education, and nationalized the nitrate fields. Clearly he was one of the most realistic and effective leaders of his era. Near the end of his term he supported an army general for the presidency, hoping such a president would be able to deal with the continuing insurrections. In 1878, at the age of forty-four, Pardo was killed by an assailant. Peru lost one of its finest leaders.

As it was, things became worse. The focus of the export economy shifted south. Nitrates began to yield profits in what is now southern Peru and northern Chile, then part of Bolivia, and Chilean investors refused to pay new taxes on nitrates claimed by Bolivian president Hilarión Daza. In retaliation, Daza ordered the seizure of a Chilean-owned nitrate operation in Antafogasta. Chile sent in troops to occupy the region. After some hesitation, the Peruvian government of Mariano Ignacio Prado decided to honor an 1873 alliance with Bolivia.

Thus began the War of the Pacific (1879–83), pitting Peru and Bolivia against Chile. It was a total disaster for Peru. Chile won a stunning military victory and occupied Lima. In the ensuing peace treaty in 1883, Chile gained outright control of the nitrate-rich province of Tarapacá, including the city of Iquique; and it was to keep control of Tacna and Arica for ten years, their subsequent fate to be decided by a plebiscite.

The War of the Pacific had far-reaching effects on all three countries. For Chile it ushered in a nitrate boom and boosted national confidence. For Bolivia it denied access to the sea. For Peru it was a humiliating defeat, which further discredited the politicians. Furthermore it increased the debt and disrupted commerce. It was a failure in every conceivable way.

In the wake of defeat came General Andrés Cáceres, who engineered a coup in late 1885 and governed from 1886 to 1890. Attempting to

pick up the pieces at home and to placate angry bondholders abroad, the new president, like so many before him, looked beyond Peru for help. Cáceres began discussions with the London bondholders. Under their plan, named for Michael Grace, the British negotiator, Peru would satisfy its creditors by ceding control of its railroads for sixty-six years, turning over all guano not needed for domestic use, and making thirty-three annual payments of 8000 pounds sterling. The bondholders, for their part, advanced a new loan for 6 million pounds and promised to invest a fair share of earnings in the railroads. Amid disagreement and controversy, the Grace Contract was finally approved in 1889. Like the Dreyfus consignment, for nationalists it has remained a symbol of their government's excessive zeal to please foreign investors.

The Grace Contract marked the end of Peru's guano age. Deposits were nearing exhaustion in the early 1880s and after 1889 Peru had relatively little surplus available for export. The cycle came to a close.

The guano boom left a deep impression on Peruvian history. It provided a lure for foreign investors; it led to extravagance and corruption within the Peruvian government; and it altered social outlooks within the elite, particularly among coastal landowners. "Popular imagination endowed fantastic proportions to the improvisation of fortunes," one writer recalled. "It was the first time that the old forms of social life, more or less static even during political upheaval, suffered a severe shake-up. For the first time money emerged as the exclusive social value." Tempted by the prospects, landowners tried to emulate merchants, traders, and financiers. But this did not mean the creation of an independent middle class or entrepreneurial middle stratum. It meant, instead, the adoption of some entrepreneurial qualities by Peru's coastal aristocracy.

Overview: Economic Growth and Social Change

Since the early nineteenth century the Peruvian economy has undergone three long cycles of export-led growth. Figure 6-1, showing the volume and value of exports, illustrates the general pattern. The first phase, corresponding to the guano age, stretched from the 1830s to the late 1870s. After a period of oscillation the economy recovered in the 1890s and began a phase of expansion that lasted to the Great Depression of the 1930s. The conclusion of World War II reopened international markets and precipitated another cycle of growth that continued to the mid-1970s, when world prices for agricultural and other commodities went into decline.

The pattern of these trends serves to illustrate several key features of Peru's economic development. First, the country has remained highly dependent on exports as a stimulus to growth. Policymakers have almost always focused on the international market, not domestic de-

mand. Second, Peru has therefore remained extremely vulnerable to price swings in the international market and thus to forces beyond its control. Third, as we shall see, each of the three cycles has reflected the rise (and fall) of one or more products, a fact which had led to boom-bust cycles for different regions of the country. Fourth, twentieth-century Peru has created an economy in which links between the capital-intensive "modern" sector—dominating the export-import process and mainly on the coast—and the labor-intensive "traditional" sector—mainly low-income and in the highlands—do not lead to a systematic increase in the latter sector's income. And fifth, partly as a result of the domestic impact of this export-oriented economy, Peru acquired a notably unequal pattern of income distribution. By 1986 the top 20 percent of the population received 51 percent of the income, while the lowest 20 percent got only 5 percent of the income.

After the guano decline, it was mining, along with sugar and cotton, two agricultural commodities both grown on the coast, that fed the 1890–1930 expansion. As in Cuba and elsewhere, sugar was by this time a capital-intensive operation. Machinery for the modern mills was expensive, and it took large amounts of land to feed sufficient cane to the mills. Along Peru's north coast, where most of the production centered, it was also a year-round activity (in contrast to most other areas, such as Cuba, where the rhythm of work is seasonal). As a result plantation owners, mostly Peruvian, developed a stationary labor force approaching 30,000 workers by the late 1920s. Some were descendants of African slaves; some were Chinese; others were Indians from the sierra who came under coercive conditions as *enganchados* (literally, "hooked ones"—indebted by cash advances).

Sugar production and exports grew especially rapidly in the mid-1890s and again during World War I. By the 1920s productive capacity reached 320,000 tons, nearly double the prewar level. Though the market collapsed by the end of the decade, most planters were able to survive. Their largest export market was the United States, however, where protectionism hurt imports, so Peru's sugar sector fell stagnant during the 1930s.

The other major source of growth was cotton, which could be grown all along the coast. Up to the late nineteenth century, Peruvian long-staple cotton could not compete with the short-staple North American variety, but technical innovations in the 1880s permitted blending of the Peruvian product with wool. World demand increased and landowners responded. Production leapt from around 400 metric tons in 1890 to more than 2000 metric tons in 1910 to around 6000 metric tons in 1930, by which time cotton was accounting for 18 percent of Peruvian exports.

Cotton cultivation in Peru is a seasonal enterprise, in contrast to sugar, and this created two kinds of labor: sharecropping (*yanaconaje*), by far the most important mode; and independent production by

FIGURE 6-1 Exports from Peru, 1830–1975: Indices of Volume and Dollar Value (1900=100)

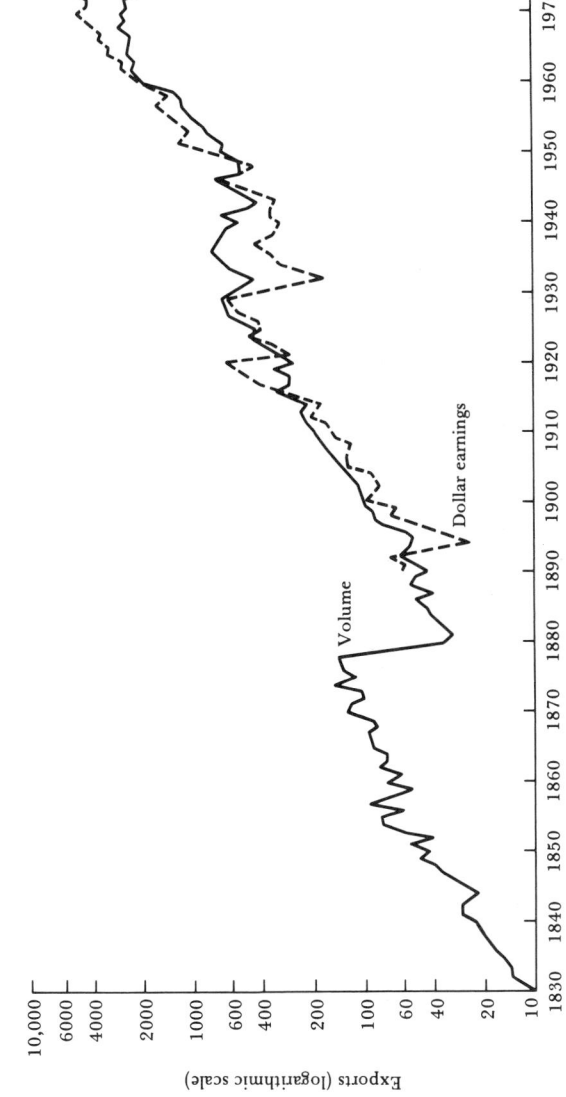

Source: Rosemary Thorp and Geoffrey Bertram, *Peru 1890–1977: Growth and Policy in an Open Economy* (New York: Columbia University Press, 1978), p. 5.

small-scale peasants. Partly for this reason ownership remained Peruvian, since foreign investors were reluctant to get involved in neofeudal (i.e., nonmarket) relations of production. By the 1920s the labor force in cotton included more than 40,000 people.

Foreign entrepreneurs made their mark not as landowners but as merchants, as they occupied crucial positions in the ginning, financing, and marketing of cotton. Up to the 1920s they also held a virtual monopoly on credit, and planters usually needed short-term loans from season to season. But most of the benefits went to Peruvian producers. In addition cotton cultivation provided stimulus for domestic investment in textiles and by-products (such as soap, candles, and cottonseed oil), but it did not lead to a wide range of investments outside the cotton sector.

The sierra also played a part in agriculture. From the grazing lands of the Andes came wool, accounting for more than 10 percent of exports in the 1890s, in 1903, and in 1918–20. The highest-grade variety, from the alpaca, was produced mainly by traditional peasants. Sheep wool came from the largest-scale haciendas, and conflicts over land led to a wave of peasant revolts early in the century. The collapse of the market in the 1920s brought on a recession, especially for sheep growers, but the coast-oriented central government did not offer any relief.

The Peruvian *montaña,* like the Brazilian Amazon, went through a short-lived rubber boom. Hard-driving domestic entrepreneurs such as Julio C. Arana amassed huge fortunes, and by the turn of the century the city of Iquitos, the key market town of the Peruvian Amazon, had grown to about 20,000 inhabitants. The boom was short-lived. Soon Peru, like Brazil, was pushed out of the market by the more efficiently grown plantation rubber from the Far East.

From the 1890s onward the once-dominant sector of the Peruvian economy, mining, underwent major change. In the late nineteenth century small-scale operators concentrated on precious metals, first silver and then gold. But Peruvian ores, located throughout the sierra, are unusually complex, there often being more than one metal in a single mine. Industrial development in the United States created demand for nonprecious metals like lead, zinc, iron, and copper.

Copper became the most important mineral product up to 1930. Technological advances and railway extensions made it feasible to tap the Peruvian veins. Large-scale U.S. investment arrived in 1901, with the purchase of the complex at Cerro de Pasco, and production quickly accelerated. By 1930 nearly half the value of all metal production came from copper, with silver and gold now mined as a by-product of copper. Migrant *serranos* provided the labor force, and ownership fell almost completely under the control of foreigners, especially the Cerro de Pasco Copper Corporation and Northern Peru Mining, a subsidiary of the American Smelting and Refining Company.

Silver refining at Cerro de Pasco in the early 1900s featured large-scale technology. (Courtesy of the Library of Congress.)

The search for, and exploitation of, petroleum also expanded in these years, especially during World War I. Coastal fields contained notably high-grade deposits, much sought after in the international market, and U.S. capitalists became active in the early stages. In 1913 the International Petroleum Company (IPC), a Canadian-registered subsidiary of Standard Oil, gained access to major fields at Negritos and La Brea-Pariñas under terms later to become a source of continuing controversy. About 10 percent of output went for domestic sales (since Peru did not have any coal); the rest was for the export market, where it brought handsome profits to the owners. By 1930 oil made up 30 percent of total Peruvian exports. That same year an informed observer guessed that 50 percent of the wholesale price abroad was clear profit for the companies.

In comparison to these export products—sugar, cotton, wool, rubber, copper, oil—domestic activity had a minor role in the 1890–1930 cycle of growth. Textile production expanded early in the century and then leveled off. Some agricultural goods were grown for domestic consumption—meat, potatoes, sweet potatoes, maize—but a fair amount of production and exchange took place within subsistence economies and not in the money economy.

The depression and World War II then altered the international picture and prompted a modest reorientation of the Peruvian economy, but—in contrast to their counterparts in Argentina, Mexico, and Brazil—national policymakers did not even try to embark on a sustained program of import-substitution industrialization. When opportunity beckoned in the late 1940s they turned instead to a tried-and-true strat-

egy: export-led growth with ample room for foreign investment and integration of Peru's economy with the international economy, above all the U.S. economy.

Most economic sectors joined in the postwar cycle, albeit with some important changes. Sugar production expanded in the 1960s, as Peru received a share of the U.S. market quota that was taken away from Cuba, and Peruvian sugar workers began to organize a militant union movement. Cotton output went up from 182,000 tons per year in 1945–49 to 381,000 tons in 1960–64 and then declined, partly because of oversupply in the world market and partly because of competition from synthetic textiles. *Yanaconaje* was joined by paid labor and cash rentals in the production process, and sharecroppers managed to enhance their bargaining position. Sheep wool virtually disappeared from the export list, though the high-grade alpaca trade continued on a modest scale.

Mining underwent its own diversification. Copper retained its preeminence. Iron ore received attention, and was developed by the Marcona Mining Company, a conglomerate whose major participant was the Utah Construction Company. Even as mining moved into other metals, U.S. investment remained predominant. The same was true for petroleum, though its relative weight among exports declined throughout the 1960s.

One new commodity contributed importantly to the overseas balance of trade: fishmeal, which came to be widely used in the U.S. animal-feed industry (especially for poultry and pigs). From a standing start in the 1950s Peru leapt ahead to become the world's leading fishing country by volume in 1964, with 18 percent of the total world catch, and it was producing about 40 percent of the world's fishmeal supply. The leading entrepreneurs were neither foreign nor elite: like Luis Banchero Rossi, the legendary anchovy tycoon, they were from the local middle class. Nature then played a cruel trick. Changes in the offshore ocean currents, first in 1965 and again in 1972, along with overfishing, led to the disappearance of the fishing grounds. Just as quickly as Peru had found a new source of wealth, it vanished.

Manufacturing added another dimension to the country's economic picture. Some industrial growth occurred in the 1950s but it tended to be export-supporting rather than import-substituting: it focused on by-products of export goods, often for the overseas market, rather than on the replacement of imported goods for the Peruvian market. This pattern started shifting around 1960, as industrial output began to include such items as cement, chemicals, and consumer durables. Because of liberal legal provisions it was the foreign sector, not local investors, that took advantage of this opportunity. During the 1960s at least 164 major foreign corporations, many of them multinational, came into Peru. By 1968 manufacturing accounted for 20 percent of the gross national product (compared to 15 percent for agriculture, 14

TABLE 6-1   Composition of Peruvian Exports, 1890–1976

Products	*Percentage Share of Exports, by Value**		
	1890	*1930*	*1976*
Sugar	28	11	7
Cotton	9	18	6
Coffee	—	0.3	9
Fishmeal	—	—	13
Silver	33	4	11
Copper	1	10	17
Lead	—}	7	4
Zinc	—}		11
Rubber	13	—	—
Oil	—	30	—

*Columns do not add up to 100 because of exclusion of some export items.

Sources: Rosemary Thorp and Geoffrey Bertram, *Peru 1890–1977: Growth and Policy in an Open Economy* (New York: Columbia University Press, 1978), pp. 40, 153; James W. Wilkie and Peter Reich, eds., *Statistical Abstract of Latin America*, 19 (Los Angeles: UCLA Latin American Center, 1978), Table 2732.

percent for commerce, 6 percent for mining), and in 1970 the industrial sector included 14.5 percent of the labor force.

Exports nonetheless continued to be paramount and, as shown in Table 6-1, copper was the country's leading export by the mid-1970s. Metals made up 44 percent of the total. Fishmeal, having recovered somewhat from the setback of the early 1970s, earned 13 percent. Sugar and cotton together came to 13 percent, and coffee came on the list with 9 percent.

Indeed, the table demonstrates that Peru has had a diversity of export products since the 1890s. The relative importance of products has changed—sometimes dramatically, as in the cases of rubber and oil—but Peru has not relied on a one-product export economy since the guano age. To this extent the country has been fortunate.

By the mid-1960s processes of economic change had led to fundamental changes in the country's social structure. At the top was the traditional elite, the so-called forty-four families, but it was by no means monolithic. The coastal segment of the aristocracy had since the 1890s become involved in commercial agriculture and in export-import transactions. It was a cosmopolitan group, shrewd and well-educated, with its center of action in the capital city of Lima. It took a flexible, pragmatic approach to matters at hand, often collaborating with foreign investors and frequently permitting newly rich investors and military officers to join its social circles.

The *serrano* elite was in general more traditional, both in its attitudes and in its insularity. The provincial *patrón* was bound to the land and maintained an intimate (though hierarchical) relationship with the working *peón*. At the same time, many of these landowners had carried out major innovations in agriculture, including dairy farming in the north and livestock raising in the center and south. This gave the sierra a new economic impulse.

In terms of political power, the *serrano* elite had since the 1890s relied on the support of the coast to bolster its position and help put down rebellions. By the 1960s the *serrano* oligarchy found its local authority eroding badly. For one thing, many peons had left the haciendas, looking for jobs either in the sugar or cotton plantations on the coast or in the mining camps of the mountains. Second, the rural gendarmerie was replaced by the national *Guardia Civil*. And in addition, Indians and Indian communities acquired increasing legal protection during the twentieth century. By the 1960s, as if in admission of defeat, frustrated landlords gave up dealing with recalcitrant peons and simply turned their haciendas over to pasture.

The lower class—probably 80 percent of the population, if not more—was heterogeneous. It included the rural proletariat on the sugar plantations, tenant farmers and hired hands in the cotton fields, and peasants and subsistence farmers in the sierra. It included wage earners on fishing boats, miners in the mountains, and organized workers in the cities. It included domestic servants in Lima, peons in the countryside, and residents of suburban squatter settlements. It included speakers of Aymará and Quechua, on the fringes of national society, as well as workers taking part in downtown strikes. It was a large-scale stratum, divided along three dimensions: between workers and peasants, between coast and sierra, between non-Indian and Indian. Yet in practice, network and family lines often bridged these divisions, and migration helped to reduce once-major geographical gaps.

And there were, of course, people in between, middle sectors if not yet a middle class (in the sense of a cohesive, self-aware social class that emerged in nineteenth-century West Europe), mostly living in urban areas. Significantly, according to one estimate for the 1960s, nearly one-half the income-earners in the middle sectors had positions in government—including the military. Bureaucracy, it seems, as well as commerce, had turned into the ultimate middle-sector occupation.

Through these transformations Peruvian society became more urban. The national population grew from 3 million in 1900 to 17.7 million in 1980, and the urban proportion increased in that same period from 6 percent to 45 percent (as of 1975). Lima maintained its preeminent position, with 4.5 million inhabitants in the metropolitan area, but other cities grew as well. Although not as urbanized as Argentina or Chile, Peru was on a level equal to Brazil. It was a complex and changing society, one that clearly held the potential for political conflict. Let

us now turn back to Peruvian political history, to see how that conflict emerged.

Oligarchic Rule

The military continued to dominate Peruvian politics in the early 1890s. General Andrés Cáceres and Colonel Remigio Morales Bermúdez exchanged power between 1885 and 1895, as the Grace Contract received approval from the Congress, but then politics changed under the influence of new ideas. Just as Argentina gave birth to its Generation of 1880 and Mexico was fostering its *científicos*, Peru produced a similar class of civilian leaders. Highly educated, neopositivistic by training, and liberal by outlook, they comprised a curious breed: for lack of a better phrase, they might be classified as aristocratic technocrats.

Their initial spokesman was Nicolás de Piérola, who, as José Balta's brash young treasury minister, had negotiated the questionable Dreyfus contract of 1869. Now more experienced, and as both a mystical Catholic and a scientific revolutionist, he sought to reconcile idealism and materialism. To advance his position and gain supporters he created the Democratic Party—the country's second political party, the first being the *Civilistas*—and in 1895 he became president.

Eager to promote the export-led expansion starting in the 1890s, Piérola moved to strengthen Peru's credit rating. He tightened tax legislation and increased duties on commerce, which led to a doubling of government income during his four-year term. The resumption of payments on the foreign debt restored the country's standing with merchants and creditors abroad. Piérola also established a ministry of development *(ministerio de fomento)* to assist local entrepreneurs and to establish a means for governmental participation in economic growth.

In politics Piérola put through a direct suffrage law and strengthened municipal governments. Perhaps hoping to diminish the prospect of military intervention, he undertook to professionalize the armed forces and in 1898, just as Chile and Argentina were bringing in German missions, he invited a French military advisory team to help modernize the Peruvian forces. A clever politician, intent on civilian rule and quick to seize advantages, Piérola earned an appropriately ambiguous sobriquet: "the democratic *caudillo*."

After Piérola stepped down, Peruvian politics entered an era of "bossism" known as *gamonalismo*. Effective competition for power was restricted to the upper-class elite. Elections took place but ballots were not secret, so landowners could herd their workers and peons to the polls and be sure they voted correctly. *Hacendados* from the sierra had themselves elected to the national Congress, where they customarily supported the president—in exchange for unchecked powers in their own local arenas. And the coastal elite, consolidating its control of eco-

nomic policy, pursued the path of export-led growth. The careful alliances Piérola had formed with landowners were functioning well.

The untimely death of President Manuel Candamo in 1904 precipitated a brief succession crisis that was resolved by the election of José Pardo. An activist, Pardo supported education, increasing its share of government expenditure in his presidency from 9.6 percent to 17.2 percent, and he put forward proposals for social and labor legislation (which Congress greeted with skepticism). Through his able treasury minister, Augusto B. Leguía, he set up a series of public agencies to define and strengthen the state's role in the promotion of economic development. The most prominent of these was the Peruvian Steamship Company *(Compañía Peruana de Vapores)*.

As though to ratify this economic policy, Leguía assumed the presidency after uncontested elections in 1908. He himself had been a highly successful businessman, with interests in insurance and exports, and he had served as president of the National Bank of Peru. Once in office he established a state-run agency for guano, now channeled entirely to domestic consumption. Confronted by a haphazard revolt in 1909, he jailed his opponents. Throughout his career, in fact, Leguía showed much more concern with the substance of policies than with the procedures of consultation. Disdaining politics, absorbed in administration, Leguía staked his claim for law and order.

Ironically enough, Leguía was succeeded by an erratic populist, Guillermo Billinghurst, who won the 1912 elections. Organized labor had made a timorous appearance in Peru in 1904, with strikes in textile mills and other factories, and 1911 witnessed some serious protests against inflation. A prosperous businessman and journalist, the grandson of a British naval officer, Billinghurst had been a popular mayor of Lima. Gaining the support of the masses—and of a segment of the ruling elite—Billinghurst campaigned on a platform that included the promise for a bigger loaf of bread for five cents. Hence his nickname, "Big Bread" Billinghurst.

Billinghurst's hope was to reconcile the interests of workers and owners under the mantle of enlightened capitalism. Having a touch of the demagogue, he was comparable in ways to Argentina's Hipólito Yrigoyen. He proposed public housing and proclaimed support for an eight-hour day, and in 1913 he secured congressional approval for labor legislation that guaranteed collective bargaining. He studied the condition of the peasants but, sensing his limits, took no action there. Billinghurst's following was in the cities, not the countryside, and he began to encourage street demonstrations in support of his policies. Aghast at such events, the elite closed ranks against the president. In 1914 Billinghurst fell victim to a military coup.

The next year power reverted to civilian technocrats, and José Pardo, president in 1904–8, began another four-year term. Despite his own status as a practicing Catholic, he oversaw the promulgation of a law

establishing religious toleration. As World War I continued, he finally severed relations with Germany, partly in hopes of gaining U.S. support against technically neutral but pro-German Chile. Pardo put through some labor legislation, including regulations for women and children. Billinghurst was right to worry about this sector, as it was here that Pardo faced the greatest challenge. In January 1919, as labor protests erupted in Buenos Aires and Santiago and São Paulo, workers in Lima-Callao proclaimed a three-day general strike. Supported by students mobilized from the University of San Marcos, they demanded lower food prices, an eight-hour day, and enactment of other legislation. Pardo had to call out the army to disperse the workers, and in the wake of the violence he acceded in part to their demands.

The clashes of 1919 also led to the formation of Peru's most important labor organization, the Regional Peruvian Labor Federation (*Federación Obrera Regional Peruana,* or FORP). Its leadership was moderate and sought, unsuccessfully as it turned out, industrywide and nationwide collective bargaining. By mid-1919 a Socialist Party and a Workers' Party had also appeared, more Marxist in approach, but labor organization and activism were still in early stages.

Leguía: The Oncenio

In the midst of this confusion, former president Augusto B. Leguía returned from abroad, gathered widespread support, and won the 1919 election. Then he executed a masterful plan. Claiming that Pardo and the Congress were plotting to annul his victory, Leguía got military backing for a coup. On July 4 he and his comrades seized the national palace, sent Pardo off to exile, dissolved the legislature, and ensconced themselves in power. Thus began Leguía's dictatorship, a watershed remembered throughout Peru as the *oncenio,* or eleven-year rule.

After a hastily arranged inauguration, Leguía defined the goals for his regime. "I have come not only to liquidate the old state of affairs," he is reported to have said, "but also to detain the advance of communism which, because it is premature among us, would produce dreadful consequences." Reacting to the red scare of 1919, Leguía proposed to siphon off working-class discontent by building a coalition between capital and labor, an alliance that would herald the foundation of a *nueva patria* (new fatherland) for Peru. Ironically Leguía was seeking to complete the task undertaken by Billinghurst, the chief executive who had himself sent Leguía off to exile some time before.

The first step was to tighten the grip on political power. In 1920 a pliant Assembly devised a new constitution, a document that legitimized Leguía's rule and charted a course for the state's role in the economy. The government was given explicit powers to set prices, impose taxes, and monitor labor-management relations. It was to assume responsibility for the education and assimilation of the Indians. And it

was to frame its policies without the tangled executive-congressional relations that had bedeviled previous administrations: every five years there were to be simultaneous elections of the president and both legislative houses, thus virtually assuring working majorities for the chief executive.

Leguía was able to consolidate such power in part because of the fragility and decline of the old-line political parties—*Civilista*, Democratic, Liberal, and Constitutionalist. In fact, they were never strong institutions. They had been personal vehicles, creations of individual *caudillos*, not stable and self-perpetuating organizations. By 1920 they were unable to meet current challenges, either from the masses or Leguía, and they quietly withered away. Indeed with one exception, described below, Peruvian politics has seen a relative absence of strong and durable parties.

Leguía aggressively silenced his critics. The minister of the interior, Germán Leguía y Martínez, a cousin of his known as "The Tiger," had a free hand in forcing opponents out of the country. The regime exiled Antonio Miró Quesada, the publisher of *El Comercio,* and later seized *La Prensa* and turned it into a pro-government newspaper. The dictator dismissed dissident professors from university chairs, and turned as well against students who had once hailed him as their "mentor." In 1924 Leguía even appropriated religion to his side, when he dedicated the republic of Peru to the Sacred Heart of Jesus. The opposition demonstrators were easily repressed. He used the occasion to arrest and deport at least thirty university students, among them a rising young leader by the name of Víctor Raúl Haya de la Torre.

Other forms of discontent were dispatched with equal efficiency. In fulfillment of the 1920 constitution, Leguía created an office of Indian affairs under the activist direction of Hildebrando Castro Pozo. As the bureau began investigating land titles and other ticklish matters, however, the *gamonales* reacted, and Leguía took action. In 1923 he sent the army to crush an Indian uprising in the sierra. Despite public rhetoric, peasants and Indians had no genuine place in the Leguía coalition.

To construct the "new fatherland," Leguía undertook a vigorous program of public works. With earnings from the export trade he invested in transportation and infrastructure, improving roads and bridges and building 600 new miles of railroad. Some of the work force came from the infamous *conscripción vial* (highway conscription), which forced poor rural males into unpaid construction work. And he encouraged foreign investment. By 1925 the Cerro de Pasco Copper Corporation holdings came to about 50 million dollars, and IPC was preparing to exploit the La Brea-Pariñas oil fields.

The cornerstone of Leguía's economic policy was, of course, promotion of the export-import model of growth. A post–World War I slump in foreign-exchange receipts put pressure on the system, but it did not lead to support for industrial development. Even before World War I,

Peru had begun to buy most of its imports from the United States rather than Europe, so the wartime cutoff in supply was not as severe as for some other Latin American countries. The postwar slowdown in trade reduced the amount of capital available for investment, of course, and entrepreneurs put what they had into relatively safe ventures—including areas favored by Leguía's public-works program, such as construction or real estate. Consequently all major partners in Peru's capitalist elite—merchants, bankers, and importers—expressed alarm when the terms of trade began leading to the depreciation of the nation's currency. (Even though devaluation boosts exports and is often favored by commodity producers, the Peruvian elite on this occasion was more interested in retaining its capacity to import goods from abroad, or to invest in the international market.) The Leguía administration listened. In 1922 the central bank sold off a large share of gold and foreign reserves, and in the mid-1920s a large-scale loan was raised for the express purpose of maintaining a high exchange rate for the Peruvian sol.

Leguía pursued an active foreign policy with bordering states. In 1927 his government settled a longstanding boundary dispute with Colombia. Two years later, in 1929, he reached an agreement with Chile: the northern province of Tacna would go to Peru, and Arica would remain under Chile's control. The War of the Pacific, so ignominious for Peru, finally came to the end of its diplomatic coda.

Within this carefully constructed political environment, Leguía had no trouble getting reelected in 1924 and 1929. He seemed invincible. His supporters controlled the Congress, while his opponents were in exile or jail. Throughout the 1920s he functioned as the quintessential aristocratic technocrat: Like the *científicos* of prerevolutionary Mexico, Leguía and his followers made economic policy to fit the prescription of export-oriented capitalist expansion. But, as in the case of Mexico, it could not last forever.

The Reformist Critique

Peru's checkered history since independence inspired various types of internal criticism. Some critics blamed the Indians, some denounced the Spaniards. Some sought to uplift the country with new ideals, or new embodiments of spiritual traditions, while others focused on the material base of national progress. Among these intellectuals three deserve special attention: Manual González Prada, José Carlos Mariátegui, and Víctor Raúl Haya de la Torre.

González Prada, born in 1848 of aristocratic Spanish parentage, hoped to rekindle patriotic spirits after the War of the Pacific. Peru had lost its moral bearing, he thought, and the false prosperity of the guano age had led to degeneration. "Riches served as an element of corruption," according to González Prada, "not of material progress.

. . . No means of acquisition seemed illicit. The people would have thrown themselves into a sewer if at the bottom they had glimpsed a golden sol. Husbands sold their wives, fathers their daughters, brothers their sisters. . . ." Influenced by European socialism in the 1890s, González Prada excoriated both capitalism and Catholicism. Peru should seek renewal by joining forces with its Indians to topple, by violence if necessary, the prevailing system. The time to act was now. "Old people to the grave," he declared in a famous rallying cry, "the youth to work."

One of Leguía's sharpest critics was José Carlos Mariátegui, a writer and journalist of humble origin. At heart a socialist, he became a spokesman for pro-Indian *indigenismo*. "Socialism preaches solidarity with and the redemption of the working classes," he characteristically reasoned. "Four-fifths of Peru's working classes consist of Andean Indians. Therefore, socialism means the redemption of these Indians." How? Land was the answer. "The Indian question derives from our economy. It has its roots in the system of land tenure. Every effort to solve it with administrative or protective measures, with educational methods or road-building projects, represents a superficial labor as long as the feudalism of the great landowners exists." As for those who agonized over Peru's supposed racial decadence, Mariátegui had a tart response: "The degeneracy of the Peruvian Indian is a cheap invention of legal lickspittles at the feudal table." In sum, he wanted to create a utopian society by drawing on and adapting the collectivism of the Incan empire, especially the communal *ayllu* form of organization.

In 1926 Mariátegui launched an influential journal, *Amauta*, a wide-ranging review of art and politics. His most famous collection of writings appeared in book form as *Seven Interpretive Essays on Peruvian Reality*. In 1929 he helped found a Socialist Party which soon became an affiliate of the Communist International. In 1930, at the age of thirty-five, he died of a chronic illness. Peru, and Latin America as a whole, lost one of its most creative political thinkers.

A colleague and rival of his was Víctor Raúl Haya de la Torre, who as a student had been exiled by Leguía in 1924. That same year, in Mexico City, Haya de la Torre founded what he called the American Revolutionary Popular Alliance (*Alianza Popular Revolucionaria Americana*, or APRA). Sharing some of Mariátegui's insights, Haya intended to create a practical instrument for action. A rather grandiose scheme, APRA proclaimed a five-point program for the redemption of "Indo-America": resistance to Yankee imperialism, political unity of Latin America (Indo-America), nationalization of land and industry, internationalization of the Panama Canal and, most generally, solidarity with the oppressed around the world. For Peru, he eventually declared a more specific program: state control of the economy, long-run nationalization of key sectors, and the protection of political freedoms and human rights.

A brilliant organizer and electrifying speaker, Haya de la Torre later

founded the *Aprista* (APRA) party in Peru. Underneath its populist rhetoric lay a concern for the welfare of the middle sectors in society. As Haya once put it: "It is this middle group that is being pushed to ruination by the process of imperialism. . . . The great foreign firms extract our wealth and then sell it outside our country. Consequently, there is no opportunity for our middle class. This, then, is the abused class that will lead the revolution."

Revolution or not, because or in spite of its middle-sector base, APRA was to become the most enduring political party in the history of Peru. Until its inglorious demise in the late 1980s, it was to have a significant impact on the course of national politics.

Flirting with Alternatives

The Great Depression took an immediate toll on the Peruvian economy. Export earnings plummeted (as shown in Figure 6-1). Profits shrank and discontent grew. In the face of crisis, the helplessness and decadence of the Leguía government prompted anger and contempt. In August 1930 a young military officer, Luis M. Sánchez Cerro, led an uprising in Arequipa that ended by ousting Leguía (who later died in prison).

Sánchez Cerro became the presiding officer of a military junta, and Peru embarked on a new political era. Of modest background, unmistakably *cholo*, Sánchez Cerro brought a distinct touch to the executive office. As he campaigned throughout the country, both to gain legitimacy for the junta and to advance his personal ambitions, he built a strong popular following. He did not offer a coherent ideology, but he was clearly attempting to forge a populist coalition between selected aristocratic elements and the working masses. In March 1931, in a fit of frustration, he resigned his post and sailed off to Europe. A few months later he returned, this time as the presidential candidate of the Revolutionary Union. His main opponent was Haya de la Torre and the *Aprista* party.

The 1931 campaign proved to be one of the most fateful and controversial events in Peruvian history. Sánchez Cerro called for agrarian reform, rural extension programs, and assimilation of the Indians: "So long as we do not consider all of our people, Indian and non-Indian, to be Peruvians, with the same rights and duties, there will never emerge the unity which is the indispensable element of the nation." Haya de la Torre countered by stressing the evils of imperialism as well as the existence of social inequities. It was an intense campaign, marked by violence and mutual accusations. *Aprismo* (as the APRA movement was known) revealed an electoral stronghold in the northern coast region, particularly in the sugar-growing areas around Trujillo, where economic modernization had led to social dislocation and popular frustration over the destruction of the traditional society; the party received ample support in Lima and elsewhere as well. But in the end it

An *Aprista* election poster expresses both the party's outlook and the intensity of the 1931 campaign; the slogans say "Only *Aprismo* will save Peru . . . vote for Haya de la Torre." (Private collection.)

was Sánchez Cerro who won with an official tally of 152,062 votes to 106,007 for Haya de la Torre.

Polarization ensued. The *Aprista* movement, with its Marxist premises, offered a left-wing critique of Peruvian society. There was also a formidable movement on the right. As in Argentina, some conservatives openly sympathized with fascism. José de la Riva Agüero was eloquent on the subject. "Up with Catholicism," he declared, "up with the corporate state and fascism, with order, hierarchy, and authoritarianism." It was time, he said, for decisive action:

> There can be no middle ground. Either to the right or to the left. Democracy, capitalism, the liberal tradition, all represent a middle ground which is really disguised communism or else the certain road to it. . . . The only solution is to return to the medieval, Catholic, Hispanic tradition as now embodied in fascism.

Not surprisingly, Francisco Franco's right-wing movement in Spain had numerous adherents in Peru during the Spanish Civil War.

Tension continued to mount. In early 1932 a young *Aprista* tried to assassinate Sánchez Cerro. The president was now determined to crack down on the *Apristas*. But they had their own plan—for an insurrection in the provincial city of Trujillo. The rebels succeeded in taking the entire city, including the army garrison. A powerful military column soon surrounded the *Apristas,* who decided to flee. In the ensuing panic, and perhaps even by top *Aprista* orders, the insurgents executed some ten military officers, fifteen policemen, and twenty-five civilians, all hostages. When the invading government troops discovered this atrocity, they summarily executed at least 1000 (the estimates ran as high as 6000) of the local residents suspected of giving armed support to the *Apristas.* This ugly exchange of murders set the tone for APRA-army relations thereafter. For the army it took on the character of a blood feud, and convinced many officers that they must never let APRA come to power.

Meanwhile, Sánchez Cerro was attempting to consolidate his power and pushed through a new constitution. In April 1933 another *Aprista* gunman succeeded in assassinating the president. Beset by yet another crisis, Congress elected General Oscar R. Benavides to serve out the remainder of the presidential term.

As Benavides took office Peru entered a phase of economic transition, a period that held out the possibility of reducing the country's dependence on international markets and investments. Led by cotton, exports began to recover after 1933. But because of the depression, and later on World War II, foreign capital beat a steady retreat: the proportion of exports produced by foreign-owned firms fell from around 60 percent at the end of the 1920s to 30 percent or less in the late 1940s. (By 1967 it would climb up to 50 percent.) Local entrepreneurs, sometimes with government help, gained increasing control of lead and zinc, and of some gold and silver as well. Petroleum output went up throughout the 1930s. Industrial capacity was modest but it nonetheless appeared that Peru was becoming able to redirect its economy if it so chose.

Benavides showed some desire to seize the chance. In 1934 his government began the state-directed development of petroleum. Trade doubled from 1933 to 1936, and the sol remained stable against the dollar. He used a fair share of governmental receipts for public works and social projects, including road construction, working-class housing, and a compulsory social security system. He supported an agricultural bank founded in 1931 to give credit to cotton planters and other landowners, thus reducing the role of foreign merchant houses.

In the political arena Benavides tried to reduce polarization and to achieve a national consensus. Settlement of a border dispute with Colombia earned him both applause and time, but the going proved to

be rough. At first he offered amnesty to imprisoned *Apristas* but then, as conflict intensified, he appointed the pro-fascist Riva Agüero as his prime minister. Crackdowns followed. In 1936 Benavides annulled elections won by an *Aprista*-backed candidate and held on to power for three more years.

In 1939 the presidency passed to Manuel Prado, a moderate civilian who adopted a conciliatory posture toward the *Apristas*. All major parties expressed support for Prado's pro-Allied stance in World War II. Peru also won a brief military clash and diplomatic triumph in a boundary conflict with Ecuador. Foreign affairs seemed to be having a healing effect.

Electoral triumph in 1945 went to José Luis Bustamante y Rivero, who ran as the candidate of the National Democratic Front—with APRA's support. He soon faced a series of political struggles, as *Aprista* delegates to Congress sought to curtail executive authority. Inflation and food shortages presented serious socioeconomic challenges. And amidst public controversy, Bustamante approved a contract giving IPC permission to search for oil in the Sechura desert.

Together, Prado and Bustamante furthered the modest reorientation of the Peruvian economy. They vastly increased government spending (which Benavides had tried to keep in check). They introduced a fixed exchange-rate system, and they established import controls along with the rationing of foreign exchange. And they launched a scheme for a state-controlled iron and steel plant. In short, they modified the path of Peru's economic development in at least two crucial respects: they strengthened the role of the state, and they reduced the reliance on exports and imports. All of which prompted the wrath of export producers, particularly the sugar planters.

A central thrust of the Prado-Bustamante policy was the diversification of agriculture, and this entailed a series of measures which the sugar barons correctly saw as threatening: increased export taxes, rationing of guano, pressure to cultivate food crops, and, in 1941, an insistence that sugar producers satisfy domestic-market quotas before exporting to overseas markets. Taking advantage of the hullabaloo over the Sechura contract, which aroused strong nationalist sentiment, the coastal elite supported a military move against the Bustamante government. In 1948 General Manuel A. Odría seized power.

Economic Liberalism and Political Vacillation

The Odría coup of 1948 constitutes another watershed in contemporary Peru. In economics it led to restoration of the open, export-led model of growth. In politics it ushered in an era of uncertainty, of dictatorial rule and electoral confusion. The contradictions in this system would eventually come forth with startling clarity.

Under the watchful eye of the elite, the Odría regime adopted pro-

planter measures. The exchange-rate was freed, import restrictions were lifted, foreign investment was encouraged in mining and oil, advice was sought from U.S. experts on currency stabilization. All in all, as Rosemary Thorp and Geoffrey Bertram observe, the Odría government undertook to implement "that dream of orthodox development economists," a vision that would be pursued till 1968: "an export-led system in which cyclical balance-of-payments difficulties were handled by domestic demand restraint and exchange devaluation, in which the entry of foreign capital and the repatriation of profits were virtually unrestricted and in which government intervention and participation were kept to a minimum." Particularly in this last respect, Peru seemed out of step with the rest of Latin America. Notes another economist, Shane Hunt: "In a continent that was witnessing ever-increasing state intervention in economic life in country after country, Peru . . . turned around to begin a march in the other direction. . . ."

To tighten his hold on authority, Odría immediately took after his opponents, especially *Apristas*. Haya de la Torre managed to escape imprisonment by seeking refuge in the Colombian embassy, where he remained for more than five years, awaiting safe conduct from the military. Dissent was stifled, and civil liberties became precarious.

Odría consolidated his position by winning the 1950 elections—there was no opposition—and proceeded to impose his regime. Rather consciously, he began to emulate the style and manner of Argentina's Juan Perón. He courted working-class masses, more so in the coastal cities than in the rural sierra. He lavished funds on ostentatious public works, most notably in Lima. He developed a personalistic following. And with the aid of his wife, María Delgado de Odría, he sought to mobilize women in support of the regime, extending suffrage to females in 1955.

A decline in export earnings after the end of the Korean War in 1953 brought Odría's heyday to a close. Unemployment mounted, inflation increased, and workers went on strike. Civilian oligarchs expressed apprehension over Odría's capricious form of rule. Under pressure, Odría finally consented to allow free elections in 1956.

The leading contenders that year were former president Manuel Prado, supported by the *Apristas*, and a political newcomer named Fernando Belaúnde Terry, candidate of the National Front of Democratic Youth. Belaúnde, a University of Texas–trained architect from a prominent family, began to articulate the hopes and frustrations of the country's educated middle sectors. But his time had not yet arrived. The winner was Prado, with 568,000 votes; Belaúnde, with 458,000, came in a respectable second.

Prado brought in a period of political liberalization, permitting trade-union organization and allowing Communists as well as *Apristas* to operate freely. By the end of his term, organized labor could claim about 330,000 members. Through his economics minister, the aristo-

cratic Pedro Beltrán, he continued Odría's policies, emphasizing exports and foreign enterprise. A program for "shelter and land" (*techo y tierra*) was proclaimed in the name of the peasants, but little was done about it.

The next presidential election, in 1962, offered a clear picture of political forces in Peru. Presidential candidates were Haya de la Torre, at last able to run on the *Aprista* platform; Belaúnde, who had created a new organization called Popular Action; and the always hopeful Odría, at the head of the *Unión Nacional Odriísta*. The only genuinely institutional party was APRA; the others were personal vehicles for their respective leaders. Partly for this reason Haya won the largest number of votes (557,000), but his scant plurality of 33 percent meant the outcome would have to be decided by Congress. Ever the opportunist, Haya proceeded to reach an accord with Odría. The armed forces, who had never forgotten or forgiven the *Aprista*-inspired murder of their colleagues at Trujillo in 1932, refused to accept the prospect of an *Aprista* president, however, and the military suspended Congress and announced that new elections would be held.

The new elections came in 1963. Having obtained the support of Christian Democrats, Belaúnde emerged on top with 39 percent. APRA was next with 34 percent and the *Odriístas* garnered 26 percent. With the *Apristas* safely defeated, the military accepted the results and let Belaúnde enter the presidency.

Belaúnde revealed himself to be an appealing, sometimes even charismatic, politician. Relishing the role of the visionary, he proposed building a trans-Amazon highway that would open up the lush Peruvian *montaña* region to development. He called on the memories of the Incas and urged his countrymen to aspire to greatness again. He traveled throughout the country, talking to peasants and workers as well as to planters and managers, tirelessly seeking to generate a national consensus.

On a more practical level Belaúnde recognized the need for adjustments in the Peruvian economic model. In particular, he began to increase the role of the state and to expand social services. He started to provide incentives for manufacturing. And he urged the need for agrarian reform.

Yet his efforts foundered. By the time his bill for land reform got through Congress, it had been virtually emasculated. It emphasized technical improvement rather than land distribution, apparently in hopes that *hacendados* would increase production. And because of objections from *Apristas* and *Odriístas*, it exempted sugar estates on the coast—where the *Apristas* had built up solid unions, and where *Odriístas* had their right-wing support.

So land reform was more promise than performance, and peasants in the sierra finally reacted angrily. They started to invade the haciendas, seizing cultivated lands as well as fallow fields, and conflict raged

throughout the Andes. Coinciding with the peasant protest was a Cuban-style guerrilla movement bent on establishing a rural *foco,* or nucleus. Ché Guevara and the young French intellectual Régis Debray had argued that a successful *foco* could be the spark to set off a country-wide revolution.

Within a year or so Belaúnde decided he had to suppress the peasant guerrillas. He sent the regular army on this mission, and by 1966 the movement was crushed. According to one account the repression left 8000 peasants dead, 19,000 homeless, 3500 in jail, and 14,000 hectares of land destroyed by fire and napalm. It was a traumatic experience—for both the 300,000 peasants who had risen up and for the soldiers who had put them down.

One of Belaúnde's other promises was to resolve the longlasting dispute with IPC, and here, too, he faced a no-win situation. After five years of protracted negotiations, during which the United States revealed continuing hostility to Peru's nationalistic inclinations, the Belaúnde government gave in. IPC surrendered its claim to the La Brea-Pariñas oil fields, by now exhausted anyway, while Peru gave up its claims to back taxes; IPC gained access to new fields in the Amazon; and the government agreed to sell crude oil to IPC at a fixed price for refining at the company's Talara complex. Amid a chorus of denunciation the Belaúnde administration published the agreement, all but the page containing the price for state-produced crude. The scandal created an uproar, as opponents accused the government of selling crude oil too cheaply to the foreign firm.

The economy was grinding to a halt. In August 1967 the government devalued the sol by 47 percent. Import controls and export taxes improved the balance of payments. As inflation declined so did growth. Belaúnde's optimistic vision, the dream of a united and prosperous Peru, was proving an illusory dream. In October 1968 the military once again decided to send their tanks to the presidential palace. This time, however, it would be no caretaker government they installed.

The Military Revolution

The military *golpe,* or coup, paved the way for one of Latin America's most ambitious military governments. Led by General Juan Velasco Alvarado, the junta immediately declared its intention to bring far-reaching changes in the structure of Peruvian society. A manifesto, issued the day of the coup, indicted the "unjust social and economic order which places the usufruct of the national wealth solely within the reach of the privileged, while the majority suffer the consequences of a marginalization injurious to human dignity." What Peru needed, the officers proclaimed, was a new economic order, "neither capitalist nor communist," a system that would abolish prevailing inequities and create the material foundations for harmony, justice, and dignity.

A combination of three qualities set this regime apart from earlier military governments in Peru—and, for that matter, in the rest of Latin America. One was its social and political autonomy. This time the Peruvian armed forces acted alone, rather than in collusion with civilian power groups; the middle-class military was beholden to no one but itself, and it consequently had an exceptionally free hand. Second, the leaders of the regime implicitly adopted the outlook and premises of the "dependency" school of analysis. One of their principal purposes, therefore, was to end what they called "the subordination of the Peruvian economy to foreign centers of decision, where actions originate which fundamentally affect the economic life of the nation and prevent an autonomous development process geared to the achievement of national objectives." And third, largely as a result of its antiguerrilla campaign in the sierra, the Peruvian military had genuine sympathy with the plight of the long-oppressed peasantry.

The result was a revolutionary military regime. The prospect seemed startling, if not contradictory—in Fidel Castro's phrase, "as if a fire had started in the firehouse."

A key to the military government's program was agrarian reform. In mid-1969 the Velasco government announced the enactment of the most sweeping land reform program in Latin America since the Cuban Revolution. "Peasants," declared General Velasco, "the landlords will no longer eat from your poverty." All large estates, regardless of productivity, were subject to expropriation. The ax fell first on the highly mechanized sugar plantations of the coast, which were placed under the administration of worker-run cooperatives, called CAPS (Agricultural Production Cooperatives). For the sierra the idea was to create small- or medium-sized farms, but before long the government gave in to peasant demands for cooperative forms of organization. The principal form was the SAIS (Agricultural Society for Social Interest), the government-recognized unit that could combine former hacienda peons and villagers from surrounding communities. By the mid-1970s it was estimated that three-quarters of the country's productive land was under cooperative management of one sort or another. Reported one observer in 1974: "Practically speaking, the agrarian elite has disappeared as a power group." By 1979 half the 21 million hectares of agro-grazing land on the coast and in the highlands had been expropriated from the *hacendados* and delivered to the peasants. No land remained in the huge estates that had dominated the Peruvian agrarian sector.

The Velasco regime took firm steps to institutionalize these fundamental changes. The traditional association of the planter elite, the National Agrarian Society, was abolished by decree in 1972. In its stead appeared the National Agrarian Confederation (CNA), a peak organization for regional collectives.

In 1971 the Velasco regime created one of its most characteristic in-

stitutions, the National System for Support of Social Mobilization (*Sistema Nacional en Apoyo de la Mobilización Social,* or SINAMOS). Sometimes written as two words—*sin amos,* "without masters"—it was to serve as the integrating institution for peasant and working-class groups. It would thus link the regime with the masses, identifying the government with its constituent groups and promoting a harmonious set of leader-follower relationships.

This desire to organize and mobilize the peasantry became one of the hallmarks of the new military government. It reflected the extent to which Peru's new rulers intended to restructure the distribution of power—not only at the upper reaches of authority but also at the bases of society.

A second area that captured Velasco's attention was the squatter shanty towns, the sprawling *barriadas* around Lima and other cities. By the late 1960s it was estimated that about 750,000 recently arrived migrants lived in the environs of Lima alone. Peru's military officers concluded that this was an explosive situation. So they set about organizing the settlements, now renamed "young towns" *(pueblos jóvenes),* and enlisted the aid of the church in their efforts. Part of the solution was simple enough: the granting of property titles to the migrant occupants.

The other tactic was to bring the *pueblos jóvenes* under the umbrella structure of SINAMOS. The purpose was to create conditions for stability. As one official put it: "We want participation but it should be organized participation. We want to make as many people as possible homeowners, then they will act responsibly towards their community and have a stake in it." By 1974 the majority of urban squatters had been reorganized from above into state-chartered *pueblos jóvenes.*

This top-down pattern of organization and mobilization illustrates a crucial feature of the Velasco regime. It was not attempting to construct a socialist society, as did Allende in Chile or Castro in Cuba. Nor was it seeking to exclude and repress already organized working-class movements, as did military governments in Chile, Brazil, and Argentina. Instead, the Peruvian regime was intent on integrating marginal urban and rural masses into the national society in order to lay the groundwork for industrialization and autonomous development. The chief strategy toward this end was to incorporate such groups under the guidance and tutelage of the state, which would in turn regulate civil disputes. This would permit achievement of a major goal, one that has frequently appeared in Latin American settings: the reduction of class conflict.

In its stress on eliminating class struggle and establishing social harmony, the Peruvian regime emerged as a typical corporate state. In its emphasis on organizing and integrating lower-class groups, it was reminiscent of the Cárdenas government in Mexico (1934–40) and the early part of the first Perón rule in Argentina (1946–50).

The same basic principle applied to the manufacturing sector, for which the government promulgated an "industrial community" law. According to this scheme employees in every firm with six or more workers—members of the "industrial community"—were gradually to acquire 50 percent of the stock in each company and to gain representation on the board of directors. Workers would become co-owners with management and, in principle at least, class conflict would disappear. Employers found various ways to get around this legislation, but by late 1974 there were approximately 3500 industrial communities with 200,000 members in control of 13 percent of all the shares in their firms.

To accelerate this process the government created the "social property" sector, through which firms would be controlled and managed exclusively by workers—with proceeds to go to the sector as a whole. Part of the income went to wages and housing and services, and part was reserved for reinvestment in other social-property industries.

While granting workers participation in the Peruvian industrialization, the military regime embarked on a series of measures to reduce the role of foreign capital. At the time of the 1968 takeover, 242 firms with significant foreign investment were responsible for 44 percent of the country's industrial production. This was unacceptable to the government. Declared Velasco in 1970: "The moment has now passed when we judged the process of industrialization in the abstract as a panacea for all our problems. Now it is imperative to determine the type of industrialization. . . ." To curtail the role of foreign capital the government began to require approval of new investment by a regulation board, to prohibit the purchase of viable, locally owned firms, and to exclude foreign participation from sensitive areas. Peru also assumed a conspicuous role in the promotion of regional economic cooperation through the Andean Pact.

But the most decisive steps involved expropriation. Soon after the coup, the regime announced the nationalization of IPC, to the delirious approval of the public, and the establishment of the state-supported *Petroperú*. In time the government took over other prominent foreign-owned firms: ITT (1969), Chase Manhattan Bank (1970), Cerro de Pasco (1974), and Marcona Mining (1975)—the latter two replaced by *Minoperú*. These actions met with hostility from the United States, but in February 1974 the two governments reached an accord through which Peru would pay $150 million as full settlement of all outstanding claims by North American businesses (including IPC) and Washington would withdraw its opposition to international loans for Peru.

Despite its populist stance, the Velasco government met with considerable resistance at home. Preexisting labor unions, such as the *Aprista*-dominated organizations among sugar workers, resented the inroads on their terrain. Peasants often found the top-down institutions unresponsive to their demands and began to stage protests in SINAMOS

offices. The traditional elite expressed nothing but horror at the policies of the regime, of course; in response the generals seized control of the media. Four newspapers were taken over in 1970, TV and radio followed the next year, and six other Lima dailies were put under progovernment management in 1974. This only compounded the government's problem, as journalists and middle-sector spokesmen denounced these restrictions on dissent.

Obviously the military government was an authoritarian regime. Opponents were harassed, intimidated, exiled, and jailed. It was furthermore staffed by a cadre of technocrats intent on achieving economic growth and development. So it was a "bureaucratic-authoritarian" regime to this extent, but it differed from the prototypical forms in Argentina and Brazil in at least three respects: first, it began as an extremely autonomous military government that did not involve a coalition with foreign investors and domestic capitalists; second, it was attempting to build support through the inclusion of lower-class groups; third, it did not engage in the campaigns of systematic terror that were occurring in the southern cone.

Economic conditions added to Velasco's woes. Export earnings declined, as shown in Figure 6-1. The fishmeal industry disappeared, petroleum explorations yielded no new oil deposits, world prices for sugar and copper dropped. The balance of payments deteriorated, the foreign debt swelled, and inflation struck. Workers began to demonstrate their discontent. In 1975 there were 779 strikes, compared to 414 in 1967.

As these problems first loomed on the horizon, Velasco himself succumbed to ill health in 1972: the diagnosis was serious circulatory problems. His grasp on power weakened, if slowly, and his colleagues eventually concluded that he had to go. In August 1975 Peru's joint chiefs replaced him with General Francisco Morales Bermúdez (who happened to be the descendant of a former president). An era thus came to an end.

In effect, Morales Bermúdez presided over the modification—if not the dismantling—of the 1968–75 experiment. SINAMOS was permitted to wither away. Under pressure from the IMF, the government imposed an economic austerity program that reduced the proportion of adults with adequate employment to 42 percent. The real income of the urban working class declined by 40 percent between 1973 and 1978. In February 1977 Morales Bermúdez unveiled Plan Túpac Amaru, a program rhetorically committed to "fully participatory social democracy" but actually designed to undo much of the Velasco scheme. It called for, among other things, economic decentralization and austerity, the encouragement of foreign investment, and transfer of the state-controlled press to private hands. And it sounded the death knell of the regime. There were to be a constituent assembly in 1978 and general elections in 1980. The officers were getting out.

What this revealed, in retrospect, was the regime's inability to gain solid support from any social class or grouping and, thereby, establish institutional foundations for its authority. By reaching into so many areas of Peruvian society, the military government succeeded in alienating almost everyone: No group felt safe from intervention or control, no stratum offered its unconditional adherence. Ironically, the feature which had given Peru's revolutionary military government so much freedom of action—its autonomy—also led to its eventual demise.

Back to Normal?

The 1978 elections for the constituent assembly produced several surprises. Not surprising was APRA's 35 percent of the vote, evidence of the party's continuing appeal. But no one had expected the five main parties on the left to win 33 percent. The left was apparently benefiting from the wave of popular mobilization in 1976–78. There was also a new party on the right, the Popular Christians (PPC), which won 24 percent. Would the left and right be able to hold to these gains in the 1980 presidential elections?

The top two parties were APRA, whose ticket was now led by Armando Villanueva (Haya de la Torre having died), and Popular Action, still under Fernando Belaúnde Terry, which had boycotted the 1978 elections. After an arduous and open campaign, the sixty-eight-year-old Belaúnde captured a commanding plurality with 42 percent of the vote, partly because of his reassuringly statesman-like image. APRA took only 28 percent, a substantial setback; the right, represented by the PPC, received 11 percent; an assortment of left-wing parties amassed about 16 percent. Thus the left and the right each dropped to less than half their percentage of two years earlier. Belaúnde had clearly picked up votes from both. Villanueva declined Belaúnde's offer to take part in a coalition government, preferring to maintain APRA's time-honored role in the congressional opposition.

So Belaúnde became president again. The scene had a familiar touch, but there were some novel features, too. In contrast to 1963–68, Belaúnde had virtual control of the legislature, with a clear majority for Popular Action in the Chamber of Deputies (95 seats out of 180) and room for maneuver in the Senate (26 of 60 seats). APRA began to fall apart and no longer resorted to purely obstructionist tactics. The right was relatively isolated. And the parliamentary left, though growing in strength, was divided into multiple factions. Belaúnde could claim a mandate almost by default. His reelection was also a vindication—the military that had deposed him in 1968 now escorted him back into the presidency.

Belaúnde was still an old-style electoral politician, promising progress through new public works. His long-time dream was to build a perimeter road deep into Peru's Amazon jungle, thereby opening new land

for settlement. At the same time, Belaúnde proposed to reduce the state role in the economy, strengthen private enterprise, and encourage new foreign investment, especially in the petroleum sector.

Belaúnde's economic team was led by individuals with close ties to North American and European banking circles. Their pro-free-market views (and their willingness to adopt an IMF stabilization program) helped in renegotiating the foreign debt and in attracting foreign capital. By 1982 the Belaúnde regime seemed well on its way to diversifying its exports (mostly commodities) and adopting a more open stance toward the world economy.

Such was not to be. Although GDP growth was a respectable 3.1 percent in 1981, it fell to less than 1 percent in 1982. In 1983 the GDP dropped a staggering 12 percent. Partially responsible for this decline was the worldwide debt crisis triggered by Mexico's near default in 1982. That shock was amplified by the 1981–83 world recession. By December 1982 Belaúnde had to turn to patchwork foreign bank financing to keep Peru solvent. Even if it had worked, the Belaúnde program would have done little to reduce the gap between the desperately poor highlands and the more prosperous coast. Worry over that gap as a source of social conflict had in part prompted the military to depose Belaúnde in 1968. Now he was back and was following an austerity policy hitting those on the bottom hardest.

The Belaúnde government faced another formidable challenge: the emergence of *Sendero Luminoso* (Shining Path), Maoist guerrillas based in the highland peasant communities of Ayacucho province. They burst onto the scene by brutally assassinating any village leaders who resisted their call to smash authority and establish an egalitarian utopia. Many local police crumpled in the face of Sendero, who issued no manifestos and maintained absolute silence about their structure and leadership. Mounting Sendero violence in the highlands forced Belaúnde to authorize a military offensive, which left its own wake of brutal repression. Nor did these tactics eliminate Sendero. On the contrary, the movement spread to other highland provinces and to Lima.

Economic growth revived to 4.7 percent in 1984, while inflation, another recurrent problem, hit 110 percent in both 1983 and 1984. Servicing the foreign debt was stretching Peru's dollar earning power to the limit. Unfortunately, neither Belaúnde's political style nor his economic formulae had measured up. But it could at least be said that he served out his five-year term, no small feat in modern Peruvian history.

The electorate chose as Belaúnde's successor in 1985 a thirty-six-year-old political newcomer from APRA, Alan García. An eloquent and forceful orator, he electrified crowds while promising dynamic leadership, social reform, and a new economic nationalism. García had risen rapidly in APRA, winning the prize that had escaped the party's founder and long-time presidential hopeful, Haya de la Torre. García won 46 percent of the vote, decisively defeating the candidates repre-

senting the United Left, the right, and Belaúnde's Popular Action. APRA also won control of both houses of Congress, giving García a strong political base.

The new president moved on the economic front first. He increased real wages, cut payroll and sales taxes, reduced interest rates, froze prices, and devalued the sol. The net effect was greater demand, which the García economic team hoped would activate Peru's underused industrial capacity. The government also announced investment programs to aid small-scale agricultural development in the long-neglected highlands. García's "heterodox" economic policies produced a boom in 1986–87, accompanied by special programs for the needy. But the return to populism proved short-lived. Peruvian policymakers had forgotten that their country, like the rest of Latin America, was facing a far more hostile world economy. They defied their foreign creditors by defaulting on all external debt. A mushrooming trade deficit, fueled by the consumer boom, rapidly exhausted the meager foreign exchange reserves.

In 1988 Peru suddenly turned into Latin America's economic basketcase. The gross domestic product plunged over 30 percent in three years, as inflation soared above 3000 percent. Massive unemployment drove millions into the "informal economy," where they paid no taxes, further swelling the government deficit. Peru had become the most bankrupt country on a bankrupt continent.

More massacres in the sierra and blackouts (from dynamited power lines) in greater Lima demonstrated growing Sendero strength. Some now compared Sendero's followers to the Khmer Rouge. Branding García as just another fascist puppet, Sendero forced the government into increasing reliance on the police and military. Government forces killed indiscriminately, prompting García to fire numerous commanders for atrocities in the field and for the slaughter of prisoners who had surrendered after a massive prison riot. In the end, however, García had no choice but to rely on the men in uniform against Sendero.

The new savior ready to rescue Peru in the 1990 presidential election was the internationally acclaimed novelist Mario Vargas Llosa. But the handsome conservative, who promised neo-liberal economic policies to shrink the state and promote private enterprise, blew an early lead in the polls, losing to Alberto Fujimori, a hitherto little-known agrarian economist born of Japanese immigrant parents.

After a year in office, Fujimori's technocrats launched a radical restructuring program (slashing tariffs, welcoming foreign investment, weakening labor laws) much like the shock treatment of Pinochet's "Chicago boys" in Chile fifteen years earlier. Hyperinflation was controlled and payments on the foreign debt resumed. But this economic success was soon overshadowed by political regression. In April 1992 Fujimori closed Congress and announced a radical restructuring of the

judiciary. It was an "auto-coup," possible only because of solid military backing. Peru thus became the first South American country of the 1990s to slip back into authoritarianism. Its fragile democracy had collapsed under the pressure of economic chaos, deep social divisions, corruption, terrorism, and the discrediting of the traditional political parties, especially APRA. To the consternation of Fujimori's critics at home and abroad, the coup enjoyed wide public support.

Sendero initially welcomed the news of the coup, assuming that an authoritarian regime would hasten the revolution. On the contrary, it proved a disaster for the guerrillas. In September 1992 the movement's founder and maximum leader, Abimael Guzmán, was captured, imprisoned, and theatrically displayed to the press. He soon cooperated with his captors, urging his fellow guerrillas to give up. More arrests of high-level Sendero leaders followed. The movement quickly began to disintegrate, although diehard *Senderistas* continued sporadic attacks.

Like García before him, Fujimori had to rely on the police and army in fighting Sendero. The result was a shocking record of human rights violations. Although their revelation caused delays in foreign aid, the Peruvian government eventually convinced the foreign lenders that its human rights violations would be rapidly corrected and that democracy would soon return. The creditors proved less interested in such promises than in Peru's austerity and privatization programs. The credits and new investments were soon forthcoming.

Fujimori again confounded his detractors by winning reelection in April 1995 with an electoral margin of 64 percent. He ran especially strong in the Lima shantytowns and in the highlands. His success, both political and economic, could not disguise the fact that Peru remained a deeply divided and profoundly unequal society. Not least of the scars were the 22,000 lives lost in the battle with Sendero.

SEVEN

MEXICO
The Taming of a Revolution

The history of Mexico offers a study in contrast. Rich in natural resources, the country has known both prosperity (if only for the elite) and poverty. For several decades after independence the nation's political life was a prototype of chronic instability. National governments came and went at gunpoint, threatening the new nation's territorial integrity. By mid-nineteenth century Mexico was heading toward a liberal government, which would have greatly reduced church power and the corresponding burdens of its colonial legacy. Political liberalism, however, gave way to the dictatorship of Porfirio Díaz (1876–80 and 1884–1911) and thence to the Mexican Revolution—the first of the world's great twentieth-century revolutions. Out of the Revolution came a political system which produced, since the 1930s, a political stability unmatched in Latin America.

Mexico's emergence from its colonial past has been conditioned by one factor no other Latin American nation shares: a 2000-mile border with the United States. That proximity had produced benefits and liabilities (as a Mexican president once exclaimed, "Poor Mexico! So far from God, and so close to the United States!"). Having tasted bitter defeat on the battlefield, the people of Mexico have retained their dignity and pride—and now, having both discovered massive deposits of oil and been faced with repeated economic crises, the country has encountered the risks involved in becoming a leading member of the international community. Mexico's future, like its past, arouses emotions of anxiety and hope.

Mexico after Independence

The Wars for Independence left Mexico in disorder and decay. Conditions were far worse in Mexico than in Argentina or Brazil, because the actual fighting had been so much more widespread and protracted in Mexico. The economy was in shambles. Spaniards had taken their capital out of the country. The gold and silver mines, once the pride of

Spain's overseas empire, had fallen into disrepair. Insurgents and royalists had both made a point of killing technicians while thousands of
miners had gone off to war; without sufficient supervision, the mines
had flooded and machinery became utterly useless. Production plummeted to one-third its prewar level. Mining communities languished:
Valenciana, for example, had 22,000 residents in 1810 and only 4000
in 1820. It would take another generation—and considerable sums of
foreign investment—to restore the precious mines to full production.

The textile industry had also fallen on hard times. The scars of battle
were visible throughout the country, especially the central valley. As
one traveler recalled, there were "ruins everywhere—here a viceroy's
palace serving as a tavern, where the mules stop to rest, and the drivers
to drink pulque—there, a whole village crumbling to pieces; roofless
houses, broken down walls and arches, an old church—the remains of
a convent." Noting the toll on agriculture, too, another observer described how he and his companions "saw the houses roofless and in
ruins blackened by fire, and had ridden over the plains still bearing
faint traces of the plow; but the ranchers who tilled the ground had
been murdered with their whole families during the war."

Roads had been neglected as well, so the country lacked a workable
system of transportation and communication. Having ruled for 300
years, the Spaniards had managed to construct only three highways
worthy of the name. Travel by stagecoach was difficult and hazardous,
and transport—often by pack saddle—was costly and slow. This was a
serious obstacle to economic integration.

Economic disorder meant there were very few jobs, and much unemployment. According to one estimate, about 300,000 men, most of
whom had fought in the wars, had no job or income when the battles
came to an end. This represented 15 to 30 percent of the entire adult
male population. They were eager, often angry, and usually armed.
They posed not only an economic problem but a social threat as well.

Some of these veterans did find work. Others turned to crime (highway robbery being a particular favorite). Others stayed on in the army.
Others drifted into unofficial, quasi-military units that provided support for local political bosses, generally known as *caudillos,* who were
soon to play a dominant role in the Mexican political scene.

The wars also had a direct effect on Mexico's social structure. In the
late 1820s the new government issued a decree expelling all Spaniards
from Mexico. This ruling not only allowed the public to vent its hatred
for the Spaniards; it also deprived the economy of an important source
of capital. And it eliminated, at a single stroke, a leading segment of
the nation's upper class or aristocracy. Now creole landowners, not
Spanish-born, made up the upper echelons of Mexican society.

Economic transformations dating back to the Bourbon era, together
with gradual recovery in the 1830s and 1840s, had made it possible for
new groups to acquire wealth and status. Centered mainly in Mexico

City, these aspirants, like most nouveaux riches, were ostentatious, putting on elaborate displays. As Fanny Calderón de la Barca, the English-speaking wife of a Spanish diplomat, later recalled, the new rich did not understand all the rules of social propriety: "But although there are cabinets inlaid with gold, fine paintings, and hundreds of rich and curious things," she observed in her memoirs, "our European eyes are struck with numerous inconsistencies in dress, servants, &c., in all of which there is a want of keeping very remarkable." In sum, early nineteenth-century Mexico had a creole upper class with two parts: one consisted of old, traditional families who for the most part kept to their land; the other was new, drawn from commerce and the professions as well as land. And it was the new segment, the recently arrived, who became active in politics.

Poverty persisted among the vast majority of the population. Especially in the center and the south, Mexico had a classic peasantry—large masses of *campesinos,* or country people, who scratched out meager livings from the land. Largely of Indian origin, sometimes mixed-blood or *mestizo,* Mexico's peasants furnished labor for the agricultural sector. Many worked on haciendas, where they lived in virtual serfdom, and some went begging in the cities.

The existence of this underemployed peasantry also guaranteed Mexico a large surplus labor force. Partly for this reason, and partly because of antiforeign sentiment, Mexican authorities did not encourage immigration from abroad. Unlike Argentina, Mexico never acquired a predominantly European-born working class. Nor did it undergo rapid population growth at any point in the nineteenth century. Starting with about 6 million residents in 1800, the country had about 7.6 million people in 1850; by 1900 the figure had climbed to 13.6 million, but even this represents a modest annual average growth rate of less than 1.2 percent over the fifty-year period. Mexico's population explosion would not come until the twentieth century.

There were two institutional bases of power in Mexico after independence—the church and the military. The church had come through the independence wars with most of its immense wealth intact. According to at least one observer, the church may have controlled nearly one-half the nation's land. The church earned regular income from rents on its vast real estate holdings, its investments were everywhere, and it was by far the largest banking operation in all Mexico. Its generous loans to large landowners not only guaranteed a steady income but also created a firm alliance with the upper echelons of Mexican society. Small wonder that the church and its economic holdings would eventually become a target of opposition, particularly among those who failed to benefit from ecclesiastical largess.

The second power base was the military, which dominated national politics. During the forty-year period from 1821 to 1860, Mexico had

at least fifty separate presidencies, each lasting for an average of less than one year; thirty-five of these ill-starred regimes were led by army officers. The basic means of winning presidential office was through a military coup. And looming throughout this period was the tragicomic figure of Antonio López de Santa Anna, who held the presidency on nine separate occasions and who installed figureheads at other times.

Santa Anna was the most famous of Mexico's *caudillos*. These strong-men assembled their armed followers—miniature armies—who were primarily seeking wealth. Once they fought their way into national power, however, they often found that the treasury was running out (usually from previous military spending). Eventually the reigning *caudillo* band would break up and a new leader, with new followers, would seize power. The *caudillos* themselves did not bother with the arts of governance. That was left to a cadre of lawyers and professionals, many from Mexico City, who staffed the ministries (and in this, the same faces often reappeared: there were nearly 600 separate cabinet appointments between 1820 and 1860, but they went to only 207 individuals). Thus did *caudillo* politics entail continuity as well as change.

The North American Invasion

Crippled by the Wars of Independence, Mexico was a weak and vulnerable new nation. To the north lay another new nation, which had thrown off its English master fifty years earlier. Now the fledgling United States was rolling westward and southward, headed for the vast, virtually unpopulated northern domains of what was formerly the Viceroyalty of New Spain.

Spaniards had never found the resources to settle the north—the huge territories of California, the entire Colorado River valley, and Texas. The best they could do was to create a network of religious missions, manned above all by the resourceful and loyal Jesuits. These sprawling lands became an obvious magnet for the restless North Americans. In 1821 Stephen Austin and a group of settlers moved into Texas, then a part of Mexico. Eventually chafing under central rule from Mexico City, the Texans revolted in 1835 and declared independence the following year. Attempting to crush the rebellion, Santa Anna led Mexican troops against the Alamo, killing the Texan defenders to the last man, but he later suffered defeat at San Jacinto and Texas remained independent. In 1845 the U.S. Congress voted to annex Texas, whose leaders promptly agreed.

The Mexicans saw the annexation of Texas as equivalent to an act of war by the United States, and disputes over financial claims continued to complicate U.S.-Mexican relations. President James K. Polk sent American troops into a disputed border area, a step that Mexicans saw as invasion. When Mexicans counterattacked, Polk called it war. By

consent of Congress—but with the opposition of such prominent legislators as John C. Calhoun and Abraham Lincoln—Polk had the war he
and his supporters sought.

It was a total mismatch. Santa Anna was the hapless Mexican commander and at first his forces managed to resist American troops under Zachary Taylor, but in 1847 Winfield Scott led his columns directly
from Veracruz to Mexico City. Ordinary Mexicans joined in the effort
to fight off the U.S. army, and young military cadets—since remembered as the "boy heroes of Chapultepec"—chose death rather than to
surrender their national flag. But it was to no avail. Mexico lost. The
price it paid was heavy.

The treaty of Guadalupe Hidalgo brought a formal end to the war in
February 1848. By the treaty, the United States paid Mexico a modest
settlement of $15 million and took the entire expanse of territory from
Texas to California—about half of Mexico's national domain. This was
a galling defeat, and its painful memory has never died in Mexico. Just
as Americans are taught to "Remember the Alamo," Mexicans learn
tales of valiant struggle against overpowering odds. The official name
of the dispute offers a clue to sensibilities. In the United States it is
called the "Mexican-American War," but in Mexico they call it the "War
of the North American Invasion."

The Mexican creole elite was deeply shaken by the loss. Their frustration erupted in a bitter attack by the Conservatives on the Liberals,
the predominant faction in government since independence. The Conservatives, led by Lucas Alamán, maintained that Mexico had been
humiliated because it had foolishly tried to adopt the values of the
Anglo-Saxons to the north. What their nation needed, argued the Conservatives, was a return to its Hispanic tradition. Specifically, it needed
to promote aristocratic ideals, protect the legal privileges of the military
and the church, and create a constitutional monarch (perhaps by importing a European prince).

The Liberals, who continued to enjoy the advantage of power,
fought back: in the mid-1850s, a flurry of new reforms, which caused
the period to be labeled *La Reforma,* stripped power from the army
and the church, and culminated in a new, notably liberal, constitution.
Unfortunately, the controversy was not restricted to speeches and laws.
In 1858 the Conservatives counterattacked, opening almost twenty
years of destructive civil war. Those years included the ill-fated empire
(1863–67) ruled by Maximilian von Hapsburg of Austria. It literally
enacted the Conservative formula. Not suprisingly, importing a foreign
monarch aroused nationalist sentiments on which the Liberals quickly
capitalized. Benito Juárez, the Liberal leader, led a successful overthrow of the empire and ordered the execution of the luckless Maximilian. The Liberals consolidated power under President Juárez until his
death in 1872. A subsequent interval of instability was ended with one

of the president's former associates seizing power: General Porfirio Díaz.

The Díaz Era: Progress at a Price

For the thirty-five years from 1876 to 1911, Díaz proved himself to be a master of politics. He began with his military colleagues and followers and from there went on to create a broad coalition. He gave the regional *caudillos* room to maneuver, encouraging them to fight among themselves. As his presidency matured, he steadily built up the army. In order to maintain control of the countryside, where the vast majority of Mexicans lived, Díaz relied heavily on the feared *guardias rurales*, or rural police. In short, Díaz patiently built up the power of the federal government where it counted—in military and police power.

At first Díaz did not seem to represent anything new in politics. He was, after all, a product of the liberal movement. As time passed it became clear that Díaz was a Liberal with a difference. He cultivated neutrality on the crucial question of the church, neither attacking it (like most Liberals) or defending it. He conspicuously allowed his devoutly Catholic second wife to serve as a symbol of reconciliation toward the institution the Liberals had pilloried.

In other respects Díaz stuck to liberal principles. In one of his most important and far-ranging measures, he ruled that the ban on corporate landholdings, a liberal measure of the 1850s aimed primarily at the church, should apply to Indian villages. This opened vast new areas to speculators, ranchers, and political favorites. In 1894 Díaz helped the landowners even more by decreeing that unused lands, or *terrenos baldíos*, could be taken over for private exploitation. The crucial source of new capital was to come from abroad. Díaz and his leading ministers sought out prospective foreign investors, especially U.S. and British, and offered them generous concessions. All this was an obvious application of the principles of economic liberalism that had captured most Latin American elites in the closing decades of the nineteenth century. In Mexico the writers, technocrats, and intellectually inclined politicians who articulated these doctrines earned the label of the *científicos*, underlining their supposed link to Positivist philosophy.

Díaz proved his mastery of politics in that most fundamental of ways: he stayed in power far longer than any would have dared to predict. For three and a half decades he held the presidency, with only one interruption (Manuel González: 1880–84). He believed that he was giving Mexico the precious gift of political stability, which he saw as indispensable for economic growth. If that required some repression, it was for a good cause. A shrewd politician, Díaz avoided ever presenting himself as a dictator. He simply had the constitution amended, time and again, so that he could be reelected to the presidency. Díaz knew

how to appeal to the privileged sectors, how to make them loyal, how to orchestrate their support for the economic schemes that would raise their country to a "civilized" level.

Economic development was impressive. Railroads were a striking example. Díaz first tried to build them with public funds, but by late 1880 he was granting concessions to foreigners. In only four years the track in operation grew from 750 miles to 3600 miles. Mexico reached 12,000 miles of track by 1900. (On the other hand, paying interest and dividends on this foreign investment was a burden on the balance of payments.) Originally foreign-built, most railroads were taken over by the state in 1907.

As elsewhere in Latin America, foreign trade rocketed: ninefold between 1877 and 1910. The United States became Mexico's leading trade partner, as mineral exports expanded to copper and zinc, as well as silver and gold. Modest industrialization occurred, centered in textiles, cement, iron, and light consumer goods. Díaz set great store by the need to pursue economic policies that would maintain Mexico's creditworthiness in the United States and Europe. In 1895 the federal government produced a budget surplus, and for the rest of Díaz's regime all budgets were balanced. As celebrations for the independence centennial of 1910 approached, Díaz and his lieutenants could claim that they had realized in Mexico the Positivist ideal of "order and progress."

Economic activity varied in character from region to region, and this led to differing social structures. The north was primarily a mining and ranching area, where the workers were hired laborers—miners, for instance, and cowboys. The central valley, by contrast, produced wheat and grain on both medium- and large-sized farms. Sugar was raised in the south-central region, particularly in the state of Morelos, where traditional peasant lands were being seized for use by the mills. Vast henequen plantations prospered in the Yucatán, where local natives were compelled to work as peons.

Under Díaz, Mexico never developed a strong entrepreneurial class. Concessions and favors came from the state, and capital came from abroad—England, France, and, of course, the United States. The middle sectors were extremely weak as well.

These social factors bore deep political significance. Elsewhere in Latin America middle-class professionals provided pressure and leadership for reformist movements, as in Argentina, and on occasion they drew support from fledgling industrialists, as in Chile. Not so in Mexico. It appears that turn-of-the-century Mexico had the social ingredients for a revolution, but relatively little material for reform.

The economic progress of the Díaz years also had its cost. While the wealthy prospered and duly copied the ways of European aristocracy, the vast majority of Mexicans faced grinding poverty. Given its labor surplus, Mexico's wage rates remained very low. Indeed, one estimate

(doubtless exaggerated) showed that average purchasing power in 1910 was only one-quarter the 1810 level. Mexico exported agricultural products, while production of most Mexicans' dietary staples—corn and beans (*frijoles*)—barely kept up with population growth. There could be no improvement in the notoriously low per capita consumption levels prevailing at the outset of the Díaz era. Vital statistics were alarming. In 1900, 29 percent of all male children died within their first year, and many of the survivors ended up working twelve hours a day in a sweatshop. Only a quarter of the population was literate.

This highly unequal economic "progress" drew repeated protests from workers, both urban and rural. There were strikes, sometimes fierce, especially where wage labor worked under industrial-type conditions. Between 1906 and 1908, for example, Mexican workers at the Cananea Copper Company repeatedly protested the higher wages given to U.S. laborers. Significant strikes occurred also among the railroad workers and at the Río Blanco textile mills. Labor protest was intensified by the international financial crisis of 1906–8. In the rural sector, peasants in the Morelos area bitterly resented losing their land to commercial cultivation of sugar and other market crops. In the north there was a similar reaction to the loss of land for railway construction.

Díaz and his advisers could pursue a consistent economic policy because they had created the most effectively centralized government Mexico had seen since independence. Decision making was centralized in Mexico City, at the expense of local or regional *caudillos*. Political office, especially at the federal level, was sought after by the higher level of society. Those who made it were envied, since economic gain so often required contact with the government. Díaz himself knew full well the kind of system he had promoted. Near the end of his regime he explained: "We were harsh. Sometimes we were harsh to the point of cruelty. But it was necessary then to the life and progress of the nation. If there was cruelty, results have justified it. . . . Education and industry have carried on the task begun by the army." Many of Díaz' opponents agreed on the need for national power, but denounced the way Díaz used it. Pressure was mounting, as frustration grew among the younger elite who were excluded from the Díaz coterie. Time was working against Díaz, but who could have predicted how his carefully constructed house would come tumbling down?

The Mexican Revolution

Few revolutions are precipitated by the oppressed. Far more often they begin with a split within the dominant elite. Disgruntled dissidents, frequently young, become angry enough to attack the system. So it was in Mexico in 1910.

One of the leading critics was Francisco I. Madero, scion of a family

that had made a fortune in cattle and mining. It was also linked to Díaz' political machine. Evaristo Madero, Francisco's grandfather, had been governor of the state of Coahuila from 1880 to 1884, and the Madero family had cultivated a close friendship with José Y. Limantour, Díaz' long-time finance minister. Francisco got the best of a foreign education, studying in Paris and at the University of California. He returned to apply his skills in commercial agriculture, especially on the family's cotton plantation. He was a strong liberal in economics, which fit the Díaz era, but also in politics, which did not. His belief in political democracy soon alienated him from the rigidities of the late Díaz regime. He became an outspoken opponent, arguing that Mexico was ready for liberal democracy and that if Díaz chose to run for reelection in 1910 (as everyone expected), then the vice presidential candidate must come from outside the presidential clique.

Díaz was by now the captive of his own success. Why should he take seriously the lamentations of an ambitious and spoiled young oligarch? When the president failed to heed his message, Madero did the unthinkable: he entered the 1910 campaign as the candidate of the Anti-Reelectionist Party. Díaz now faced greater opposition than at any time in decades. His machine produced another victory, but it was far from effortless. The police had to jail 5000 of the opposition, including Madero. The young rebel, now emboldened, refused to recognize the legitimacy of Díaz' reelection. Instead, he issued (while in jail—which suggests Díaz hardly had an iron grip) his famous Plan of San Luis Potosí, and called for armed resistance. The rebel movement grew rapidly, as its troops took Ciudad Juárez (across the border from El Paso). Díaz now dropped the mask of the infinitely resourceful autocrat. In a surprising show of weakness, he capitulated and left the country in May 1911. A new presidential election was held and Madero triumphed. In 1912, he became the nation's president before delirious crowds in Mexico City. Democracy, it seemed, was on its way.

Francisco Madero and his fellow dissidents may have started the Mexican Revolution, but they did not long control it. Other rebels had larger goals: Emiliano Zapata, for example. Zapata had emerged as the rock-hard leader of landless peasants in the southwestern state of Morelos. They were the country dwellers who had seen their traditional land rights taken away by the smooth-talking lawyers and speculators using the new laws of "liberal" inspiration. These *zapatistas* (as they inevitably became known) saw the rebellion as a chance to restore justice. That meant regaining their lands.

The *zapatistas* quickly became disillusioned with Madero, and with reason. Why should this son of a great landholding family sympathize with their cause any more than the Díaz gang? In November 1911 Zapata and his followers in Morelos angrily attacked Madero in their *Plan de Ayala*. "Having no intentions other than to satisfy his personal ambitions, his boundless instincts as a tyrant, and his profound disrespect"

Emiliano Zapata gave determined leadership to the revolutionary peasant movement that began in the state of Morelos.

for the Constitution of 1857, they said, Madero "did not carry to a happy end the revolution which gloriously he initiated with the help of God and the people." Instead, he let the Porfirian political machine continue, thereby showing his indifference to the plight of the people. The rural dwellers now had only one option: direct action. "We give notice [that] . . . the *pueblos* or citizens who have the titles corresponding to those properties will immediately enter into possession of that real estate of which they have been despoiled by the bad faith of their oppressors, maintaining at any cost with arms in hand the mentioned possession." The *zapatistas* were as good as their word. These rural smallholders had believed the Revolution would help them regain their lands. When Madero failed to deliver, they contemptuously dismissed him and declared their own revolution.

Madero was hardly a true revolutionary. He was a would-be parliamentarian who thought Díaz' abdication would open the way to true democracy. Madero belonged in England or Scandinavia, not Mexico. He flinched at the thought—suggested to him by less squeamish rebels—that he should strike at his opposition before they struck at him. The mistake cost him his life in 1913. His killer was his own military chief of staff, Victoriano Huerta, a high-ranking general under Díaz. Huerta dragged the indiscreet U.S. Ambassador Henry Lane Wilson into his plot, thereby insuring that the United States would continue its notorious role in Mexican politics.

In the north Pancho Villa created a powerful military juggernaut, but his personal flamboyance earned him a dubious reputation in Mexico and the United States (Courtesy of the Library of Congress.)

Huerta was a crude figure, who thought he could reestablish a version of the Porfirian regime. He tried to impose his authority across the aroused country, but soon met resistance. Many Mexicans who had been caught up in the revolt against Díaz now saw Huerta as the usurper. Opposition began to build, and as it gathered force it coalesced into the genuinely "revolutionary" phase of the Mexican Revolution.

One of the most powerful centers of resistance to Huerta was the northern state of Chihuahua, where Pancho Villa gained control. Villa was a rough-hewn ex–cattle rustler who had mobilized a small army. Unlike Zapata, with whom he was often compared, he led no peasant rebellion. Villa's supporters, at least initially, were small ranchers, unemployed workers, and cowboys: men who wanted jobs, not small plots of land. So it was not surprising that when Villa pronounced an agrarian reform, in December 1913, he called for confiscation of large haciendas, but not for their subdivision into plots. The state would administer the haciendas, and their commercial crops would help finance Villa's military machine.

Villa quickly put this idea into practice. It may have created administrative problems, but it achieved its goal. Money was produced and supplies were obtained (mainly from the United States, which remained the great arms supplier for all Mexican revolutionaries). Villa's army was well fed and well equipped. Indeed, Villa's followers now had a sure source of employment in his army, which emerged as a well-paid professional mercenary outfit.

There were other challenges to Huerta's bloody accession to power. One was in Madero's home state of Coahuila, where governor Venustiano Carranza mounted a strong resistance movement. Carranza, like Madero, was a dissident member of the elite, having risen to the level of senator during the regime of Díaz. A wealthy landowner, he had also been an interim governor. As the anti-Díaz forces slowly gathered strength in 1910, Carranza first cast his lot with Bernardo Reyes, another opposition candidate for president. During the campaign, however, he joined the "Anti-Reelectionist" group. Once in power, Madero rewarded Carranza by naming him governor of their home state, Coahuila.

Carranza contested Huerta's usurpation with little more than a counterclaim. Carranza's *Plan de Guadalupe* (March 1913) simply declared that Huerta held power illegitimately and that he, Carranza, should be recognized as "First Chief of the Constitutionalist Army." Once established, the new president would then convoke new elections. The *Plan* included no attempt to discuss larger socioeconomic or ideological questions. The *carrancista* movement looked like another *caudillo*-type rumbling. Support was scattered, mostly rural, obviously limited to the north.

All attention now centered on Huerta: Could he hold power? The opposition hammered away from the southwest (Zapata and the agrarian rebels) and the far north (Villa and his roaming army). Huerta's most dangerous enemy, however, was Carranza, the ultra-respectable elite politician. Mexico was now plunged into a bloody civil war that saw the federal army swell to more than ten times what it had been at the end of Díaz' rule. The *zapatistas* drew off Huerta's forces by their stubborn rebellion in Morelos, while the Constitutionalists in the north kept up their pressure. Eventually it was foreign intervention, not Mexican arms, that doomed Huerta. U.S. President Woodrow Wilson, determined not to recognize Huerta's government, had sent marines to occupy Veracruz after an incident involving the arrest of U.S. sailors. To counter the U.S. marines Huerta had to pull troops out of the civil war. Soon he saw his situation was hopeless. In early July 1914 he resigned, accusing the United States of having overthrown him.

By mid-1914 the Revolution was up for grabs. All of the forces that had overthrown Huerta gathered to discuss a possible coalition government. Carranza was immediately suspicious of the common agrarian origins of the Zapata and Villa forces. He withdrew from the negotiations, attacking the legitimacy of that putative "government," and set up his own regime in the eastern seaport of Veracruz.

The social fissures in the Revolution were now becoming painfully obvious. Villa, and especially Zapata, represented claims for radical social change. Carranza sensed that he would have to offer more than the liberal rhetoric that had sustained Madero. In a December statement Carranza began to edge leftward. He promised, without details, "legis-

lation for the improvement of the condition of the rural peon, the worker, the miner, and in general the proletarian classes." The following month he pronounced an agrarian reform, calling for the restoration or creation of agricultural communities *(ejidos)*, requesting procedures for restoring legal titles, and establishing a national agrarian commission. In February 1915 Carranza made his move toward labor: he got the anarcho-syndicalists—the best organized of the small urban labor movements—to agree that in return for favorable labor laws their Red Batallions would back the *carrancista* cause.

During 1915 the issue was joined. Villa, the most formidable of Carranza's military enemies, pressed forward for a quick victory. He met his match in Álvaro Obregón, Carranza's brilliant army commander from the northern state of Sonora. In mid-1915 Obregón decisively defeated Villa, who retreated to the hills of Chihuahua to continue a guerrilla war but no longer to offer a national threat. The *zapatistas* could not mount a sustained challenge to Mexico City, and withdrew into their native Morelos to hold out against federal incursions.

With his principal enemies safely at bay, Carranza could afford to call a constitutional convention in late 1916. In May 1917 he formally assumed the presidency. The stage was now set for the writing of the Mexican Constitution of 1917, a premier document of the Mexican Revolution.

Carranza himself had no radical ideas. He drafted a pale imitation of the Constitution of 1857, little more than a restatement of principles of classical liberalism. The convention delegates had other thoughts. They promptly took control and wrote a charter that was startlingly radical for this pre-Bolshevik era. Article 27 empowered the government to redistribute land. Article 123 announced rights for labor that had certainly never been heard of in North America. Article 3 subjected the church to new restrictions, which imposed a virtual straightjacket. Socialist overtones permeated the constitution. Suddenly it became obvious that what had started as a mere revolt of dissident elitists against Díaz was threatening to become a social revolution, to change significantly the power and property relationships in Mexico. After 1917 every aspiring political leader had to adopt at least a rhetorical posture in favor of Mexico's workers and the peasants.

The agrarian rebels—Villa and Zapata—continued to hold their strongholds and represent a possible threat to Carranza. Zapata was taken care of in 1919, murdered by *carrancista* troops in an ambush. The following year Carranza faced his own problem: he wanted to impose a little-known politician, Ignacio Bonillas, as his successor. In this Carranza was short-sighted. The "no reelection" slogan of the 1910 campaign had been its most powerful rallying call and it found explicit expression in the new constitution. Now Carranza was violating that rule in spirit by imposing a successor who would be his stooge. The Revolution reverted to its bloody practice: the valiant Obregón, the ar-

chitect of victory over Villa, led an uprising. Carranza was forced to flee and, while on the run, was assassinated by one of his own guards, probably acting on behalf of Obregón. The succession problem, which had led to Díaz' fall, was still far from solved.

Institutionalizing the Revolution

Obregón succeeded to the spoils of the presidency. The need was for reconstruction after years of civil war, but the world recession after World War I sharply reduced Mexico's export earnings and deepened a domestic economic slump. Nonetheless, the government launched an ambitious rural education campaign under the leadership of the noted intellectual José Vasconcelos. In the area of labor, the Obregón government bet heavily on the newly founded *Confederación Regional Obrera Mexicana* (CROM), which Obregón soon co-opted, while at the same time harassing the communist- and anarchist-led unions. On land distribution Obregón was cautious, fearing a loss of production. The last of the original popular rebels, Pancho Villa, succumbed to a fusillade of bullets in 1923, and the era of effective demands for fundamental social reform was over for the moment. Obregón did make two important contributions to the stability of the Revolution. First, he achieved an understanding with Washington—an agreement on how U.S. oil firms would be treated, in return for U.S. diplomatic recognition. Second, Obregón managed to transfer power peacefully to his successor, something no Mexican president had done since 1880.

The new president was another general from Sonora, Plutarco Elías Calles. This solid officer-politician soon proved to be the man who would put the revolutionary political system on a solid footing. For Calles, however, the threat was from the right. Calling themselves the *cristeros* ("Christers"), Catholic militants presented the revolutionaries with the first broad-based, ideologically committed opponents of the secularizing Revolution. The *cristeros* were by no means limited to the wealthy defenders of the old economic order; they included many simple folk who saw the Revolution as the work of the devil, to be stopped only by the sword. This pious belief was reinforced by reactionary clergy, especially in the state of Jalisco, where they desperately needed foot soldiers in their crusade against the anticlerical Revolution.

When the presidential term of Calles expired in 1928, Obregón, never politically reticent, presented himself for election anew. It was not a reelection, Obregón reassured Mexico, because he was not the incumbent. He won easily but did not live to enjoy his power play: before the inauguration he was assassinated by a religious fanatic.

Into the vacuum stepped the lame-duck Calles. He got the political leaders to agree on a new election and on the creation of a new party, the *Partido Nacional Revolucionário* (PNR). During the subsequent short-term presidencies of Emilio Portes Gil (1928–30), Pascual Ortiz Rubio

(1930–32), and Abelardo L. Rodríguez (1932–34), Calles continued to be the power behind the scenes.

Most observers expected Calles to continue that role in the presidency of Lázaro Cárdenas, elected in 1934. Cárdenas was a relatively obscure army officer and politician from Michoacán who surprised everyone, promptly sending the stunned Calles into exile. It was the first of many moves that proved Cárdenas was going to be his own man.

Many peasants had grown cynical about the "revolutionary" goals of their rulers. Where was the land they had been so often promised? Cárdenas decided to make good on those promises. During his term (1934–40) he presided over the distribution of 44 million acres of land to landless Mexicans, almost twice as much as that distributed by all his predecessors combined. Cárdenas knew the dangers in simply distributing land without the necessary supporting services. All too often that led to subsistence agriculture, with the farmer able to feed his family but unable to produce a surplus for the market. That would create grave problems in the food supply to the cities, as well as for the export markets.

Cárdenas' solution was to rely heavily on the communal system of the *ejido*. It had the advantage of being genuinely Mexican, while being neither capitalist nor socialist. The land distribution was made to the *ejido*, which was then the owner, even if plots were subsequently apportioned for individual use. *Ejidos* could include hundreds, even thousands, of families. The plans called for schools, hospitals, and financing, which was to be provided by the newly founded *Banco de Crédito Ejidal*. Not all the land distribution was made to *ejidos*. Individual peasants and families got plots as well.

The huge distribution created an initial euphoria, as over 800,000 recipients saw a life-long dream realized. But the longer-term results were not uniformly happy. Agricultural production for the market fell in many areas, as had been feared. The social and financial services promised by the government often never materialized in the volume needed, despite some successes. The result was low productivity and disorganization on many communal units and an insufficient integration into the market for many smaller units. Notwithstanding these problems, Cárdenas earned enormous popularity among the peasants for his boldness in distributing so much land. He had deeply reinforced the agrarian character of the Revolution.

Cárdenas also reorganized the party structure. Calles had led the way by creating a stronger machine than he found upon entering office in 1924. In 1938 Cárdenas reorganized the official party and renamed it the *Partido de la Revolución Mexicana* (PRM). It was now to be built around four functional groups: the agricultural (peasant) sector, the labor sector, the military sector, and the "popular" sector, which was a residual category including primarily the middle class. In applying this concept of functionalist representation, Cárdenas and his political ad-

visers were borrowing from corporatism, the political doctrine then in vogue in Mediterranean Europe, especially Italy, Spain, and Portugal.

In this fashion Cárdenas devised a strategy for dealing with the lower classes: mobilize and organize *both* the workers and the peasants, but keep them apart from each other. Thus the creation of separate (and competing) sectors for each group within the official party. This way the government could maintain control of popular movements and prevent the possible appearance of a worker-peasant coalition. (As of the 1990s, the regime appears to have achieved its goal.)

Cárdenas also took a more radical line in relations with the United States. The toughest issue was oil. In the early twentieth century, Mexico possessed a significant percentage of the world's confirmed oil reserves. By the 1930s, foreign oil firms, mostly U.S. but some British, had huge investments in Mexico.The companies inevitably got into a wage dispute with their Mexican employees, and it was finally carried to the Mexican Supreme Court, which ruled in favor of the employees. The foreign companies disregarded the court decision, assuming that now, as before, there must be a way around such legal problems in backward Mexico. To everyone's surprise, the president intervened and announced the expropriation of the companies. The precipitating factor cited by Cárdenas was the companies' refusal to obey the Supreme Court decision. The legal basis given for expropriation was Article 27 of the 1917 constitution, in turn based on the long-standing principle in Spanish law that all subsoil rights belong to the state (crown), not to the owner of the surface rights. The oil companies were infuriated. The U.S. firms demanded that President Franklin Roosevelt intervene on their behalf. Right-wing propagandists in the United States had a field day at the expense of the "atheistic" Mexican revolutionaries who had first attacked religion and were now attacking property.

In Mexico the news of expropriation provoked an ecstatic response. Mexican nationalist sentiment, never far below the surface, poured forth; Cárdenas was now an authentic hero for standing up to the gringos.

At first Roosevelt issued some angry demands to the Mexicans, but cooler heads prevailed in Washington. After all, Roosevelt's much ballyhooed "Good Neighbor" policy meant, at a minimum, no more U.S. invasions of Latin America. In fact, the Mexican government had already said it would compensate the companies. Dispute then centered on the value of the expropriated properties. The companies filed enormous claims, including the future value of all the oil in the ground they had owned. The long negotiations which followed favored the Mexican government, since the Roosevelt administration had early on ruled out intervention on behalf of the investors.

The companies were paid and the Mexicans created a state oil monopoly, *Petróleos Mexicanos* (PEMEX). It has since remained a high sym-

bol of nationalism—above all, because its target had been the United States. The oil companies and their friends in the U.S. government did not forget either. For another thirty years they enforced a world boycott against all Mexican oil and effectively obstructed the development of PEMEX's refining operations by getting it blacklisted with all leading foreign equipment suppliers. One reason the companies and the U.S. government thought they had to punish the Mexicans for their nationalist boldness was to prevent other Latin American governments from being tempted to similar expropriations. Mexico paid a price for standing up to Uncle Sam.

Cárdenas succeeded in giving the Revolution new life for many Mexicans. On the other hand, the wealthy sector—landowners and large merchants—hated him, much as their counterparts in the United States hated Franklin Roosevelt. In both countries the right saw these charismatic reformers as only preparing the way for radical leftism. Cárdenas gave his enemies ammunition with highly publicized gestures, such as granting asylum to Leon Trotsky in 1937 and welcoming Spanish Republican refugees fleeing Franco's vengeance at the end of the Civil War. But these were gestures on the world stage, not policies for Mexico.

Stability, Growth—and Rigidity

Cárdenas would have been difficult for any politician to follow. The choice of his successor followed a pattern which has been repeated at the end of every six-year presidency since 1940: endless speculation, mostly ill-informed, over the likely nominee. The mechanics of choosing have varied, but it has remained essentially a negotiation among top-level political bosses, carried out in private, with frequent trial balloons to test public opinion. In short, perfect material for political rumor mills.

In 1940 the choice rested with Cárdenas, who had accumulated overwhelming influence in the official party. He chose neither of the two much-discussed front-runners (one radical and one conservative), but turned instead to his little-known minister of war, General Manuel Ávila Camacho. Clearly there was a consensus on steering the Revolution onto a moderate course.

In his campaign, Ávila Camacho made it clear that he was not anticlerical; he even declared himself a believer. And he actually faced an opponent: Juan Andreu Almazán, candidate of the *Partido de Acción Nacional* (PAN), a fledging pro-clericalist party on the right. The official PRM candidate easily prevailed.

In several key policy areas Ávila Camacho soon proved more moderate than Cárdenas. One was land redistribution. Cárdenas had endeared himself to the Mexican peasantry by his much-publicized land grants, given almost invariably to the collective groups who were to

form *ejidos*. Ávila Camacho targeted his distribution at individual families, rather than the *ejidos,* since he favored small-scale, single-family ownership. There was also a contrast in the total amount of land involved. Ávila Camacho distributed about 11 million acres, whereas Cárdenas had distributed 44 million acres.

In the labor field Ávila Camacho made another move away from the left. He replaced the official leader of the party's labor sector with Fidel Velázquez, who was openly hostile to the more militant union leaders and helped to make strikes more difficult. While autonomous union action was being discouraged, the government moved on another front: creating the *Instituto Mexicano de Seguro Social* (IMSS), a social security agency which provided workers with medical care through a network of clinics and hospitals. The coverage was limited to a few hundred thousand workers by the mid-1940s, but it was the precedent for a fringe benefit system which would be steadily extended to the best organized elements of labor.

In addition, Ávila Camacho faced the challenge of a spreading world war. Mexicans felt a strong sympathy for the Allied cause, but an almost equally strong suspicion of an automatic alliance with the United States. After Pearl Harbor the Mexican government broke off diplomatic relations with Japan, Germany, and Italy, but stopped short of declaring war. It was only the repeated sinking of Mexican ships by German U-boats that led the Ávila Camacho government to obtain a declaration of war from the national Congress in May 1942.

Mexico, along with Brazil, was one of the only two Latin American countries to supply combat forces to fight the Axis. The Mexican contribution was an air force squadron, trained in the United States and subsequently attached to the Fifth U.S. Air Corps in the Philippines. The squadron performed valiantly, and brought Mexico honor in the distant conflict with Japan. Mexico also supplied vital raw materials to the United States at government-controlled low prices.

Another step would have grave importance for the future. After an explicit agreement between Presidents Franklin Roosevelt and Ávila Camacho, Mexico began sending agricultural workers north, to fill the gap left in the U.S. fields by the military draft. Spontaneous Mexican migration north had long been under way. As the war continued, the Mexican laborers (known as *braceros*) began to fill nonagricultural jobs as well—a development that aroused the opposition of U.S. organized labor. The war ended with an important precedent established: the officially endorsed northward movement of Mexican workers to perform jobs for which no Americans could be found. Yet there were enormous problems. The Mexicans, used to far lower levels of pay at home, were often willing to be cajoled (or forced) into conditions of employment inferior to what had been officially agreed upon. When the war ended, some 300,000 Mexicans had undergone the experience of working in the United States. Although many had encountered prejudice and dis-

crimination, most had earned much higher wages than was possible in Mexico. The promise of a higher income across the border, however tarnished, remained a constant attraction to impoverished Mexicans after the war.

With the end of World War II, Mexico saw industrialization as a way out of persistent poverty. The man to lead the way was Miguel Alemán, the first civilian president since the Revolution. One of Alemán's first acts was to reorganize and rename the official party, now called the *Partido Revolucionario Institucional* (PRI). Adding the word "institutional" signaled a turn toward pragmatism. The party was made up of three sectors: peasant, worker, and popular, the form it has since retained. It emerged as an utterly dominant official party, different from any other in Latin America.

The new president's hallmark was to be economic development. What Mexico most needed was infrastructure—roads, dams, communications, port facilities. Alemán therefore launched an ambitious program of public works, stressing irrigation and hydroelectric projects. There was also highway and hotel construction to facilitate the tourist trade from the United States. This investment paid off, as tourism became an all-important foreign exchange earner for Mexico, although with cultural and social implications that Mexican nationalists found distasteful.

The Mexican economy did show significant growth. The foundations were laid by sharply increasing protection against imports. The short-run justification was to ease Mexico's severe balance-of-payments deficit, but the net effect was to provide a guaranteed market for domestic production—which made sense in a market as large as Mexico's. Domestic manufacturing responded with a spurt of growth, averaging 9.2 percent a year between 1948 and 1951. Agricultural production did even better in those years, averaging 10.4 percent. Inflation and a balance-of-payments deficit slowed the rate of growth in 1952. An additional cloud over Alemán's economic record was the constant charge of corruption.

The bosses of the PRI knew, when it came time to choose Alemán's successor in 1952, that they had a serious problem in improving the government's image. The man they chose was at least a partial answer. Adolfo Ruiz Cortines had been governor of Veracruz and later secretary of the interior in the Alemán presidency, yet he had managed to earn a reputation for honesty. Once elected president, Ruiz Cortines made good on a campaign pledge to root out grafters by firing a series of suspect officials.

The most important policies of Ruiz Cortines came in the economic sphere. Since the war, Mexico had been experiencing an inflation rate which was high for Latin America. The Mexican economic managers made a crucial decision. They opted for a "hard-money," low-inflation strategy, which meant setting an exchange rate (peso/dollar) and then

managing their economy (by conservative fiscal and monetary policy) so as to maintain that exchange rate. The first step was to devalue the overvalued peso from 8.65 pesos to the dollar to 12.5 pesos to the dollar in 1954. This devaluation was larger than almost anyone expected. It gave an immediate stimulus to Mexican exports, now cheaper in U.S. dollars, and made Mexico cheaper for foreign tourists. Mexico quickly became known as a promising target for international investors.

When Ruiz Cortines left office at the age of sixty-seven, the king makers chose a successor two decades younger. He was Adolfo López Mateos, secretary of labor in the previous government. Since López Mateos had the reputation of being pro-labor, some thought the pendulum might be swinging back toward the center or even the moderate left.

Somewhat cryptically, López Mateos himself declared that his administration would be "on the extreme left, within the Constitution." Mexico was not highly unionized. The vast majority of lower-class citizens, especially the *campesinos,* had no organized means of protecting or promoting their own interests. The unions that did exist were closely tied to the regime itself. This contrasted sharply with Argentina, where Peronist trade unions had represented a base of political opposition since the mid-1950s, and with Chile, where worker movements identified with one or another political party. In Mexico, unions functioned as part and parcel of the political system.

Notwithstanding this pattern, López Mateos soon found himself facing an exception. He was quickly challenged by militant railworkers, who staged a major strike in 1959. Their leader, Demetrio Vallejo, was contesting the government-dominated structure of labor relations, not least since the railroads were government owned. He was demanding the right to genuinely independent union action. The workers followed the strike order and braced themselves for a long siege. López Mateos applied an old-fashioned remedy: he arrested the leaders and ordered the workers back to work. The strike was broken and Vallejo remained in jail for years, an object lesson to other would-be militants.

The López Mateos government did not rely only on the stick in dealing with labor. It also instituted a profit-sharing plan under which many workers increased their take-home pay by 5 to 10 percent a year. But this measure was typical of the PRI style of social policy: a beneficence granted on government initiative, not conceded under worker pressure. Given the fact that Mexico still had surplus labor, workers had little economic leverage. If they tried to organize independently, the apparatus was at hand to co-opt or repress them.

López Mateos nonetheless saw a need to change the course of the Revolution. He and his aides sought to distance his presidency from the pro-business administrations since 1940. The obvious starting point was land ownership. A chance to acquire land remained the greatest dream for Mexico's poorest rural dwellers. López Mateos ordered the

distribution of approximately 30 million acres of land, giving him a land reform record second only to Cárdenas. Furnishing basic services (and credit) for these new landowners was much more difficult, and too seldom achieved. Nonetheless, revolutionary momentum had been resumed in a crucial realm.

In economic policy López Mateos continued the hard-money policies implicit in the 1954 devaluation. Investment remained high, and Mexico began raising capital abroad, above all in the New York bond market. The attraction was high interest rates, guaranteed convertibility (into dollars), and apparent political stability. The government succeeded in achieving extraordinarily low inflation, thereby making it possible to stick with its fixed exchange rate of 12.5 pesos to the dollar. Yet Mexico was by no means a 100 percent free market economy. Indeed, state intervention in the economy increased in the years of López Mateos. U.S.- and Canadian-owned electric companies were nationalized, for example, as was the motion picture industry, which had been largely U.S. controlled.

The López Mateos years (1958–64) brought some significant changes in other areas. In foreign affairs, the Mexican government succeeded in finding a definitive solution to the longstanding Chamizal border dispute with the United States. A 1964 formal agreement between López Mateos and U.S. President Lyndon Johnson gave Mexico sovereignty over a long-disputed river bank territory in the area of El Paso. At the same time, López Mateos preserved independence on another issue: Fidel Castro's Cuba. After 1960 the United States was pushing incessantly for anti-Cuban votes in the Organization of American States. Mexico was the only Latin American country never to break relations with Cuba. It took pride in its refusal to bow to the U.S. call for a uniform response from its Latin American allies.

The official candidate to succeed López Mateos in 1964 was Gustavo Díaz Ordaz, whom many thought would swing the PRI back toward the right. He was from the state of Puebla, Mexico's Catholic stronghold. As the incumbent secretary of the interior he had earlier ordered the arrest of certain "radicals," including the world-famous artist David Alfaro Siqueiros.

Díaz Ordaz countered this expectation by pledging to continue the policies of his predecessor; he soon failed the first test. López Mateos had taken seriously the criticisms of the PRI's one-party system and pushed through a constitutional amendment that guaranteed opposition parties a minimum of congressional seats if they won a minimum national vote. Applying this principle in the 1964 elections, both the PAN (a right-oriented party) and the PPS (a left-wing party) had won seats in Congress, although still overwhelmingly outweighed by the PRI representation.

Díaz Ordaz began by honoring this reformist thrust. But the en-

The student movement of 1968 began as a limited protest with an eclectic ideology, as suggested by the declaration of solidarity with Ché Guevara during this peaceful march along the Paseo de la Reforma in Mexico City. It eventually became a tragic crisis for the nation's political system. (United Press International.)

trenched PRI soon made known their fury at the newly appointed leader of the party, Carlos Madrazo, who was attempting to open up the nomination procedures—always the critical link in a one-party electoral system. Responding to the party machine complaints, Díaz Ordaz fired Madrazo. The new hard line was further evident when the federal government annulled mayoral elections in two cities in the state of Baja California Norte which PAN candidates had won. The democratization of the one-party system had overreached its limit.

Díaz Ordaz would have been lucky if mayoral elections had been his only political worry. But it was his fate to govern in the era of student protest that shook the Western world in the late 1960s. The precipitating factor was Mexico's hosting of the summer Olympic games in 1968. The government went all out to "sell" Mexico to the world. The Mexican left, always strong among students in Mexico City, was upset at the idea that the government might succeed in this public relations venture. There began a test of wills. A secondary school clash in Mexico City in July 1968 was met by brutal force from the riot police. Protest spread to the national university in August, culminating in a strike. The government thought it was a "subversive conspiracy," bent on disrupting the Olympic games. President Díaz Ordaz responded by send-

ing army troops onto the campus, thereby violating its historic sanctuary status. The battle was joined. Could the student left stop the Olympic games?

The tragic rhythm of confrontation between students and troops continued. On October 2, 1968, a rally of students in the Mexico City section of Tlatelolco drew an unusually heavy contingent of security forces. An order to disperse was allegedly not observed and the police and paramilitary forces moved in. Later they claimed to have taken sniper fire from surrounding buildings. They began shooting and the crowd was caught in a murderous cross fire, as hundreds fell dead and many more wounded. The massacre at Tlatelolco sent a shudder through Mexico. There was no inquiry, no convincing explanation from the military or civilian authorities responsible for the slaughter. A chorus of critics said the massacre had proved the bankruptcy of the PRI monopoly on power. By the same token, the brutal show of force convinced virtually everyone that mass challenges to authority would only bring more wailing ambulances. The effect was chilling.

Despite the turmoil on the political front, the Mexican economy continued to boom. The gross national product grew at 6 percent a year, although the distribution of income remained troublingly unequal. Between 1950 and 1969 the income share going to the poorest tenth of the population dropped from 2.4 percent to 2.0 percent. Meanwhile, the richest tenth increased its share from 49 percent to 51 percent. The top two-tenths widened their share at the expense of the bottom segments. By a standard measure of overall inequality (the Gini coefficient), Mexico's "miraculous" growth had only increased the maldistribution of income.

When the time came for the PRI bosses to nominate Díaz Ordaz' successor, they settled on Luis Echeverría, the secretary of the interior responsible for the security forces at Tlatelolco. It was hardly a choice likely to reunite embittered Mexicans. Echeverría tried to show a new face in his energetic campaign, and, after the usual landslide victory, plunged into his new duties. The sphere in which the new president sought to make his greatest mark was the one where he was soon most criticized: management of the economy.

Echeverría and his advisers wanted economic growth, but also better distribution of its benefits at the same time. An obvious place to begin, as always in Mexico, was the rural sector. Effort centered on infrastructure, such as rural electrification and the road system. In order to pacify consumers in the cities, the Echeverría government tightened the existing price controls on basic foodstuffs. In effect, the federal government was committing itself to an escalating subsidy on food for the urban masses. This could be financed only by draining the federal treasury or paying farmers below-cost prices for their goods. The latter would inevitably discourage production and the former would tend to

be inflationary. As Echeverría's term continued, he resorted increasingly to short-term measures that would channel resources (wages, land, social services) to the poor.

At the same time the state was increasing its general control over the economy. In addition to direct spending through federal departments and ministries, the government allocated a large share of the budget— well over half in recent years—to dozens of special agencies and state-supported companies. The leading lending institutions, most conspicuously the *Nacional Financiera,* were operated by the government, and the manipulation of credit regulations endowed the state with considerable influence over the economy. As of 1970, for instance, the government controlled principal shares in nine of the country's top ten firms, in thirteen out of the top twenty-five, and in sixteen out of the top fifty. Most of the leading state-dominated firms were involved in credit banking, public services (telephone and electricity), or in high-cost infrastructural activities (such as steel or oil), so they did not always compete directly with the private sector.

While the Mexican state took an active part in the country's capitalist economy, it retained considerable independence from the private sector. Much of this autonomy stems from the fact that Mexico's public leaders were, for the most part, professional politicians. They did not come from wealthy families, and after finishing school or university, they moved directly into political careers. In contrast to the United States, there was very little crossover of personnel between private corporations and public office. Consequently the Mexican state was not captive to any social group or interest. In tended to collaborate with the private sector, to be sure, but this was not always the case—a situation that gave the government considerable freedom of action.

While this process continued, the Mexican government faced a new problem: a guerrilla movement. Mexican politicians had long reassured themselves that their country was "different" from the rest of Latin America, where guerrillas were rife. After all, Mexico had already had its revolution. But Mexico was not immune. Guerrillas appeared, calling for violent action against the PRI and all its works. Beginning in 1971, they staged a series of bank robberies and kidnappings. The latter reached into the diplomatic corps: their victims included the U.S. consul general in Guadalajara and the daughter of the Belgian ambassador. In 1974 the father-in-law of the president was seized and held for ransom by militant guerrillas. In the state of Guerrero an ex-schoolteacher, Lucio Cabañas, led a guerrilla army that began to strike at will. They kidnapped the official (PRI) candidate for governor and defied the army by direct attacks on isolated outposts. It took a 10,000-man army more than a year to hunt down and kill the rebels and their leader. Despite predictions on the left, Cabañas had no successor in Guerrero or elsewhere, as the guerrilla threat faded. Why? Was it the

genius of the co-optive system of the PRI? Or was it the repressive network developed over the decades as the government's counterpart to its participatory electoral machine?

But Echeverría's major problem was not with the guerrillas. It was with the economy. The weak point in Mexico's economic strategy was inflation. In crude terms, Mexico could not expect to guarantee the peso's convertibility at a fixed rate unless its inflation was no higher than the U.S. level. By 1973 Mexican inflation was running 20 percent and remained at that level in 1974. Mexico's goods, based on the 1954 exchange rate, were growing uncompetitive on the world market. Yet the government stuck with the fixed rate, which had been the bedrock of Mexican development and a powerful political symbol.

Above all, Mexico had to continue attracting foreign capital. It had become crucial in financing Mexican investment as well as helping the balance of payments. Since the Revolution began there had been significant change in the level and allocation of foreign investment in Mexico. Total direct foreign investment in 1911, on the eve of the Revolution, was about 1.5 billion dollars (in 1970 value). By 1940 the level was less than a third that amount. The decline could be traced to revolutionary turmoil, deliberate government policy, and the Great Depression. In the postwar era, however, the level soared to 2.8 billion dollars, with 80 percent from the United States. And in sharp contrast to previous eras, when mining and communication-transportation were the dominant activities for foreigners, nearly three-quarters (73.8 percent) of this investment was in the manufacturing sector, mostly in critical industries: chemicals, petrochemicals, rubber, machinery, and industrial equipment. In this way Mexico obtained a considerable share of the financial resources for economic growth from abroad—and foreigners, notably Americans, assumed substantial if indirect influence on the direction of economic policy.

Commerce and tourism provided an additional type of dependence on the U.S. economy. About 60 percent of Mexico's international commercial transactions—imports as well as exports—were with the United States. About 85 percent of the income from tourism, needed so badly to offset balance-of-payments deficits, came from the United States. However much Mexicans wanted to achieve economic sovereignty, in fact they retained intimate ties to their northern neighbor. Nothing could have made this plainer than the economic crisis at the end of the Echeverría presidency.

Why was inflation plaguing Mexico? Many Latin Americans might have reversed the question: How had Mexico avoided it for so long? The answer was that the Mexican government, trying to please so many constituencies, was running large deficits, and financing them in an inflationary manner. There was also pressure from the balance of payments, which went into serious deficit by the middle of Echeverría's term of office. Mexico's continuing industrialization required heavy

capital goods. But a relatively new import was even more worrisome: food. The economy's failure was in agriculture. Production had grown for selected foods (tomatoes, strawberries) for export, especially to the United States, but output of basic foodstuffs, especially cereals, was falling short. Imports to meet this demand put an enormous burden on the balance of payments.

The reckoning came in Echeverría's last year as president. The drama centered on the greatly overvalued peso. With the government stubbornly maintaining its fixed rate of 12.5 to the dollar, every Mexican of means tried to convert pesos into U.S. currency. The government's ever more frequent denials of devaluation rang hollow. In September 1976, after capital flight had reached panic proportions, the government gave way. The peso was devalued by 60 percent. Government credibility was so low that a month later another devaluation of 40 percent was needed to settle the market. Could this incompetently managed devaluation convince investors (including Mexicans) to make new commitments in pesos? Although Mexico at last had a realistic exchange rate, the Echevarría government had failed to attack the rising public sector deficit—an essential step if future overvaluation of the peso, and thus future balance of payments crises, were to be prevented.

Echeverría ended his term in a flurry of histrionic gestures. Only eleven days before the end of his presidency, he expropriated rich farmlands in the north for redistribution to landless peasants. Panic spread among landowners. For the first time in years, Mexicans talked seriously about the possibility of a military coup.

Apparently Echeverría was motivated in part by a desire to win the secretary generalship of the United Nations, then up for election. Mexican politics suddenly seemed hostage to the ambition of one man. But the system—both in the UN and in Mexico—proved stronger than Echeverría. His term ended peacefully and on schedule. In retrospect, his presidency appears as merely another swing of the pendulum. (See Figure 7-1 for a schematic representation of the political positions of the presidents since 1934; question mark signifies the uncertainty of the current president's location on the political spectrum.)

The new president was José López Portillo, a leading moderate in Echeverría's cabinet. López Portillo made the moves to be expected of a new president shifting government policy back toward center. As the finance minister under Echeverría, he had presided over an economy that seemed to be wildly out of control. Mexico had growing deficits, both in its federal budget and in its balance of payments. Inflation had reached 30 percent. Although modest by Latin American standards, it was enough to erode confidence in the Mexican growth model, which had been based on guaranteed peso convertibility and free capital movement. López Portillo therefore gave first priority to that eternal task of restoring foreign confidence in his economy. Within weeks after his inauguration in December 1976, the new Mexican president trav-

FIGURE 7-1   Presidencies and the Political Spectrum in Mexico since 1934

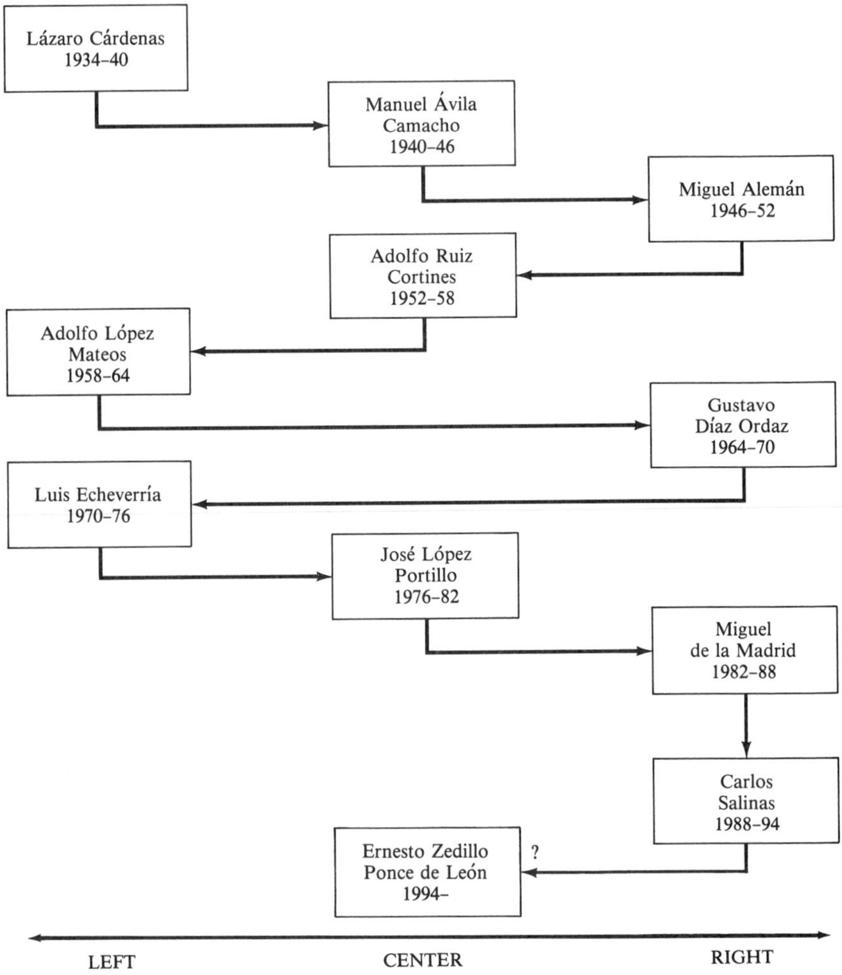

Lázaro Cárdenas 1934–40	
Manuel Ávila Camacho 1940–46	
Miguel Alemán 1946–52	
Adolfo Ruiz Cortines 1952–58	
Adolfo López Mateos 1958–64	
Gustavo Díaz Ordaz 1964–70	
Luis Echeverría 1970–76	
José López Portillo 1976–82	
Miguel de la Madrid 1982–88	
Carlos Salinas 1988–94	
Ernesto Zedillo Ponce de León 1994–	?

LEFT   CENTER   RIGHT

eled to Washington for a highly publicized visit with outgoing President Gerald Ford and an address to a joint session of the U.S. Congress. It was a powerful reminder that the Mexican elite still saw its fate closely linked to U.S. opinion.

López Portillo's presidency came to be dominated by economic issues. Just as he took office Mexico began discovering vast quantities of oil, and by 1980 López Portillo could announce that the country possessed proven reserves of 70 billion barrels and potential reserves of more than 200 billion. In a world apparently beset by chronic shortages and soaring costs for energy, Mexico had suddenly acquired new international clout. Declared an ebullient López Portillo: "There are two kinds of countries in the world today—those that don't have oil and those that do. We have it."

FIGURE 7-2 Mexican Exports, 1970–89

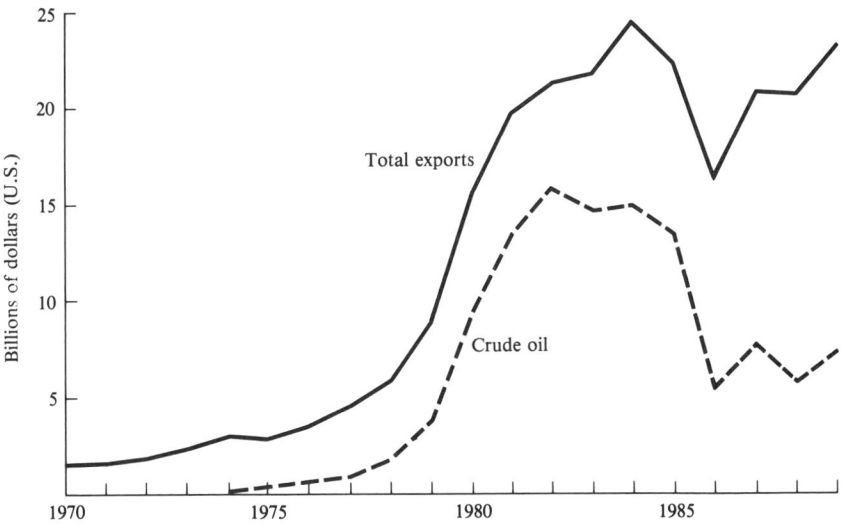

Sources: International Monetary Fund, *International Financial Statistics* (April 1977), 250–51; (April 1980), 268–69; (April 1983), 288–91; (December 1990), 78, 366; IMF, *1985 Yearbook*, 111, 450–51; and IMF, *1986 Yearbook*, 11, 460–61.

Optimism and pride surged through the nation. Government officials declared their intentions to increase production only gradually, not rapidly, in order to avoid the sad experiences of Venezuela and Iran—where the influx of petrodollars spurred inflation and exacerbated social inequities. Exports grew and world prices mounted, however, and Mexico's petroleum earnings jumped from $500 million in 1976 to more than $13 billion in 1981. As shown in Figure 7-2, the dollar value of nonpetroleum exports grew at a much less rapid rate. Somewhat in spite of itself, Mexico was becoming excessively dependent on oil revenues.

Economic problems persisted. Mexico was finding that the hard-money strategy which had worked so well between the mid-1950s and the late 1970s was no longer possible. The government could not get inflation below 20 percent, except for one year (1978), and by 1982 it shot up to almost 60 percent, an unprecedented rate for postwar Mexico. Another painful devaluation became inevitable in early 1982. Mexico had simply not adapted its financial system to inflation (as the Brazilians, for example, had managed to do).

Mexico had hoped to avoid all this by cashing in on its huge oil reserves, but the world slump in oil prices after 1981 reduced dramatically the projected foreign exchange earnings. The López Portillo government was therefore driven to heavy foreign borrowing, which raised the foreign public debt to $57 billion by the end of 1981. Most worrisome was the fact that the Mexican economy was still not produc-

ing jobs at a rate fast enough to absorb all the Mexicans entering the work force.

When faced by opposition, the Mexican regime's most frequent response was to bring its critics into the system—by offering a voice, a job, or a policy concession. As one observer neatly summarized the government's approach: two carrots, maybe even three or four, then a stick if necessary. By embracing (and defusing) the opposition, the Mexican state managed to strengthen its support.

In keeping with this tradition, López Portillo sponsored a program of political reforms. These included two innovations that seemed particularly far-reaching: first, the rules for registration of political parties were made easier, so much so that the Communist Party gained official recognition; and second, opposition parties were guaranteed a total of at least 100 seats in an expanded, 400-member Chamber of Deputies. Such alterations seemed unlikely to lead to a fundamental change in the locus of power, but they at least provided an outlet—within the system—for the opposition. The official presidential nominee was Miguel de la Madrid, a Harvard-trained technocrat and the minister of budget and planning under López Portillo, and he won a predictable victory in the elections of July 1982.

Before de la Madrid could take office on December 1, however, the Mexican economy was shaken by another and much larger, financial crisis. Mexico had run out of dollars with which to make payments on its foreign debt—now over $80 billion. Near panic ensued in Washington, New York, Frankfurt, and London, where it was feared that other Latin American debtors might follow Mexico's example and declare a de facto default. If that were the case, U.S., European, and Japanese banks would face huge losses, posing a formidable threat to world financial markets. The causes of the crisis were obvious. The price of Mexico's prime export (oil) had nosedived, interest rates had spiraled upward, and rich Mexicans had transferred billions of dollars out of the country. The U.S. government, the IMF, and the commercial banks rushed a "rescue" loan package to Mexico. These new loans enabled Mexico to continue paying interest but did not allow for amortization.

The rescue had its price: Mexico had to adopt an IMF-approved austerity plan. A key goal was to reduce the inflationary public deficit, which was at a dangerously high 15 percent of the GDP. This meant phasing out government subsidies on food and public utilities. Mexico also had to reduce its tariff barriers, thereby stimulating greater industrial efficiency and thus greater competitiveness in world export markets.

President de la Madrid dutifully followed the IMF prescription but at the price of inducing a deep recession. By 1985 real wages had fallen by 40 percent from their 1982 level; living standards fell even further as subsidies for such staples as corn tortillas were ended. In September 1985 a severe earthquake in Mexico City compounded the economic

disaster. The 1985–86 drop in oil prices depressed export earnings, further weakening the economy.

Amid these difficulties de la Madrid and his advisers decided to adopt a dramatic shift in economic policy, a new emphasis that came to be characterized as "liberalization." There were two main pillars to the program. One was to reduce and—recast the economic role of the state. This was to be done through continued cuts in public spending and through a program of "privatization" of state-owned companies. Of the 1115 publicly owned companies that his government inherited in late 1982, de la Madrid managed to sell off nearly 100 and to close down 279 by late 1986.

The second component of the new policy was commercial liberalization and "opening up" of the economy. This was most dramatically demonstrated by Mexico's accession to the General Agreement on Tariffs and Trade (GATT) in September 1986, which meant a long-term commitment to the reduction of barriers to imports from abroad. Mexico promptly began lowering and phasing out its tariffs and promoting its exports, especially nonpetroleum exports. For all practical intents and purposes, these changes amounted to a near-complete abandonment of the postwar policies of import-substitution.

In July 1986 Mexico needed another emergency loan package from its foreign creditors. Once again Mexico was told to bear down on its public deficit (down to 8 percent of the GDP in 1984 but nearing 15 percent again in 1986) and further reduce its protectionism. Mexican nationalists angrily charged that reducing protection would destroy their industrial base, and benefit foreign producers.

By early 1988 the de la Madrid government could see little prospect for relief. Inflation had accelerated to an annual rate of 143 percent, the public sector deficit was approaching 19 percent of the GDP, and the domestic capital market had been shaken by a 75 percent drop in the Mexican stock market. Yet another U.S.-engineered capital infusion came in December 1987. In a complex scheme Mexico would buy U.S. bonds to post as collateral against commercial bank loans. The move offered no prospect for large-scale relief from the debt, which had clearly become unpayable.

Despite these agreements, there would be continuing friction with the United States. A dramatic example was the 1985 case of an agent on assignment in Mexico for the U.S. Drug Enforcement Agency. Apparently his investigation had gone too well. He was kidnapped, tortured, and murdered, allegedly on orders from one of Mexico's multimillionaire drug kings. The pace of Mexican justice infuriated U.S. officials, who in retaliation ordered slowdowns at U.S. customs checkpoints on the Mexican border. This act in turn infuriated the many thousands of Mexicans who legally cross the border daily. An additional ongoing cause for bilateral tension was the U.S. policy toward Mexicans working (legally and illegally) in the United States. The Simp-

son-Rodino Act, passed in 1986, laid down tough penalties for employ-ers who hire "undocumented aliens." The prospect of its implementa-tion sent shudders through northern and central Mexico, whose younger generations have long seen jobs in the United States (usually temporary) as their main hope for a decent life. By 1991 the law ap-peared to have had only a minimal impact on actual migration flows, but Mexicans remained wary.

The debt crisis and economic stagnation in the late 1980s intensified social inequality and popular pressures. Investment plummeted, unem-ployment increased, and per capita income declined by more than 9 percent during the 1980s. In contrast to the southern cone countries in the 1960s and 1970s, however, Mexico did not resort to pervasive, large-scale authoritarian repression. Key attributes of the Mexican po-litical system—its restricted competition, its control of working-class movements, its autonomy from private interests, and its tactical flexi-bility—help explain why Mexico has managed to avoid the violent trauma that has afflicted Chile and Argentina.

Aware of their sagging credibility, PRI leaders made the process of choosing the official nominee to succeed de la Madrid more visible (if not more genuinely open) than the ritual had ever been. The choice was another U.S.-trained economist, Carlos Salinas de Gortari, only thirty-nine years old, who as the incumbent budget and planning min-ister had authored the highly unpopular austerity policies of the 1980s.

The election of 1988 brought surprises—and possible portents of meaningful change. For the first time in its history, the PRI faced seri-ous opposition from both the right and the left (as Cuauhtémoc Cárde-nas, son of the revered ex-president, led a breakaway faction from the PRI itself). Organized labor also showed its displeasure with the PRI candidate. Salinas de Gortari won with a bare 50.3 percent majority, according to official returns, and in claiming victory he declared an end to an era of "what was practically[!] one-party rule. . . ." Opponents nonetheless accused the regime of electoral fraud. The youthful Salinas took office in December 1988 under exceedingly difficult conditions. Would he be up to the challenge?

Apparently so. The first task for Salinas was to demonstrate political authority. He began by naming a cabinet dominated by his personal associates, instead of mending political fences. In January 1989 he mas-terminded a spectacular raid on the headquarters of the independent-minded and financially corrupt head of the oil workers' union, who was promptly placed under arrest (for illegal possession of firearms). Shortly thereafter he dismissed the longstanding chief of the large and powerful teachers' union. Unwilling to tolerate flagrant corruption within top governmental ranks, in 1990 he dismissed the naval secre-tary from his cabinet post, an unusual move in view of the delicate balance of civil-military relations in Mexico.

In keeping with his campaign promises, Salinas de Gortari promoted

a modest political opening. He commanded PRI officials to recognize a gubernatorial triumph for the PAN in the important state of Baja California (just south of the California border). He oversaw reforms of the electoral system and of the internal workings of the PRI. But there were limits to this *apertura*. The PRI claimed unrealistic victories in key elections in the state of México, near Mexico City, an area that had shown itself to be a left-wing opposition stronghold in the presidential election of 1988. The government also harassed and intimidated Cuauhtémoc Cárdenas and his followers, who found it extremely difficult to organize their forces into a coherent and durable political party. The opening, such as it was, was biased toward the right (and the PAN); it did not include the left.

Indeed, for the first time in memory the question of human rights appeared on the national agenda. Critics called attention to a number of abuses committed by Mexico's national police force in alleged pursuit of drug dealers. They reported the assassination or "disappearance" of at least sixty pro-Cárdenas sympathizers in 1990 alone. They expressed outrage at the murder of a prominent human-rights activist. To assuage the criticism Salinas appointed a National Commission on Human Rights, led by former university rector José Carpizo, but did not give it genuine authority.

It was in the economic arena that Salinas sought his most lasting achievements. In hopes of completing Mexico's structural adjustments, he continued and extended the "liberalization" strategy initiated under de la Madrid. Salinas and his team kept lowering trade barriers. They aggressively promoted privatization of state-owned industries, even putting up for sale such sacred cows as the telephone company and the banking industry (nationalized by López Portillo in 1982). With the support of the U.S. government, Salinas negotiated a new debt restructuring agreement that promised to reduce the net outflow of funds by $2 billion a year until the mid-1990s. The government also sought to assist local development by establishing a "program for national solidarity" to provide seed money for self-help projects throughout the country. Perhaps in response to these measures, the national economy showed signs of picking up: annual inflation moved down to the 20–30 percent range while annual growth rates for the GDP rose to 3.1 percent for 1989 and 3.9 percent for 1990.

North American Free Trade

The crowning achievement of the Salinas *sexenio* was the North American Free Trade Agreement (NAFTA). Unable to attract large-scale investment from Europe or Japan, the Salinas administration in June 1990 announced its intent to negotiate a free-trade compact with the United States. The proposal entailed a total repudiation of the protectionist strategies of import-substituting industrialization, and it dis-

carded the national tradition of keeping a suspicious distance from the "colossus of the north." Small-scale industrialists and grain farmers expressed fear that they might be destroyed by U.S. competition, and some intellectuals mourned the imminent demise of the nation's economic sovereignty and cultural pride. Salinas persisted nonetheless.

Unveiled in August 1992, the NAFTA accord envisioned the creation of a three-nation partnership (including Canada as well as Mexico and the United States) that would forge one of the largest trading blocs in the world—with a population of 370 million and combined economic production of approximately $6 trillion. It would promote the free flow of goods among the member countries by eliminating duties, tariffs, and trade barriers over a period of fifteen years. Sixty-five percent of U.S. goods gained duty-free status immediately or within five years; half of U.S. farm goods exported to Mexico immediately became duty-free. There were special exceptions for certain "highly sensitive" products in agriculture, typically one of the sectors most resistant to economic integration; phase-outs on tariffs for corn and dry beans in Mexico and orange juice and sugar in the United States would extend to the year 2009. Tariffs on all automobiles within North America would be phased out over ten years, but rules of origin stipulated that local content would have to be at least 62.5 percent for vehicles to qualify. Not surprisingly, spokespersons for Asian governments regarded this clause as a thinly disguised effort to exclude their industries and products from the North American market.

NAFTA opened Mexico to U.S. investments in various ways. Under the treaty U.S. banks and securities firms could establish branch offices in Mexico, and U.S. citizens could invest in Mexico's banking and insurance industries. While Mexico continued to prohibit foreign ownership of oil fields, in accordance with its constitution, U.S. firms became eligible to compete for contracts with *Petróleos Mexicanos* (PEMEX) and operate, in general, under the same provisions as Mexican companies. One item was most conspicuous by its absence: beyond a narrowly written provision for movement of corporate executives and selected professionals, the treaty made no reference at all to large-scale migration of labor.

NAFTA precipitated strenuous debate within the United States. In the heat of the 1992 presidential campaign, Democratic candidate Bill Clinton pledged to support NAFTA on condition that there be effective safeguards for environmental protection and worker rights; by September 1993 the governments reached "supplemental" or side agreements on labor and environment. As the U.S. Congress prepared to vote on ratification, Texas billionaire (and erstwhile presidential hopeful) Ross Perot led the charge against the treaty, claiming that NAFTA would entice business to seek low-wage Mexican labor and thus lose jobs for millions of American workers. Proponents insisted that NAFTA would stimulate U.S. exports, achieve economies of scale,

and enhance U.S. competitiveness. Disregarding vociferous opposition from unionized labor, a historic bastion of support for Democrats, Clinton lobbied tirelessly on behalf of the treaty. And after Perot stumbled badly during a memorable television debate with Vice President Al Gore, the House of Representatives finally approved the NAFTA accord by the surprisingly lopsided margin of 234–200; the Senate followed with a vote of 61–38.

In final form, the NAFTA accord had several outstanding characteristics. One was its implicit commitment to regional economic integration. Despite its title, NAFTA was not primarily concerned with "free trade." By 1990 tariff and even nontariff barriers to U.S.-Mexican commerce were already low. NAFTA was primarily concerned with investment. By obtaining preferential access to U.S. markets and a formal "seal of approval" through NAFTA, Mexico was hoping to attract sizable flows of direct foreign investment—from Japan and Europe as well as from the United States. By obtaining untrammeled access to low-wage (but highly skilled) Mexican labor, the United States was hoping to create an export platform for manufactured goods and thus improve its competitive position in the global economy. It was for these reasons that the NAFTA treaty contained extensive chapters about investment, competition, telecommunications, and financial services. Implicitly, NAFTA envisioned a substantially more profound form of integration thatn its label acknowledged.

Second, NAFTA made explicit provision for environmental protection. As originally negotiated NAFTA made only passing reference to environmental concerns. In keeping with his campaign pledge, however, President Clinton oversaw negotiations on a supplementary provisions for environmental protection; and under a separate agreement, the U.S.-Mexican border received special attention under a bilateral Integrated Environmental Plan. While some observers raised doubts about the practical significance of these agreements, the mere fact of their negotiation made one point clear: trade and environment had become inextricably intertwined. As one analyst wrote, these developments forcefully demonstrated "that the environment has become a staple of trade politics in the 1990s, for it was politically impossible to contemplate the completion of the NAFTA trade accord without a complementary agreement on the environment."

Yet another distinguishing characteristic of NAFTA was its underlying political rationale. The United States was seeking several goals. One was the preservation of stability on its southern border. The idea was that NAFTA would stimulate economic growth in Mexico, easing social pressure and sustaining the political regime. A second goal was to assure the United States of increasing access to petroleum from Mexico, one of the five leading sources of U.S. imports (Mexican shipments in the late 1980s and early 1990s were roughly half as large as those from the topmost source, Saudi Arabia). A third purpose was for the United

States to obtain an important bargaining chip in its trade negotiations with Europe, Japan, and the General Agreement on Tariffs and Trade. And fourth, the United States wanted to consolidate diplomatic support from Mexico on foreign policy in general. As demonstrated by disagreements over Central America during the 1980s, this had long been a source of bilateral tension. But with NAFTA in place, Mexico became unlikely to express serious disagreement with the United States on major issues of international diplomacy.

For its part Mexico was seeking, first and foremost, preservation of its social peace. The hope was that NAFTA would attract investment, stimulate employment, provide meaningful opportunity for the one million persons entering the job market every year—and thus reduce social tension. Second, NAFTA offered Salinas an opportunity to institutionalize his economic reforms, insulating them from the historic vagaries of presidential succession by inscribing them in an international treaty. Third, Mexico was seeking international benediction for its not-quite-democratic political regime. This was especially important because, in comparison with Argentina, Chile, Brazil, and other countries undergoing processes of "democratization," Mexico no longer looked like a paragon of political civility. Finally, Mexico believed that NAFTA would provide the country with diplomatic leverage vis-à-vis the rest of Latin America and, by extension, the Third World as a whole. Association with Canada and the United States would link Mexico with advanced industrial democracies and leaders of the First World. Consequently Mexico could serve as a "bridge" between the developing world and the developed world, as a representative and interlocutor for aspiring peoples of the South.

Technocracy in Crisis

All the optimism resulting from the NAFTA accord promptly came under assault. On January 1, 1994—the day that NAFTA went into effect—a guerrilla movement in the poverty-stricken state of Chiapas rose up to denounce the free trade accord, the *Salinista* economic model, and the undemocratic character of the political regime. With colorful and able leadership, the Zapatista National Liberation Army (EZLN) captured national and international attention during the course of highly publicized negotiations with governmental authorities. Despite a variety of governmental responses, from military pressure to political negotiation, the Zapatista movement would remain a thorn in the side of the regime.

Two months later, as public attention turned toward presidential succession, an assassin's bullet struck down Luis Donaldo Colosio, Salinas' handpicked successor and the candidate of the PRI. Salinas hastily chose another nominee, the forty-two-year-old Ernesto Zedillo Ponce de León, who scurried to develop a credible campaign for the upcom-

ing August election. These developments inflicted a devastating blow to Mexico's international image. Mexico could no longer be seen as an up-and-coming country on the brink of joining the First World; it looked, instead, like a Third-World society threatening to come apart at the seams.

Earnest and intelligent, Zedillo was a technocrat par excellence. A Ph.D. in economics from Yale University, Zedillo had spent most of his career in the central bank and the planning ministry. As a result he had very few contacts with career politicians or officials in the "political" ministries of the federal government. Despite a lackluster campaign, Zedillo won the August 1994 elections, by all accounts the cleanest in Mexican history, with 48.8 percent of the vote (compared with 26.0 percent for the rightist PAN and only 16.6 percent for Cuauhtémoc Cárdenas' populist Party of the Democratic Revolution, PRD), thus becoming the fifth man in a row to reach the presidency without ever holding prior elective office.

Inaugurated in December 1994, Zedillo faced crisis right away. Fearful of the overvaluation of the peso, investors withdrew more than $10 billion from Mexico within a week. In response the Zedillo administration had to devalue the peso, which eventually lost more than half its value against the U.S. dollar, and the government came within only a few days of insolvency. Early in 1995 the Clinton administration put together a multilateral package of nearly $50 billion, including $20 billion from the U.S. government. One major goal of this measure was to head off a potential default on $30 billion in *tesobonos* (short-term bonds issued by the Mexican treasury, payable in dollars), which would have inflicted major damage on U.S. pension funds, mutual funds, and other institutional investors. Another was to sustain the credibility of economic reform and the viability of NAFTA itself.

The financial crisis provoked a political crisis as well. As criticism mounted against Salinas' insistence on maintaining an unrealistic exchange rate throughout 1994, the ex-president publicly criticized Zedillo and his economic cabinet for mishandling the December devaluation. Zedillo reacted by sending Salinas into de facto exile in the United States, then authorizing the arrest of the former president's older brother on charges of corruption. The detention by U.S. authorities of an assistant attorney general under Salinas led to further denunciations of corruption, family intrigue, and official involvement in the assassination of a high-level PRI leader in September 1994. Serious fissures threatened to split apart the Mexican political elite.

The public promptly showed its disapproval. For the first time in decades, rumors began circulating that an elected PRI president might not be able to finish his term. One poll in early 1995 showed that nearly half the respondents believed a military coup was possible. Voters in the state of Jalisco, long a bastion of the PRI, elected an opposition-party PAN candidate as governor. Even where the PRI claimed victo-

ries, as in statewide elections in Tabasco and Yucatán, the results were sharply contested. Clearly, the PRI was losing its ability to curry and deliver votes.

By the mid-1990s it appeared that Zedillo, and Mexico, were confronting three long-term challenges. One focused on the economy, which fell into recession in the first half of 1995. The need was not only to regain investment and stimulate growth. It was also to alleviate problems of poverty and inequality. Between 1963 and 1981, according to one study, the proportion of Mexicans below the poverty line dropped from 77.5 to 48.5 percent; but from 1982 to 1992, under the pro-market reforms, it rose again to 66 percent. And despite its cooperation with international creditors, Mexico still confronted a massive external debt of more than $120 billion, with annual interest payments consuming about 15 percent of export earnings. The debt crisis of the 1980s was casting a long shadow.

A second challenge focused on law and order, especially in light of the emergence of new and powerful drug cartels. Soon after taking office, Zedillo received an official report which warned that "the power of the drug-trafficking organizations could lead to situations of ungovernability." The most dangerous of these cartels were involved not so much in marijuana or heroin, traditional products of Mexico, but in trans-shipment of cocaine from Colombia. With an estimated $7 billion in annual profits, these groups could spend as much as $500 million per year on bribery—more than twice the total budget of the attorney general's office. By the mid-1990s Mexico had about a half-dozen drug organizations of truly international scope (in Tijuana, Sinaloa, Ciudad Juárez, Guadalajara, and in the state of Tamaulipas, where traffickers operated a flourishing cocaine pipeline along the Gulf of Mexico). Drug cartels were implicated in a wave of violence that swept through Mexico, including the assassination of a Roman Catholic cardinal in 1993. Former prosecutor Eduardo Valle Espinosa proclaimed that the country had fallen under the heel of drug traffickers and, like Colombia, had become a "narco-democracy."

Third, and perhaps most difficult, was the challenge of political transition. It was clear that the old system of PRI domination was undergoing change. Between 1964 and 1994, for instance, the proportion of voter districts showing "strong" PRI hegemony declined from 52.2 percent to only 2.3 percent; by 1994 about 26 percent showed two-party competition, 55 percent revealed multiparty competition. While it seemed possible that Mexico was heading toward a de facto three-party system, the most pervasive fact about the political system was uncertainty. As novelist Carlos Fuentes observed, "The obvious truth about Mexico . . . is that one system is falling apart on us, but we have no other system to put in its place."

EIGHT

CUBA
Late Colony, First Socialist State

Cuba's historical development has been deeply affected by its location. Cuba is an island, lying astride a network of vital sealines that feed the rich Caribbean basin, extending from Florida to Guyana. Columbus discovered modern-day Cuba on his first voyage (1492), and it soon became a staging ground for the Spaniards' many expeditions to the Mexican and North American mainland. During the sixteenth and seventeenth centuries the island did not command much imperial attention, but its commercial and strategic importance grew in the eighteenth century with the expansion of the regular fleets between Spain and its American colonies.

The indigenous population, descended from immigrants from the Lesser Antilles, scarcely survived the first century of the Spanish colonization. Here, as elsewhere in Latin America, the European conquerors turned to black Africa for their labor supply. As a result, Cuba became a multiracial society: by the twentieth century, according to one estimate, the population was 40 percent black, 30 percent white, and 30 percent mixed (including Oriental and Indian).

Cuba's economy languished under the rigid mercantilist policies of the Spanish crown until the reforms of Charles III (1759–88) provided the stimuli that led to growth. The nineteenth century saw Cuba burst forth as an agricultural phenomenon. A brief coffee boom gave way to the cultivation of tobacco, which became a major crop by mid-century—a position it still holds, as Cuban cigars *(puros)* continue to be regarded as among the finest in the world.

But the most important source of wealth, the one that would shape the contours of Cuban society and history, was another product: cane sugar. Emphasis on sugar began in the eighteenth century and continued over time. By 1860 Cuba was producing nearly a third (500,000 tons) of the world's entire sugar supply. The human power to fuel this boom continued to come from the nightmarish slave trade, which delivered more than 600,000 Africans in chains to Cuba between 1800 and 1865. Slavery itself lasted until 1886, longer than anywhere else in the Americas save Brazil.

Cuba's economic development has thus been typical of tropical America: a monocultural, slave-based, export-oriented agricultural society. In another respect, however, it was atypical. When Cuba was less than a decade away from the twentieth century it was still a colony. An earlier independence effort had failed in the bitter "Ten Years War" (1868–78), when the nationalist Cubans—those who rose against the Spanish—failed to rally the elite and were slowly ground down by Spanish troops.

Spain's continued political control of the island was becoming anachronistic, however, since by the 1880s Cuba's trade and investment were almost exclusively with the United States. The U.S. economic interest in Cuba led to numerous offers to purchase the island country. The Spaniards invariably refused, but some prominent Cubans strongly favored annexation by the United States. Meanwhile, Cuba was drawn ever closer into the U.S. orbit.

A handful of Cuban nationalists, who had never accepted their defeat in 1878, fled into exile and plotted a new rebellion. The most famous was José Martí, an eloquent revolutionary poet-lawyer whose long exile in New York produced Cuba's most memorable anti-American rhetoric. A new revolt for independence broke out in 1895. Cuba was soon engulfed in another savage war, with rebels and Spanish forces both resorting to scorched-earth tactics. The war dragged on for three years. The Spaniards resorted to brutal methods, such as the use of concentration camps, to liquidate the guerrilla-style patriots.

Given its huge economic stake in Cuba, the United States was unlikely to remain out of the struggle. The U.S. public was excited by sensationalist press accounts of Spanish brutality, and business and religious leaders demanded U.S. recognition of the rebels. The expansionist urge in the United States was fed both by those who stood to gain economically and by those who preached of a U.S. mission to rescue the Cubans from Spanish misrule.

Although President McKinley resisted pressure to intervene, events overtook him. In April 1898 the *USS Maine* mysteriously exploded in Havana harbor. The blast, which has never been satisfactorily explained, swept away the last vestiges of antiwar sentiment, and Congress promptly declared war on Spain. The "splendid little war" (as Teddy Roosevelt called it) lasted only seven months. The ill-equipped Spaniards went down to humiliating defeat. They had little choice but to grant Cuba independence in December 1898.

Dubious Independence

Cuba began her new status under U.S. military occupation, hardly favorable for developing a healthy sense of national identity. The U.S. authorities immediately disbanded the rebel army, thus removing the only potential source of armed opposition to U.S. rule. The occupation

was a textbook example of what was regarded as "enlightened" intervention. The North Americans built badly needed schools, roads, sewers, and telegraph lines. But it was all in the service of integrating the now "civilized" Cubans more closely into the U.S. orbit.

U.S. government leaders saw nothing contradictory in *their* presiding over Cuba's emergence as an independent nation. They saw economic, moral, and political responsibilities all going hand in hand. The Cubans were allowed, even encouraged, to choose a constitutional convention, which produced a charter in 1901. But the U.S. government harbored doubts about the new country's ability to govern itself, so Washington forced the Cubans, under protest, to incorporate an amendment (the "Platt Amendment"), which gave the United States the right to oversee the Cuban economy, veto international commitments, and intervene in domestic politics at will. This proviso remained in force until 1934, making Cuba an American protectorate.

Cuba's first president, Tomás Estrada Palma (1902–6), favored outright annexation by the United States. He was typical of much of the Cuban elite, which saw little advantage, and certainly no future, for an independent Cuba. Their willingness to embrace the Yankee encroachment aroused the bitterness and fury of those few Cuban nationalists who kept alive the flame of José Martí's dream of a Cuba free from Yankee dominance.

Estrada Palma won a second term by electoral fraud. The ensuing revolt, led by the defeated Liberals, brought a second U.S. military occupation (1906–9). The United States imposed an interim president, Charles Magoon, who oversaw a new election. Fraud recurred, however, triggering another U.S. military intervention in 1917. All these interventions presented opportunities for U.S. economic interests to deepen their hold over the Cuban economy. The Cuban government gained a well-deserved reputation for venality and corruption; the Cuban political system was very far from having generated the democratic spirit which the U.S. idealists thought would result from U.S. occupation.

Overview: Economic Growth and Social Change

During Cuba's years as a protectorate, it underwent a great sugar boom. In the nineteenth century Cuba had rapidly emerged as one of the world's most efficient sugar producers, helped by the modern vacuum methods of refining. As output increased, sugar came to dominate Cuba's economy and, eventually, to have a lasting effect on class structure and social relationships.

By the early twentieth century, as shown in Figure 8-1, Cuba was producing several million tons of sugar per year—nearly one-quarter of the world supply around World War I, about 10 percent of the total during the depression years, close to 20 percent just after World War

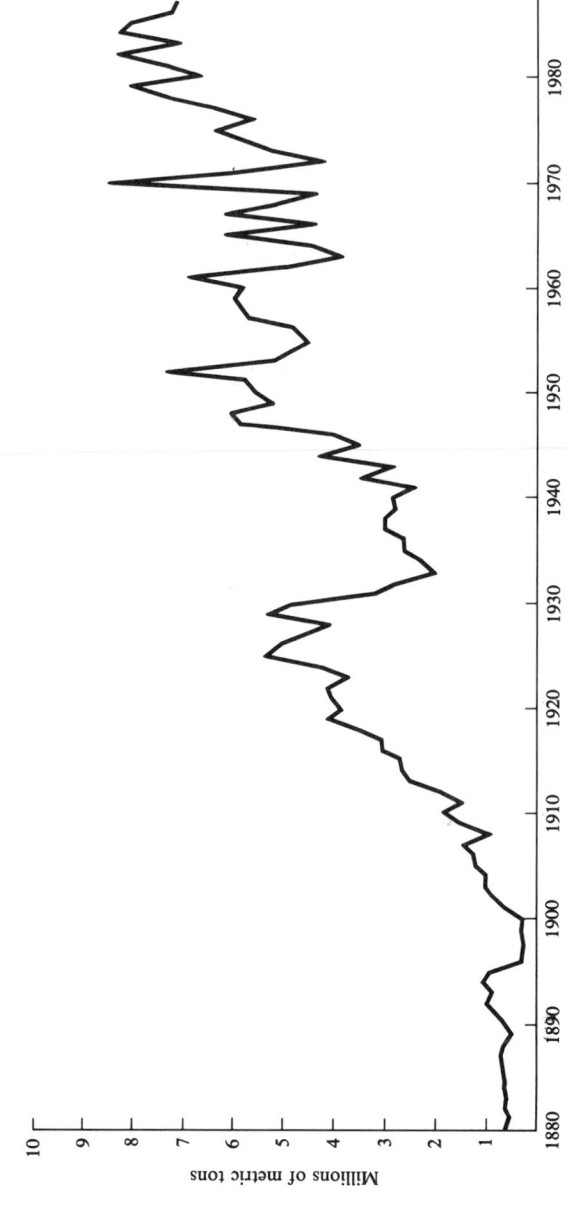

FIGURE 8-1 Sugar Production in Cuba, 1880–1988

Sources: Manuel Monreno Fraginals, *El ingenio: complejo ecoomico social cubano del azúcar* (La Habana: Editorial de Ciencias Sociales, 1978), III, Cuadro 1, pp. 37–40; Arthur MacEwan. *Revolution and Economic Development in Cuba* (New York: St. Martin's Press, 1981), p. 188; James W. Wilkie, Enrique C. Ochoa, and David E. Lorey, eds., *Statistical Abstract of Latin America*, 28 (Los Angeles: UCLA Latin American Center, 1990), Table 1714.

II. Throughout this entire period sugar exports earned approximately 80 percent of the island's foreign exchange. Such dependence on a single product obviously placed the Cuban economy in an extremely vulnerable position. If the harvest was poor (the result of weather or other conditions) or demand was low (the result of economic slowdown elsewhere) or prices were down (the result of oversupply from other exporters) the Cuban economy would suffer. The variations in production from 1920 to 1959, and even later, illustrate some of the dangers of this situation.

Another feature of the sugar boom was concentration of ownership, especially in the hands of American investors. After the 1870s, the new technology, particularly railways, stimulated a rapid reduction in the number of sugar mills (from 1190 in 1877 to only 207 in 1899), despite an increase in the cane hectarage. At the same time, the huge sugar estates began to spread. The independent growers, whose small- and medium-sized farms had produced most of the cane before the 1870s, now sold out in growing numbers to the big sugar companies. By 1912 the latter controlled more than 10 percent of all land in Cuba. By 1925 the number of sugar mills had dropped to only 184, and they controlled 17.7 percent of Cuban land.

This concentration of mill and land ownership was a natural result of the manner in which the sugar boom had proceeded. Under the shield of the protectorate, U.S. investors poured capital into the building of modern mills (*centrales*) and the consolidation of cane-growing lands. American-owned mills produced only 15 percent of Cuba's sugar in 1906, but by 1928 their share reached about 75 percent, thanks to loan defaults by Cuban owners; the figure then slacked off, and by 1950 it stood at 47 percent.

The technology of sugar production affected labor as well as ownership and management. Cultivation came to require a large-scale work force, especially at harvest time. Cane needs to be replanted only periodically, at intervals of five to twenty-five years. Therefore the principal need for labor is for the harvest, or *zafra*, a feverish three-month period of intense activity, mostly spent on the arduous cutting of cane with machetes. The rest of the year was known in Cuba as the *tiempo muerto*, the "dead season" of widespread unemployment and underemployment.

But workers had nowhere to go. Because of the enormous plantations they could not lease or purchase small-scale plots of land for their own use. Managers wanted to keep them near the mills, available for work, and for this they devised several tactics. One was to raise cane on land owned by the *centrales* themselves, usually about 10 percent of the total, thus maintaining the presence of independent growers nearby who would share the problems of labor with them. Another was to let workers go into debt, so they would remain under obligation to the ownership. A third was to encourage the formation of modest ur-

A steam-driven engine hauls wagons of sugar cane to the mill around the turn of the century. (Courtesy of the Library of Congress.)

ban settlements, called *bateyes,* that would create working-class communities.

As a result Cuba witnessed the appearance of a rural proletariat, a social group that differed greatly from a classic peasantry. To be sure, there were some isolated and self-sufficient peasant communities in Cuba, particularly in the rugged mountain regions, but they were not a predominant class. Workers in the sugar mills and in the *zafras* were laborers, not farmers. They were concerned more about wages and working conditions than the acquisition of land.

Moreover, the rural laborers had intimate contact with the working class in the cities, most notably Havana. Despite the disincentives and inhibitions, they often migrated to urban areas, living in the kind of slum that has come to characterize many of Latin America's largest metropolises: known as *colonias populares* in Mexico and *favelas* in Brazil, they acquired in Cuba the suitable name of *llega y pon* ("come and settle"). And their residents were blighted by poverty and deprivation. By the 1950s close to 40 percent of the national population lived in cities. Only 40 percent of urban lower-class dwellings had inside toilets, only 40 percent had refrigeration of any kind, and as many as a dozen people lived in a single room.

Contact and communication between urban and rural elements of the Cuban working class would eventually have a decisive effect on the course of the country's history, because it permitted the sort of unified, classwide social movement that has been found so rarely in Latin America. It is worth noting, too, that the church played only a minor role in Cuban society, and trade unions had a sporadic and precarious

existence. In other words, the outlook and behavior of the Cuban laboring classes were not conditioned or controlled by existing institutions. Workers would, in time, be available for mobilization.

Meanwhile, the United States built up more and more control over the Cuban economy. Not only did United States capital take over major ownership of plantations and mills, the United States became by far the largest customer for Cuba's sugar exports—usually purchasing 75 or 80 percent of the total. This created a complex political dimension to Cuba's economic dependency on the United States. On the one hand, U.S. investors in Cuba favored U.S. trade policies that helped Cuban sugar's competitive position in the U.S. market. On the other hand, U.S. beet sugar producers, as well as U.S. investors in non-Cuban sources of overseas sugar production, opposed favoritism for sugar imports from Cuba. To complicate matters further, U.S. sugar refiners might have wanted to favor Cuban raw sugar imports, while U.S. owners of sugar refineries in Cuba wanted favoritism only for refined imports. Through it all, Cuba was dependent upon U.S. decisions for the fate of its major industry. And U.S. sugar import policy was invariably a topic of prolonged debate in Washington.

Newly independent Cuba had originally signed a reciprocal trade treaty in 1903 which gave Cuban sugar a 20 percent reduction from the existing U.S. tariffs. In return, Cuba gave U.S. exports reductions of 20 to 40 percent in Cuban tariffs. For the next thirty years U.S.-Cuban trade relations grew ever closer, as the Cuban economy was for all intents and purposes integrated into the U.S. economy. The Cuban currency was made interchangeable with the U.S. dollar. Cuban monetary policy was actually set by the Federal Reserve Bank of Atlanta, since the Cuban authorities had in effect surrendered any control over the movement of monetary assets between Cuba and the United States.

The eager U.S. investor in Cuba might well have smiled over his good fortune, at least until the end of the First World War. The end of the war had brought a world food shortage and all food exporters, Cuba included, found themselves cashing in on near panic buying conditions for commodities. A crash then came in 1920. In a few months sugar prices fell to less than one-fifth of the record levels of May 1920, and in the following two years the value of the sugar crop declined to little more than one-quarter of the 1920 level. The decline continued through the rest of the 1920s, having a devastating effect on the economy, hitting especially those rural workers whose existence was precarious even in the best of times.

With the collapse of the world economy in 1929–30, Cuba soon suffered for its (somewhat involuntary) dependence on one trading partner. The U.S. Congress, under pressure from the domestic sugar beet producers, passed the Smoot-Hawley tariff in 1930, burdening Cuban sugar with new duties. This merely increased the pressure on the staggering Cuban sugar economy, which contracted severely. The only

bright spot came with Franklin Roosevelt's assumption of power in Washington in 1933. Roosevelt and the Democratic Congress brought lower tariffs, as the Reciprocal Trade Agreement of 1934 cut duties on Cuban sugar imports, while Cuba gave increased favor to U.S. imports. Also in 1934 Congress mandated fixed quotas among domestic and foreign suppliers of the U.S. sugar market. Cuba's quota was 28 percent, a share that endured, with modifications, until 1960. It gave Cuba a privileged access to the U.S. market. It also made Cuba constantly subject to political or economic blackmail. Most important, it tied Cuba to the will of the U.S. Congress, which could change the legislation at any time. The quota was an economic bonus and a political liability. It symbolized all the vulnerability which "independence" had brought Cuba in the era of American dominance.

In sum, the reliance on sugar produced mixed blessings for Cuba's economy and society. It brought considerable prosperity to the island, especially in good *zafra* years, but it spawned enormous social and economic inequities. It attracted investment from abroad, but placed the country in a subordinate position to the international economy and especially the United States. Also it created a volatile social structure, one in which rural and urban elements of a long-deprived working class maintained communication with each other. The top of the social pyramid was occupied not by resident landlords, as in classic haciendas, but by foreign entrepreneurs or native owners who often lived in Havana: the upper class was absentee. There was a sizable middle class, at least by Latin American standards, but it was an amorphous stratum that lacked cohesion and self-consciousness. As sociologist Maurice Zeitlin once observed, this combination of factors was bound to have its effect: "Large-scale enterprise in the countryside and the intermingling of industrial and argricultural workers in the sugar centrals permeated the country largely with capitalist, nationalistic, secular, anti-traditional values and norms of conduct. In this sense, the country was *prepared* for development—the only thing lacking being the revolution itself. . . ."

Politics: Corruption and Decay

The government the Cubans got in the 1920s and 1930s was among the most corrupt and brutal of the republic's history. Gerardo Machado gained the presidency by election in 1925 and soon used his executive powers to make himself forever unbeatable at the ballot box. Machado's repressive measures and the growth of nationalist opposition, especially among students and urban labor, brought out the uglier realities of the U.S. protectorate. When the global depression hit, Cuba's export-oriented economy suffered badly. The bottom dropped out of world sugar prices yet again, and the Cuban economy contracted even further. Total income plummeted and unemployment mushroomed.

There was no shortage of political will to exploit the economic dis-

tress. Machado's opposition included a coalition of students, labor leaders, middle-class reformers, and disgruntled politicians, held together by a common hatred for Machado and a common aspiration for a more honest and more just Cuba. Armed plots abounded. Shoot-outs regularly punctuated Havana's nighttime. Machado's police and military bore down with more repressive measures. The United States, so attentive to some other kinds of deviations from democracy in Cuba, stood by passively. Herbert Hoover's Republican administration, for all its supposed identification with business, was seeking to end the era of ruling the Caribbean via the U.S. marines.

Franklin Roosevelt's election victory brought an activist to the White House. While Washington assumed a more critical stance toward Machado, the Cubans took matters into their own hands. A successful general strike in August 1933 helped prod the army toward undercutting the dictator, who fled Havana. Now opinion began to polarize sharply. The young radicals dominant in the provisional government joined with army enlisted men, led by Sergeant Fulgencio Batista. This alliance took over the government, alarming Roosevelt's high-level envoy, Sumner Welles. The new civilian leader was Ramón Grau San Martín, a doctor-professor (the one faculty member to vote against giving Machado an honorary degree) and long-time hero to the student left, with whom he had invariably sided. "Soviets" were formed, followed by occupations of factories and farms. The new government proclaimed a socialist revolution.

Washington became deeply worried over the sharp leftward turn by its protectorate. U.S. navy ships took up stations off the Cuban coast; old style intervention seemed near. But a new strongman, eager to follow the Cuban formula for finding power and wealth, was already on stage. On signal from the United States, Batista easily ousted Grau and the radicals. A front-man president acceptable to Washington was soon arranged, and the radicals, the nationalists, and the reformers watched with bitterness as Cuban politics returned to business as usual. U.S. hegemony was so certain that Washington had no trouble agreeing to abrogate the Platt Amendment in 1934. The U.S. naval base at Guantánamo, for example, was not affected.

For the next twenty-five years Cuban politics was dominated by Fulgencio Batista. Between 1934 and 1940 Batista ran his country through puppet presidents. He ruled directly from 1940 to 1944, then went back to a behind-the-scenes role as the one-time radical Grau San Martín returned to the presidency (1944–48). There was little left of Grau the idealist, and the spectacle of his descent into the nether world of Cuban political corruption merely deepened the disgust and moral fury burning in the radicals and nationalists. Grau's successor, another Batista front man, was Carlos Prío Socorrás (1948–52). Batista himself retook the presidential reins in a coup, and thenceforth ruled with dictatorial powers (1952–59).

In reality, Cuban politics saw little change between 1934 and 1959. The futility of the electoral system was repeatedly demonstrated, as the perennial strongman (yesterday Machado, today Batista) worked his will. The honest opposition, far weaker than its true constituency, scrapped and struggled in vain. What had happened to the revolutionary fervor of 1933? Where was the coalition that had so frightened Washington? It had gone the way of all Cuban nationalist movements—rendered impotent by the unbeatable alliance of the Cuban elites, their political and military handmaidens, and Uncle Sam. If one had asked most Cubans in 1959 whether their little island had any chance of true independence, how many would have dared say yes? How many really thought Cuba could ever successfully assert its identity against the Northern Colossus? Very few. Most educated Cubans undoubtedly thought that the best their country could hope for was to win a few advantages at the margin—maximize Cuba's gain from its inevitable dependence on the United States. What else could one hope? A startling answer soon came forth.

Fidel Castro and the Making of the Revolution

Born in 1927, Fidel Castro was the son of a successful Spanish immigrant and he represented an odd Cuban tradition—the heir of a *peninsular* who had "made America," as the Spaniards put it in the sixteenth century. But this immigrant's son was not interested in enjoying the comfortable life his background and training might have promised. He wanted to make a different America.

Fidel had followed the classic path—primary and secondary education with the Jesuits, then a law degree. He plunged into the turbulent world of student politics, where every brand of nationalism, leftism, and revolutionary thought could be found. He proved to be strong-minded, articulate, and ambitious. But he was not in the most radical ranks. Passionately nationalist, he steered clear of the communists, who were the best organized of the student groups.

Soon after graduation Fidel began traveling in Latin America, meeting other radical nationalists and learning about other political realities. His most dramatic experience came in Bogotá in 1948, when the colossal urban riot of the *bogotazo* turned the city upside down for two days. The triggering event had been the assassination of the charismatic leftist Colombian politician, Jorge Eliécer Gaitán. The populace rose as one and took over a city whose authorities had abdicated in terror. Fidel was swept up into the wave of popular outrage and eagerly tried (unsuccessfully) to become a combatant. Those remarkable days must have given him a taste for the possibilities of popular mobilization.

Fidel Castro's first assault on Batista's state came straight out of the tradition of romantic Latin American revolutionaries. It was an attack on the 26th of July 1953, against the provincial army barracks at Mon-

cada in the southeastern city of Santiago. Fidel led a band of 165 youths who stormed the garrison in what afterward could only seem a suicidal attack. They had hoped for surprise, but failed. Half the attackers were killed, wounded, or arrested. Fidel and his brother Raúl were among the few who got away. The government reaction was swift and ruthless. The police began slaughtering suspects. Fidel and Raúl were captured, tried, and sentenced to fifteen years in prison. During the trial Fidel gave a long, impassioned, rambling speech ("History Will Absolve Me"), little noticed at the time but later to become a sacred text of the revolution.

The Castro brothers were lucky. They stayed in prison only eleven months before Batista granted amnesty in an attempt to court public opinion and to improve his political image. Fidel thus benefited from a tactical concession by a government he was all the more determined to destroy. Given his freedom he immediately fled to Mexico to begin organizing a new revolutionary force. At this point there was little to distinguish him from the innumerable other Caribbean revolutionaries who plotted ineffectually against the Trujillos, Somozas, and Duvaliers—tyrants cut from the same cloth as Batista.

In 1956 Fidel set out with a new band of revolutionaries in the *Granma,* an ancient yacht whose name would later be immortalized as the title of revolutionary Cuba's official newspaper. With him once again was his brother Raúl, more politically radical than Fidel. Also aboard was Ernesto ("Ché") Guevara, a twenty-seven-year-old Argentine physician who had personally witnessed the CIA-conducted overthrow of radically anti-American Guatemalan President Jacobo Arbenz in 1954.

Fidel coordinated his voyage to coincide with the mobilization of anti-Batista forces on the island. They were to rise in communities close to the planned landing site. But the navigation went awry, so did the uprising, and the *Granma* beached in a swampy area. The eighty-two men struggled to get ashore. The next few days were a nightmare of thirst, hunger, and death at the hands of army units to whom they had been betrayed by local peasants. Seventy of the men were lost, but Fidel, Raúl, and Ché were not among them. They fled into the Sierra Maestra mountains in eastern Cuba, following their contingency plan. From the mountains, Fidel rebuilt his rebel band and once more launched his war against Batista.

For the next two months, until February 1957, Fidel's column was virtually lost to the world. Most of the Cuban press, encouraged by Batista's propaganda, thought Fidel was dead. What suddenly brought the rebels to notice was not their action against the government, but their discovery by a famous U.S. journalist.

Fidel and his coconspirators knew that a key to toppling Batista would be the erosion of the dictator's foreign support, especially from the United States. Fidel's contacts found the perfect vehicle: Herbert

Matthews, a veteran foreign correspondent of the *New York Times* who had covered the Spanish Civil War and remained an unreconciled partisan of the Republican cause. Matthews was smuggled up to Fidel's mountain hideout and from there wrote a series of stories which exploded on the front page of the most prestigious newspaper in the United States. Matthews' dramatic dispatches, which extolled the discipline, bravery, and commitment of the rebels, gave Fidel international status overnight. Readers around the world wondered: How could a fifty-seven-year-old foreign journalist elude the army's cordon and spend days with the supposedly nonexistent guerrillas? Suddenly Batista was on the defensive in world public opinion. He was in that most dangerous of realms—seen to be both brutal and impotent.

The following month, March 1957, Fidel received reinforcements. Fifty-eight new recruits joined the rebels, having been led in by the underground. Most of the guerrillas now in the Sierra Maestra were newcomers to the wild country, indeed to the countryside itself. Most, like Fidel, were from the middle class. They had joined up, not because of a grand vision about transforming all Cuban society, but because they hated the brutality, corruption, and antinationalism of the politicians, who always seemed to serve the dictators. When they deserted the cities, however, they soon found another Cuba. However much they thought they knew about the "other" Cuba, they were unprepared for the realities in the mountains.

They found the peasants forced to scratch for a miserable living. The rebels took a strong interest in these people's fate because they needed peasant support to survive in the mountains. It was the first principle of the guerrilla: retain the sympathy of the local residents, not only for supplies but also so they will not betray you to the authorities.

The rebel band was still, however, primarily middle class. A few peasants joined the rebels, but they never came in large numbers and they never held positions of leadership. This is hardly surprising. Most revolutions in history have been led by members of a counterelite. This is not to say that participation and support from peasants was not important. But the Fidelista phenomenon was middle class in origin and leadership. Its later directions were another matter.

Guerrilla warfare is a lonely and dangerous business. Month after month through 1957 the rebels managed the essential—to survive. But they failed to score seriously against the enemy. By December Fidel had become discouraged. His strategy was based on waiting for the urban uprisings. But over that front he had precious little control. How long could they wait in the mountains?

There were some encouraging signs in early 1958. In February the Cuban bishops issued a pastoral letter calling for a government of national unity. In March the U.S. government, under pressure for supplying arms to the repressive Batista regime, placed an embargo on

arms shipments to both sides. This was a political slap to Batista, since it amounted to a partial withdrawal of legitimacy for the established government.

After a general strike failed to materialize in April 1958, Fidel decided to change his strategy. The guerrillas had to become more aggressive. The strike failure also convinced Batista that he should now move, and the army launched a "liquidation campaign" the following month. It was a disaster. Whole army units were captured, along with secret codes and extensive supplies for arms. By August the army had withdrawn from the mountains, defeated by their own poor leadership and faulty training and by superior intelligence and dedication on the rebel side.

Through the rest of 1958 a savage guerrilla war raged on. There were never any set battles, never any waves of peasants confronting Batista's soldiers. It was a war of hit-and-run, with bombings, sabotage, and harassment. Batista's response was counterterror. Since he could seldom catch the guerrillas, he sent his goons against the students and the middle class suspected of having links to the 26th of July Movement. In so doing, Batista was rapidly enlarging the support for Fidel. Here was the classic guerrilla tactic: goad the unpopular government into repressive measures, which then will serve to recruit new rebels against the government.

Support for Batista began to evaporate. As dictator his greatest card to play had always been his ability to keep order. Now even that was disappearing. Adding to the frustration and fury of the army and police was the fact that they could never pin down the enemy. They were unprepared for the kind of underground that could elude their network of regular informants. Torture and execution only produced new rebel adherents.

In November Batista carried out a presidential election, running a new front-man candidate in the hope that removing himself from the limelight might improve the situation. The result was a dramatic signal that the government had lost public support: most of the voters abstained. Here was the kind of political development that the rebels had been working for. Batista maneuvered desperately to hold on to his ebbing support from the Eisenhower administration. But the United States, as in the case of the dictator Machado in 1933, now saw its enormous interests in Cuba endangered by the excesses of a rapacious and brutal dictator. Batista had outlived his time.

Batista had no desire to fight a losing cause to the end. He could see that his power was shrinking daily. His army and police had become both hated and derided. He had lost the all-important support from Washington. And the country had become so convinced of his fall that the economy was increasingly disrupted as businessmen and bankers waited for the inevitable. Suddenly, on New Year's Eve, he called his

aides together, designated a successor president, and took off with a planeload of relatives for the Dominican Republic. The way was now clear for Fidel's triumphal entry into Havana.

The guerrilla war had been so savage, the repression so fierce, the buildup so long, that Batista's sudden exit took the rebels by surprise. Crowds went wild in the cities, especially Havana. The red and black flags of the 26th of July Movement were everywhere.

Defining the Revolution

Euphoria is the only word to describe Havana's mood in the early days of 1959. Fidel had achieved genuine heroic status. The question now occupying the minds of the Cuban middle class, workers, peasants, foreign investors, the U.S. embassy, and other observers was, What kind of revolution would this be?

Fidel entered a political vacuum. The civil war had not only discredited Batista; it had besmirched the entire political class, all of its members, to greater or lesser degree, compromised by the dictator. However important the urban conspirators, with their heroic tactics against the army and police in the last half of 1958, the momentum now lay with the men from the Sierra Maestra in the green fatigue uniforms. The visible power was the rebel army. It was to remain the key political institution thereafter.

Fidel's greatest asset, aside from his own formidable leadership gifts, was the desperate desire for change among his fellow Cubans. The most underprivileged, the rural poor, had never counted for anything in the electoral system. Working classes in the cities and towns had precious little more weight. The most restless and most important social sector was the middle class, which had produced the movement's leadership, such as Fidel and Raúl.

This class was ready to receive a new political message. They were first of all disgusted by the old political cadre and repulsed by the dictators (Machado, Batista) Cuba regularly produced. Second, they were moved by appeals for greater social justice. Third, they longed for a more independent Cuba. That meant a Cuba freer of the United States. Could there be a Cuban nationalism that was not anti-American? In theory, perhaps; in practice, any assertion of Cuban national dignity was bound to collide with the Yankee presence.

1959 was a year of drama for the Revolution. For all his heroism, Fidel arrived as a political unknown. The government began as a triumvirate. Manuel Urrutia was president, José Miró Cardona was prime minister, and Fidel was commander-in-chief of the armed forces. The illusion of collegiality collapsed in February when Miró Cardona resigned, protesting his lack of real power. Fidel assumed Miró Cardona's post, portending the pattern of things to come.

The first major political crisis arose over what to do with the cap-

tured Batista officials who had been responsible for the worst of the repression. The revolutionaries resorted to arbitrary procedures in trying their victims, appealing to sentiments of "ordinary justice" to legitimize their executions. In the first six months of 1959 about 550 were put to death, following trial by various revolutionary courts. These executions, punctuated by cries of *paredón!* (to the wall!), worried the liberals in Cuba and their sympathizers abroad, especially in the United States.

In April 1959 Fidel set out for New York, where he was to visit the UN headquarters. The trip was of extreme political importance, since U.S. opinion would be crucial for events in Cuba. From Fidel's standpoint, the visit was probably a success. He managed to project the image of a nationalist reformer, strongly opposed to foreign intervention, but also not a communist. He was careful to maintain only distant contact with the U.S. government (President Eisenhower refused any meeting and Vice President Richard Nixon had to receive the bearded revolutionary), while skillfully cultivating elite centers of opinion with, for example, a triumphant appearance in Harvard Stadium. Fidel hammered on the need for radical reform in Cuba, especially land reform. How could anyone who knew Cuban agriculture disagree?

Fidel returned to Cuba to carry out his most radical measure to date: the Agrarian Reform Law of May 17, 1959. The law eliminated the giant estates, expropriating farmlands over 1000 acres, with compensation to be paid in Cuban currency bonds in proportion to the declared value for taxes in 1958 (deliberately undervalued, as was the custom). No foreigners would henceforth be allowed to own agricultural land. The expropriated lands would be turned over to small private holders and cooperatives. A National Institute of Agrarian Reform (INRA) was created to implement these far-reaching measures. Critics in Cuba and abroad, especially in the United States, began to raise the alarm. Was this not the first step to communism? Hadn't Fidel appointed a communist, Núñez Jiménez, as the operating head of INRA?

Political polarization heightened in June 1959. Fidel announced the discovery of a plot against the Revolution. Noncommunists among the supporters of Batista's overthrow now became increasingly alarmed. A former president of the Senate attacked the agrarian reform and called for the elections which Fidel had promised. Later that same month the commander of the air force, Pedro Díaz Lanz, resigned in protest over the alleged communist influence in the military. Díaz Lanz then fled to the United States and pushed hard his story that Fidel was a communist. Such defections strengthened the hand of the growing anti-Castro elements in the United States.

In July Fidel staged what was to be a regular drama in revolutionary Cuba. He resigned the premiership in the midst of what he described as a political crisis brought on by the resignation of President Urrutia, whom Fidel had accused of joining Díaz Lanz in a conspiracy against the Revolution. Later in July there were huge rallies in Havana, where

the carefully primed multitudes called for Fidel to return to the premiership. He bowed to their will.

Fidel was now in a strong enough position to speak his mind on the delicate subject of elections. He promised there would be no more elections for at least four or five years.

There was now brewing a case that would for many become a hallmark of the Revolution's radicalization. Major Huberto Matos, one of Fidel's oldest political allies and a long-time revolutionary, chose to break with the Fidelista line. He resigned from the armed forces and issued a letter attacking the growth of communist influence. Fidel's response was swift. He jailed Matos and mobilized a huge propaganda campaign against him as a traitor to the Revolution. For the next decade and a half, Matos, locked away in prison, remained for the Fidelista regime the supreme symbol of revolutionary deviationism. For many foreign observers, Matos remained the quintessential victim of Stalinist-style repression.

In the waning months of 1959 Cuba's line became more stridently anti-American. There were daily charges of Yankee-backed invasion plots to restore Batista. Such charges were by no means fanciful. Exiles were already flying missions from Florida, firing the cane fields and dropping antirevolutionary leaflets. Although the White House and the State Department had not yet made up their minds about Fidel's intentions, the CIA and the Pentagon had long since ceased to have doubts. Meanwhile, Fidel had his eye on Washington—always the cockpit of decision in Cuban politics.

The year 1960 proved to be even more decisive for the course of the Cuban Revolution. By the end of Fidel's second year in power, four basic trends had taken hold: (1) the nationalization of the economy; (2) a sharp swing to the Soviet bloc; (3) the establishment of an authoritarian regime; and (4) the launching of an egalitarian socioeconomic policy.

All Cuban nationalists had been angered over the years by the degree of U.S. control of the Cuban economy. It was inevitable that any Cuban government attempting to reassert Cuban control over its economy would collide with the United States, both the investors and the government in Washington which had so often backed them up. The major clash came over oil, always an emotional economic issue in the Third World. When Fidel had discovered that he could buy crude oil cheaper from the Russians than from Venezuela, he ordered the U.S.-owned oil refineries located in Cuba to process the Russian crude. Although an old law obligated them to comply, they refused. Fidel promptly confiscated the U.S. oil companies. Partially in retaliation, President Eisenhower suspended the Cuban sugar quota in the United States.

The Cuban government now followed by seizing virtually all the rest

of U.S. property. That included electricity and telephone companies (another prime irritant to the nationalists), sugar mills, and nickel mines. Washington reacted by embargoing all trade to Cuba, except medicines and foodstuffs. Other foreign firms in Cuba were also nationalized.

The nationalization campaign was not restricted to foreign owners. In the course of 1960 all major firms in Cuba were nationalized, including textiles, tobacco, cement, banks, and department stores. Agriculture took longer. The first move, in 1959, was against the sugar plantations and mills owned by Batista or his closest collaborators. But the shape of agricultural policy, always a tough problem for command economies, had not emerged by the end of 1960.

The swing to the Soviet bloc was neither a cause nor an effect of the clash with the United States; it was part and parcel of the same process. Initially it was a question of how far the Soviets might be willing to commit themselves in Cuba, so far from Moscow and so close to the United States. The Russians proved bolder than almost anyone expected. In February 1960, well before the full economic break with the United States, the Soviets signed a trade agreement with Cuba, granting $100 million credit to buy equipment and promising to purchase 4 million tons of sugar in each of the coming four years. Fidel was now developing an alternative source of technology and equipment, and the Soviets were getting ready to integrate Cuba as a "socialist" ally in the Third World.

As 1960 continued, the Soviets added military weapons to the equipment headed for Cuba. Technical and artistic missions also arrived, to fill out the lesson in how to build a socialist society. By the end of 1960 Cuba's swing toward the East was decisive. Still, Fidel had not yet announced Cuba's total conversion to the Soviet brand of socialism, and foreign observers held conflicting views. Some, like Vice President Nixon, were convinced that Castro was a full-fledged communist. Others, sharing deep concern for social justice, hoped Fidel could find an independent path between the superpowers; if he failed, they argued, it would be because an intolerant United States pushed him into Russian arms.

Revolutionary Cuba's state was emerging in a piecemeal, ad hoc fashion. Fidel began by proclaiming his commitment to the old constitution, which Batista had repudiated by his coup of 1952. But by what institutions would or could the new Cuba be ruled? The problem was a classic, one that Salvador Allende would later encounter in Chile: How to carry out fundamental economic and social change when existing government institutions were set up to maintain the status quo.

Though the old legal system remained in place, for example, there was never any attempt to elect a new legislature. The 26th of July Movement could hardly provide an institutional base. It had never de-

veloped into a tightly knit organization and it was far from a political party. From the start, Fidel relied on the most responsive and popular institution at hand: the revolutionary army.

In the fall of 1960 the government created an important new institution: Committees for the Defense of the Revolution (CDRs). Locally based citizens' groups, they were organized primarily for civil defense. The constant threat of invasion—by exiles and/or by the United States—necessitated such a measure. Since the Revolution also had enemies at home, the CDRs also had the task of monitoring the population for counterrevolutionary opinions or behavior.

That same year Fidel moved to eliminate or neutralize the key institutions of the former "bourgeois" order. By December the press had been brought into line, often through seizure by communist-controlled labor unions. Even the satirical *Bohemia*, long a mordant anti-Batista organ, fell victim. By December Fidel had gained power to appoint new judges at will, as the independent judiciary was phased out. The universities and the unions, once centers of opposition to the government, were also brought under absolute government control. A new law gave the minister of labor the power to "intervene," i.e., assume legal control over any union. All private clubs and associations were subordinated to government direction. The church, although never strong in twentieth-century Cuba, was watched closely, as the revolutionaries launched frequent attacks on "reactionary foreign priests." In 1961 the government nationalized all private schools, thereby removing one of the church's most important prerevolutionary roles.

The Revolution set out to create new institutions in place of the old. Fidel seemed to be everywhere. Mobilization was the inexorable theme: mobilization against invaders, mobilization against social and economic problems at home. All Cubans would become guerrillas. To achieve this goal, a huge militia was created: by the end of 1960 it totaled 500,000 out of a total population of 6.7 million. It was, after all, an obvious way to organize the new Cuba. And none could doubt the identity of the commander-in-chief.

The only political party to survive the revolutionary transition was the Communist Party. Never a member, Fidel had throughout 1959 avoided any personal identification with the party. But he also made it clear that anticommunism would be considered counterrevolutionary. As 1959 continued he increasingly turned to party members to handle such areas as agrarian reform. Their increasing participation did not, however, threaten Fidel's effective control of the party.

What most Cubans cared about was not political structure but how the Revolution would change their lives. On this score Fidel and his guerrilla companions kept their eyes fixed on the poor, especially in the countryside. The revolutionaries were determined to attack the legacy of the corrupt, capitalist Cuba: illiteracy, disease, malnutrition, and dilapidated housing. A year-long crusade in 1960 cut illiteracy rates in

half (Cuba's illiteracy rate of 25 percent in 1959 was already low by Latin American standards), and it has virtually disappeared since then. Sensing the direction of the Revolution, the rich (and many from the middle class) began to flee, and the government acquired a windfall: the refugees abandoned assets—homes, offices, farms—that the state could now distribute.

In a typically populist move, Fidel began his government by freezing prices and ordering major wage increases (a step also taken by Perón in 1946 and Allende in 1970). This led to a buying spree, but the inventories soon disappeared. Batista had left $500 million in foreign exchange reserves. But that was quickly spent, especially on oil. So the era of seemingly painless redistribution was over by the end of 1959. In 1960 Cubans discovered the cost of the nationalist, egalitarian policies of the Revolution. For once in Cuban history, however, the huge inequalities of sacrifice had been eradicated.

But the number of defectors also grew. Most attacked the guerrillas for betraying the hope of rapid elections. Instead, they charged, Fidel and his clique were leading Cuba toward communist totalitarianism. Most probably were sincere. Some also thought it the best tactic to arouse the United States.

Some people in the U.S. government needed little encouragement. By late 1959 a hard-line faction in the CIA and military intelligence saw Fidel as a Soviet foil who should be dealt with directly. The CIA began formulating an endless series of often bizarre conspiracies such as slipping Fidel an exploding cigar. All were aimed at disrupting or sabotaging the new government. All involved Cuban exiles, who were flooding into Miami. Therein lay one of the CIA's greatest vulnerabilities: working with exiles made security questionable. Fidel's intelligence apparatus, soon aided by the more experienced Soviets, cultivated its Miami contacts and neutralized much of the CIA's laborious plotting.

The most obvious strategy for Washington was to support an exile invasion of Cuba. That was how José Martí had returned to the island back in 1895, and it was the standard strategy in Caribbean-exile politics. Since late 1959 the CIA had been organizing anti-Castro exiles. In July 1960 the CIA empresario of the exile invasion, Richard Bissell, convinced President Eisenhower to approve the training of an invasion force. From that moment on, Bissell, a formidable intellect and bureaucratic infighter, became a super-advocate for the invasion.

The "toughness" of U.S. policy toward revolutionary Cuba became an issue in the 1960 presidential campaign, which featured Eisenhower's vice president, Richard Nixon, and the relatively unknown senator from Massachusetts, John Fitzgerald Kennedy. In their first televised debate, Kennedy took a more aggressive stance toward Cuba than Nixon, who knew of the invasion plan, which he did not wish to compromise.

It was Kennedy, the ostensibly tougher candidate, who won the pres-

idency and inherited the "Cuban problem." Eisenhower broke diplomatic relations in January 1961, in response to Fidel's demand that the United States drastically reduce its embassy in Havana. In April, still inexperienced in foreign affairs, Kennedy found himself pressured to approve an exile invasion of Cuba. Wanting to do his anticommunist duty, but fearful of the possible effect on world opinion, the new president agonized. He finally gave the go-ahead, but demanded that there be no identifiable U.S. involvement: above all, no combat involvement of U.S. forces. It was an ironic concern, given the CIA's critical role, and it would have its effect on events.

As rumors mounted, an invasion force headed for Cuba in April 1961. The operation proved a misadventure from the beginning. After endless debate, President Kennedy reduced the exile-piloted air cover and vetoed the use of any U.S. planes. The invaders foundered in an ill-chosen bit of southern coast, on the Bay of Pigs, which Fidel happened to know well. The exiles were hopelessly disorganized. The hoped-for uprisings, which would supposedly paralyze the Cuban defenders, never materialized. The Cuban defenses proved more than adequate. The invasion brigades were quickly captured. They never had a chance to adopt their fall-back procedure—head for the mountains and mount a guerrilla operation.

The Bay of Pigs could not have been a greater triumph for Fidel and the revolutionaries. The United States had finally shown its intentions to be what Fidel had always said they were: a desire to turn the clock back in Cuba. Although the CIA had tried to screen out the more unsavory ex-Batista types, the invaders included more than a few who had served the dictator. Fidel and his supporters seized on those names to prove that the United States wanted to restore the discredited dictator.

The failed invasion marked a watershed in U.S.-Cuban relations. Washington's most obvious strategy had failed. Cuba would not be the Guatemala of the Caribbean. What options were left for the aroused United States? Precious few. Now the issue shifted to the level of the superpowers. In July 1960 Khrushchev had rattled Soviet missiles in defense of Cuban socialism. In April 1961 Cuba didn't need Soviet help to repulse the CIA protégés. But would the Americans stop there?

The Soviets decided they must back up their threat by putting missiles in Cuba itself. The decision took almost everyone by surprise. Why would the Russians want to put intermediate-range missiles in the U.S. backyard, when their long-range missiles could easily reach the United States from Soviet launching pads? Yet the Soviets went ahead, and by October 1962 they were installing intermediate-range rocket bases in Cuba. This was an unprecedented challenge to the balance of military power. The United States demanded that the Soviets withdraw their missiles from Cuba, under sanction of a naval quarantine on all Soviet military shipments to Cuba. The world seemed to balance on the edge

of nuclear war. After a fateful interval, Khrushchev complied. The missiles were withdrawn.

The superpower confrontation in the Caribbean had fateful implications for Cuba. First, Fidel himself was not consulted at any stage. The result was to make Cuba, in Latin American eyes, into a Soviet satellite in essential security matters. Second, the Soviets withdrew their missiles only because Washington (secretly) promised it would not invade Cuba. This was the least noticed and least understood result of the missile crisis: the Soviets had forced the United States to allow the socialist experiment in Cuba to proceed.

When Fidel declared himself a Marxist-Leninist, in December 1961, the statement came as an anticlimax. Whatever his ideological confessions, Fidel continued to be the overwhelmingly dominant personality in the Revolution.

Decade of Experiment

After defeating the Bay of Pigs invasion in 1961, the revolutionaries could concentrate on the economic tasks facing the new Cuba. The central fact was that the Cuban economy revolved around exporting sugar, especially to the United States. The revolutionaries were determined to change that humiliating dependence. The chief architect was Ernesto "Ché" Guevara, the Argentine physician-guerrilla who was the most creative theoretician among the revolutionaries.

Guevara drew up a Four-Year Plan which called for agricultural diversification (de-emphasis on sugar) and industrialization (the manufacture of light consumer goods). Cuba launched this ambitious plan amid great fanfare. The Revolution would break the stranglehold of a monocultural export economy.

By 1962 the results had already proved disappointing. In part, Guevara and his youthful planners were reaping the whirlwind of the shortsighted policies of 1959–60. Stocks of consumer goods were exhausted, foreign exchange reserves were gone, and shortages were everywhere. Even worse, sugar production had taken a plunge. In 1961 the Cubans had produced 6.8 million tons of sugar, the second highest harvest in Cuban history. This output merely disguised the deliberate neglect the government was showing to sugar. They seemed to take it for granted. Cane fields were plowed under, new planting was delayed, fertilizer was forgotten. In 1962 the harvest dropped to 4.8 million tons and in 1963 it was only 3.8 million tons, the smallest since 1945. The fall was disastrous for export earnings.

The industrialization drive was also going badly. Cuba lacked the raw materials and expertise to rush into industrialization, even in light goods. Since 1960 the United States had enforced a strict economic embargo against Cuba, pressuring all U.S. firms (and their Latin Amer-

ican and European subsidiaries) to cease trade with Cuba. This embargo, enforced by U.S. legal measures, forced Cuba to depend largely on the Soviets and the Eastern bloc for equipment. Direction was to come from highly centralized planning bureaucracies, modeled after Soviet and Czech patterns. The effort was ineffective and expensive. Even the Russians seemed uneasy about underwriting a socialist utopia in the Caribbean.

In mid-1963 the Soviets put their foot down. The Cubans must slow down the industrialization drive and improve their planning. They must recognize Cuba's comparative advantage: sugar. Cuban policymakers moved in this direction, not only because of Soviet pressure but also because they saw the need for change. Ché Guevara resigned, confessing his errors. Fidel, ever on the initiative, now embraced sugar, which he had so recently spurned. In 1963 he announced that in 1970 (later labeled the "Year of the Decisive Endeavor") Cuba would break all records for sugar production: it would harvest 10 million tons. Thus the famous target of 10 million tons.

Debate continued over strategies for economic development and political consolidation. Still active in the regime, Ché Guevara argued for an "idealistic" strategy, a Maoist approach that would totally eliminate the market and material incentives. The economy would be fully collectivized and directed by a centralized planning authority. A radical break with the capitalist past would require a "new man," a Cuban who would work for moral rewards (decorations, public praise) and thus reflect a new, higher level of political consciousness. Through dedication and sacrifice the "new" Cubans could contribute to the rapid construction of socialism. Here the Cuban leaders were going through the familiar dilemma of communist regimes: how to reconcile Marxist idealism with a pragmatic economic policy.

Guevara's idealists further argued that the construction of socialism at home required the aggressive promotion of revolution abroad. They wanted to prove that a guerrilla strategy could work throughout Latin America and perhaps the entire Third World. According to this voluntaristic vision, pragmatism and conventional institutions (political parties, labor unions) played only a minor role; the need was to create revolution *now.*

Guevara's main opponent in this debate was Carlos Rafael Rodríguez, an economist and long-time Communist Party member. Rodríguez took a practical approach. He favored a more measured use of central planning, partial reliance on market mechanisms, and autonomy left to the individual enterprises. He thought state firms should have to account for their expenses and earnings. In short, Rodríguez and his allies proposed a more conventional path, relying on material incentives instead of only moral ones. They favored also a strong party and a "flexible" policy toward Latin America. This meant a willingness

Fidel Castro addresses a rally in the early 1960s; the doves, frequently used as a political symbol, represent the idea of a society at peace. (Center for Cuban Studies, New York.)

to deal with regimes that Guevara saw only as targets for revolutionary opposition.

While the arguments went on, Cuba was returning to sugar. Even with the de-emphasis on industrialization, however, economic production was disappointing. 1964 yielded a 9 percent growth rate for most of the economy, and that was primarily catch-up from the declines of 1961–63. In 1965 the figure slipped to 1.5 percent, less than the rate of population growth, and in 1966 became negative again (−3.7 percent). Indecision in basic policymaking was not building a dynamic socialism.

In 1966 Fidel brought the debate to an end. He endorsed Ché Guevara's idealism. Cuba would make a gigantic collective effort accompanied by moral incentives. This immediately increased Fidel's own power, as he himself took charge of the now strengthened central planning apparatus. He and his trusted lieutenants plunged into the minutiae of economic management. Fidel selected and dropped favorite projects, often reacting impulsively on his never-ending visits to worksites around the island. The atmosphere recalled the early romantic

days of the Revolution—endless rhetoric, euphoric dreams, celebration of the selfless "new man."

Along with this idealistic mobilization at home went a stepped-up commitment to revolution abroad. Cuba sought out guerrilla movements across Latin America, offering arms, training, expertise. Ché Guevara spearheaded the drive. Always a heroic figure, Ché became the nemesis of the CIA and the Latin American military. Unfortunately for Ché, however, he chose the *altiplano* (highlands) of Bolivia to start the spread of his "many Vietnams" in South America and there met death in 1967 at the hands of U.S.-trained Bolivian Ranger troops. One important problem was his poor communication with the Bolivian Communist Party, which considered him a foreign adventurer who knew nothing of Bolivia. This rift mirrored the gap that had arisen between the Soviets and Cuba. Havana had strayed far from the prevailing Moscow line on exporting revolution.

By 1968 Fidel was pulling back from the Guevarist line. There had already been signs that Ché did not get full support from Havana during his ill-fated campaign in Bolivia. By supporting the Soviet invasion of Czechoslovakia in 1968, Fidel signaled a return to Soviet orthodoxy. He then began to downplay the export of revolution. However heroic Ché's death, the new policies suggested it might have been in vain.

On the domestic front, however, Guevarist policies continued intact. The spring of 1968 saw a "revolutionary offensive." The remainder of the private sector was nationalized, consumption was subordinated to investment, and Cubans were exhorted to give their all to reach the omnipresent target of 10 million tons of sugar in 1970.

The magic year came and all of Cuba was mobilized to cut cane. Even visiting Soviet sailors, much to their shock, were whisked off the docks to wield machetes. Everything was sacrificed to release labor for the cane fields. Sensing that the target was distant, the authorities left some of the 1969 harvest in the fields, hoping to improve the 1970 figure. At harvest time they tried to cut every available scrap. It was no use—the *zafra* reached only 8.5 million tons. It was a prodigious total, the largest in Cuban history, but it might as well have been half that amount. So much propaganda, so many promises. The entire fate of the Revolution had seemed to ride on the scales at the sugar mills. It was a mortal blow for the "voluntaristic" philosophy of Ché. The psychological toll was enormous. But Fidel, ever resourceful, was about to change policies again.

Consolidating the Regime

The failure of the 10-million-ton effort made Fidel's about-face easier. Everyone could see that the "idealistic" model had failed. On July 26, 1970, Fidel confessed all. In a marathon speech (the "Let the Shame Be Welcome" speech), Castro took on his own shoulders the responsi-

bility for the quixotic crusade for the super-harvest. He offered to resign, but the crowds cried no. The economic failure was obliterated by revolutionary theater.

Cuban policy now turned more pragmatic. First, there were to be new management and planning systems involving greater decentralization and more use of "profits" as a basis for decision making. Second, the private sector was to be given a greater role in both agriculture and services. Third, pay would now be linked to output, with premiums for needed skills. Finally, there was to be greater economic interaction with the West.

This more conventional economic policy was accompanied by a similar shift in institutional policy. The Communist Party was now strengthened, the unions and other mass organizations were reorganized and given a greater role. This move toward greater "orthodoxy," (i.e., closer resemblance to Soviet practice) affected culture as well. Central controls over education and mass media were strengthened.

Fidel began to sound like Khrushchev at his most obscurantist. In early 1971 Fidel launched furious attacks on "former friends" of the Revolution who had charged that Fidel's personalistic regime was leading Cuba toward economic defeat. One was the French agronomist René Dumont, who attributed Cuba's agricultural failures to Fidel's egomania and the helter-skelter militarization of the Cuban economy. Another critic, the Hungarian-born K.S. Karol, was more devastating because he had a deeper knowledge of Marxist thought and communist experience from which to gauge Fidel's limitations.

Also in early 1971 Fidel cracked down on the Cuban artistic scene by arresting the internationally known writer Heberto Padilla. Apparently under coercion, Padilla was forced to confess crimes against the Revolution. He later repeated his mea culpa before a writers' conference, which set the tone for a tougher standard of political loyalty now expected of all artists in revolutionary Cuba.

Part and parcel of this policy shift was an increasing approximation with the Soviet Union, which meant closer conformity to Soviet models of economic and political decision making. It had been under way since 1968, but the shift in domestic policy now made Cuba's overall stance more consistent. Radical experimentation was over. The inevitable logic of Cuba's enormous economic and military dependence on the Soviets was being played out. Fidel had now become a reliable ally of the U.S.S.R. in the Third World. Gone were the harsh attacks on the orthodox communist parties. As the decade of the 1970s began, the Cuban Revolution was approaching the Soviet model more closely than ever before.

As the Cuban Revolution entered the 1980s, there could be no doubt that the one-time guerrillas had created a new society. There had been more than two decades to educate and train new generations in the commitment to an egalitarian, communitarian ideal. They had been

able to train their own technicians, with Soviet and East European help, to replace the cadres that had fled the radicalizing regime. There had been time to make Cuba a formidable fighting force. No future would-be freedom fighter would be able to duplicate the epic of the *Granma*.

Cuba had also settled into an extreme economic dependence on the U.S.S.R., one that bore much resemblance to her one-time dependence on the United States. Soviet economic assistance to Cuba in 1989 was estimated at $4.2 billion by the U.S. government. Although the exact total was difficult to calculate, it probably equaled about one-quarter of Cuban GNP. The integration of trade, both export and import, with the Eastern bloc was close to what it had once been with the United States. Had Cuba merely traded one brand of dependency for another? In the most obvious sense the answer had to be yes. Yet the ties to the Soviet Union did not produce the direct ownership which had created such a nationalist backlash against U.S. economic penetration before 1959.

Beyond the statistics, however, there were more pressing questions. What were the consequences of this new dependency? We know that Fidel had echoed the Soviet-line denunciation of Solidarity in Poland ("the antisocialists and counter-revolutionaries") and praised the Soviet intervention against "the savage acts of provocation, subversion, and interference against the revolution" in Afghanistan. Cuba sent more than 30,000 troops and social service personnel to support pro-Soviet regimes in such African countries as Angola and Ethiopia. But what were the deeper implications for Cuban society? The profound social revolution in Cuba had been possible only because of Soviet military protection and economic aid. It remains unclear whether Cubans had more bargaining power with Moscow than they once had with Washington, since Soviet-Cuban relations occurred in far greater secrecy than had relations with the United States.

The Revolution has brought many changes to Cuba. For those who had lived with little hope in capitalist Cuba, the Revolution brought a much improved standard of living. Socialist Cuba's greatest triumphs have been in serving basic human needs. Illiteracy has been wiped out, and a comprehensive school system created. Its teaching content is, not surprisingly, highly ideological, designed to inculcate the new socialist values. Basic health care, especially preventive care, has been extended to the lower sectors. Medical training has been geared to public health. Food distribution, always one of the most shocking reflections of social inequality, has been guaranteed by rationing. Minimum nutritional standards have been established and largely met for the entire population. The result is that life expectancy rose from sixty-three years in 1960 to seventy-six in 1992 and the infant mortality rate fell by more than two-thirds in the same period. Much of this progress was of course undermined by the economic crisis that began in 1990.

The role of women has been another area of significant change. The tradition of *machismo* was especially strong in prerevolutionary Cuba and has proved a major obstacle to the feminist movement. To take a striking example, by mid-1980 only 19 percent of the Communist Party members and candidates for membership were women. Nonetheless, the Federation of Cuban Women (*Federación de Mujeres Cubanas,* or FMC) has gone a long way in changing opinion and behavior. The number of women in higher education and professional schools (especially medicine, where female students now outnumber males) has increased sharply. The FMC was instrumental in getting adopted in 1975 an egalitarian family code which obligated husbands to do half of all family chores. Any viewer of the Cuban film *Portrait of Teresa* knows that this and other feminist goals will not be easily reached in Cuba. But where is that not true? Despite the perceptible change in Cuban attitudes, married women, especially those with children, have found it difficult to enter the full-time labor force. One reason is the cost and inconvenience of child care. Another is the fact that an additional income may bring few additional benefits, since consumer goods remain scarce.

Housing was the other basic need that had been so unequally distributed before 1959. Here the revolutionaries had trouble making rapid progress. It was easy enough to expropriate the residences of the wealthy and give them to special groups (like students). But new construction was slower and more expensive. In the short run, investment in new housing was not seen as a top priority. There the Cubans were follow-ing, perhaps inadvertently, the example of the Soviets, for whom the shortage of housing had been a major social problem.

Ironically enough, one of Cuba's greatest economic failures was in agriculture. In the early years of the Revolution that was understandable. The guerrillas were eager to repudiate Cuba's long-time bondage to a single export crop. Guevara's great hope had been to diversify agriculture, as well as to industrialize. Even after the turn toward economic realism in 1963, food production lagged. According to a United Nations study, Cuba's agricultural performance for 1961–1976 was tied with that of Chile for the worst in Latin America. After 1976 farm output grew at a healthy rate, but a decade later Cuba was still seriously dependent on food imports.

As the 1970s closed, both the U.S. and the Cuban governments sought to improve their relations with one another. Fidel decided to let U.S. relatives of Cubans visit the island—the first time since early in the Revolution. And 100,000 relatives arrived in Cuba in 1979, loaded down with electronics and other consumer goods. Since these goods did not exist or were available only at high prices on the black market, many Cubans realized how limited their consumer goods were after two decades of the Revolution.

This frustration undoubtedly contributed to what became a dramatic exodus from Cuba in 1980. The trigger was the Cuban government's decision to withdraw its guard at the Peruvian embassy, in response to a violent incident involving Cubans who charged the embassy barrier to gain asylum and, ideally, safe conduct out of Cuba. Word suddenly spread that the embassy was unguarded, and within twenty-four hours 10,800 Cubans rushed onto the embassy grounds, jammed in like cattle. The government, taken aback by the embarrassing surge of dissidents, announced they would all be allowed to emigrate, along with anyone else who cared to inform the authorities. The total eventually reached 125,000. Most went from the port of Mariel via small craft—many hardly seaworthy—provided by the Cuban community in Florida.

These 125,000 followed the previous waves of exiles, including the 160,000 who had left in the officially coordinated U.S.-Cuban program between 1965 and 1973. Why the exodus? Across the newspaper front pages and television screens in the United States, West Europe, and the rest of Latin America were the images of almost 11,000 desperate Cubans crammed into the Peruvian embassy grounds, without food or water.

To refute the image of the desperate would-be exiles, hundreds of thousands of Cubans staged huge marches through Havana. But diplomats in Havana estimated that if Mariel had remained open, perhaps one million would have chosen to go to Florida. Whatever the exact potential, it was far larger than any foreign diplomat would have guessed before the embassy incident. The discontent could be partly explained as the frustration of Cubans who were weary of waiting for the higher standard of living promised for so long. The Cuban government was well aware of this discontent, and in the wake of Mariel it supplemented the food rationing system with "free farmers' markets." But the "Rectification Program" beginning in 1986 abolished small-scale private businesses and reenshrined moral incentives, attempting to make a virtue of the country's deepening economic crisis. The latter was exacerbated by a severe hard currency shortage, caused in part by low world sugar prices. The process of "rectification" took Cuba in precisely the opposite direction from the *perestroika* then being launched by Fidel's mentors in the Soviet Union.

Although domestic economic performance, if measured by production (rather than by income, as in Western economies), had held up well, averaging 7.3 percent growth from 1981 to 1985, growth dropped sharply after 1985, even turning negative for some years. Unlike in capitalist Latin America, however, Cuba's population was not put at risk by low economic growth, thanks to the food rationing and comprehensive health services that have given Cuba indicators of morbidity and mortality equal to those of the industrialized world.

The Struggle for Survival

Although Cuba had reached mid-1991 without the signs of mass discontent that doomed communism in East Europe, the strain had begun to show in recent years. In May 1987 the deputy chief of the Cuban air force and a hero of the Bay of Pigs climbed into a plane and defected to Florida. In June 1989 a heavier blow fell. The army's most respected leader, General Arnaldo Ochoa Sánchez, architect of brilliant battlefield victories over South African forces when Cuba fought to consolidate the communist regime in Angola, was tried and executed, along with three other high officers. The charges were drug-running and embezzlement. Their dramatic show trial, conducted in great haste, aroused memories of Stalin's infamous purges. Many asked how officers who had enjoyed Fidel's closest confidence could have organized such a vast conspiracy without the knowledge of a leader who possessed a legendary appetite for administrative detail. Or was this a way of eliminating a potential rival for ultimate power?

A key to the Revolution's survival would be the ability to institutionalize the revolutionary process. In the 1960s Fidel drifted into reliance on groups that had been born in the insurrection or were created to protect the new regime: the army, the militia, and the Committees for the Defense of the Revolution. The Communist Party was given a greater or lesser role at the initiative of Fidel and the top revolutionary leadership. With the turn toward orthodoxy in the early 1970s, the party assumed new importance.

The basic challenge for the revolutionaries was to transform leadership from a tiny elite of guerrilla veterans and party faithful to a growing base of loyal supporters. The most obvious means—and the one that had been used in the Soviet model—was to broaden the base of the Communist Party. In 1975 this process began. Under the banner of "popular participation," grass-roots elections for regional assemblies were held. The idea was to build a representational structure at the local level. Yet by the mid-1990s Cubans were still complaining about a highly centralized, bureaucratized, inefficient state apparatus. Power was still concentrated at the top in a structure like the one the East Europeans and the Russians had already dismantled.

In 1991–92 Cuba underwent a painful reality check, as the foreign underpinning of its economy vanished. The collapse of the U.S.S.R. and of Comecon (the foreign trade authority for the U.S.S.R. and East Europe) brutally exposed Cuba's economic vulnerability. By 1992 all Russian economic and military aid was gone. Oil shipments fell 86 percent from 1989 to 1992, while food imports dropped 42 percent in almost the same period. Vital equipment, such as buses, once supplied by East Europe, now languished for lack of replacement parts. General economic activity fell by 29 percent between 1989 and 1993. Other estimates put the decline at twice as great. Cuba had suffered an economic

blow greater than any (including the Great Depression) experienced in Latin America in the twentieth century.

Why? Because Cuba had excessively concentrated its trade and finance with one market (84 percent with Comecon). Secure in the Soviet connection, it had seen little need for a major effort to diversify its exports or its markets. It became the ultimate example of dependency, although one must acknowledge also the central role of the U.S. embargo in forcing Cuba eastward. With the collapse of the Soviet Union, Cuba confronted the equivalent of a "double embargo."

The domestic effects of the Soviet withdrawal were soon felt, as Cubans suffered a drastic drop in their living standards. The monthly rationing quotas now covered only one to two weeks, with the rest obtainable only on the black market. Havana had electricity only four to eight hours a day. Bus service virtually disappeared because of fuel shortages. Cubans were told to use bicycles, hurriedly ordered from the People's Republic of China. Fidel called the collapse of the U.S.S.R. a "disaster" and proclaimed that Cuba would now enter "a special period in peace time." The order of the day would be to save socialism in one country—Cuba.

To meet this objective, Cuba cannot do without trade with the capitalist West. Castro desperately needs capital goods and technology which only the West can supply. To buy, he needs hard currency, 70 percent of which Cuba earns from sugar sales on the open market. Cuba had run up a hard currency debt of $6 billion by 1990, and Castro's decision to suspend principal and interest payments crippled his efforts to gain further financing from capitalist sources. There were some features of the world economy Cuba could not escape.

Many outside observers predicted that under these circumstances, Fidel would be forced to move toward the market. In fact, such movement has been minimal, aside from aggressive but only modestly successful efforts to attract foreign investment from the capitalist world. Fidel and his lieutenants continue to defend the planned state economy and one-party rule. They have often discussed but not yet adopted, for example, the Chinese model of economic liberalization with continued political authoritarianism. There have been some timid liberalization moves, such as allowing limited self-employment and the holding of dollars. But in the all-important and underproductive rural sector, any return to peasant markets (an experiment abandoned in 1986) has been ruled out.

The reaction of ordinary Cubans to the economic calamity has been more stoical than might have been predicted. In April 1994, for example, Jorge Mas Canosa, the conservative Miami exile leader, confidently predicted that he and his colleagues would soon govern Cuba. Their expectations were not fulfilled. The population is clearly discontented in Cuba, as shown by the dramatic surge in the number of rafters intercepted by the U.S. Coast Guard (35,000 in the first eight months of

1994 alone). Within the island, however, there has been little organized opposition aside from a few courageous dissidents who periodically end up in prison. Power remains highly concentrated under Fidel, who more and more resembles an old-fashioned Latin American *caudillo*. Ironically, he is greatly helped by the tough U.S. stand, which Congress further tightened in 1992 and again in 1996. Without the specter of Uncle Sam, Fidel would be left without an explanation for Cuba's misfortunes. When Cuba was the only country in the hemisphere not invited to the December 1994 Summit of the Americas in Miami, Castro pronounced it "a great honor." This odd distinction offered scant consolation to the proud and patriotic citizens of Cuba.

NINE

THE CARIBBEAN
Colonies and Mini-States

Smallness is a basic fact of life throughout the Caribbean, an area stretching from the tip of Florida to the coast of Venezuela (see Map 3). With the exception of Cuba and Hispaniola (comparable to the state of Maine), the islands tend to be modest in size; the Grenadines, the Bahamas, and the Cayman Islands are absolutely minuscule. Topographies vary from the flat plains of Barbados to the rugged coasts of Martinique and Guadeloupe. A few of the islands, like Cuba and Jamaica, have rolling hills and substantial mountain ranges. In general the climate is mild, rainfall is abundant, and soil is adequate. Here the power of nature is clearly and constantly evident, from beauteous scenery to destructive forces—hurricanes, volcanoes, earthquakes, and floods.

The northernmost and larger islands of Cuba, Jamaica, Puerto Rico, and Hispaniola (now shared by Haiti and the Dominican Republic) are known as the Greater Antilles. The smaller islands to the east are known as the Lesser Antilles—also as the Leeward and Windward Islands.

Although the Caribbean was inhabited well before the arrival of Europeans, the size of the indigenous population is still subject to debate. By most accounts there were probably around 750,000 inhabitants, about two-thirds of them on the island of Hispaniola. There were three different groups: Ciboney or Guanahuatebey, Taino Arawak, and Carib (from which the region gets its name). These pre-Columbian peoples were the first victims of the crises and transitions resulting from conquest and colonization by European powers.

The arrival of Columbus and his three little ships in 1492 signaled the inexorable doom of the area's native inhabitants and the insertion of the Caribbean into the world arena of European competition. But this entrance was by no means abrupt. Spain exercised a monopoly over the region for more than a century, until rival European powers were prepared to issue a challenge.

MAP 3 Central America and the Caribbean

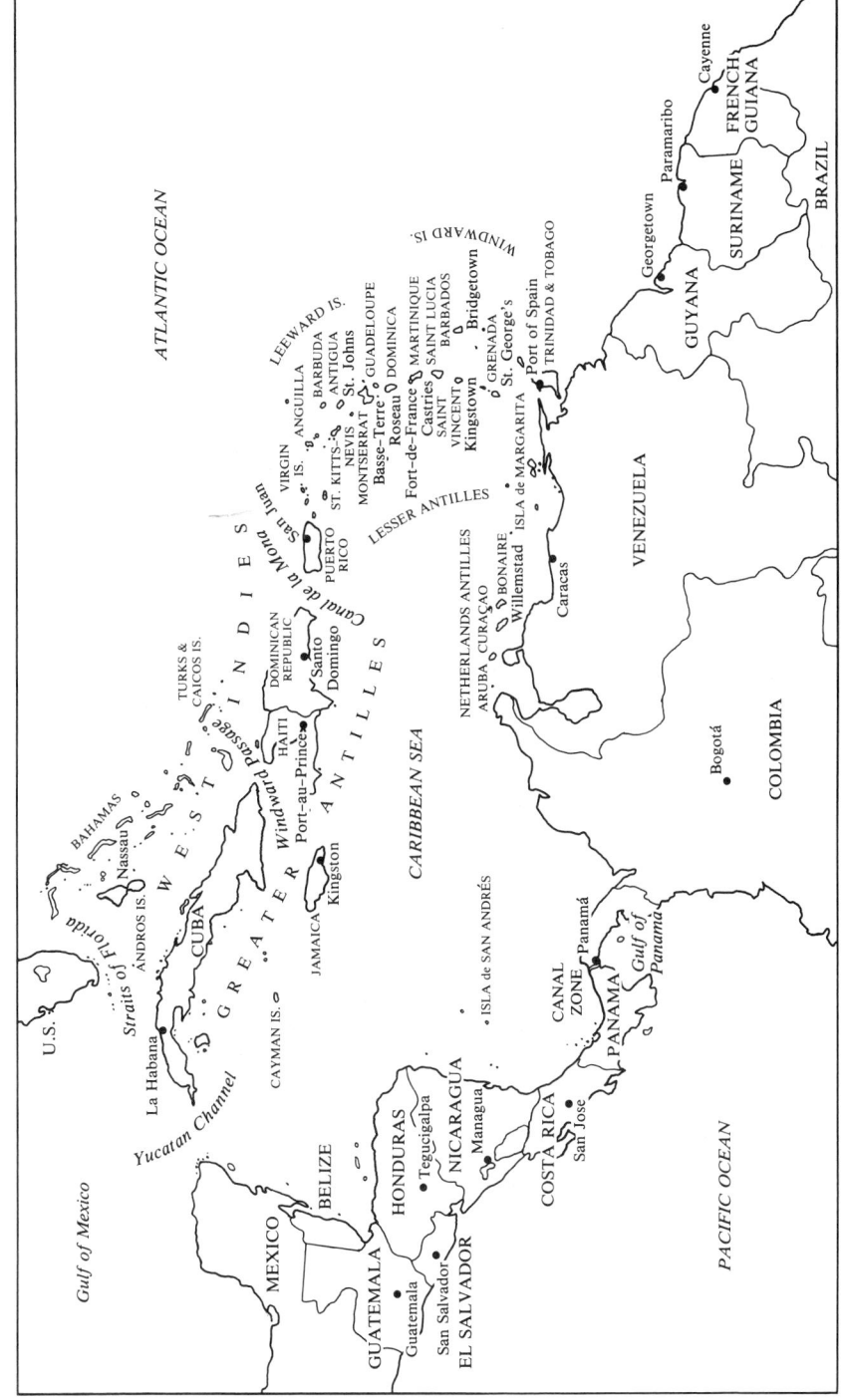

295

The Colonial Period: Conquest and Competition

Columbus landed on the island of Hispaniola in December 1492 and christened it *La Española*. The original intention of his voyage had been to replicate the advantageous commercial linkage which the Portuguese had established with the Far East. Expansion to the New World would thus yield exchange in new commodities, and Spain would become a first-rate trading nation. However, the absence of great civilizations and the prevailing conditions of production soon convinced the newcomers that they would have to change their strategy. And as Columbus observed, the local inhabitants were of generous disposition—and well equipped to serve the Spanish crown.

Unable to develop significant trade, the Spaniards chose to exploit the island as a source of land and labor through the *encomienda* system. Semifeudal institutions were imposed upon the native society. Indians were put to work in mines and fields. Harsh labor conditions and physical contact with Spaniards led to their decimation: disease and debilitation took a staggering toll. Realizing what fate held in store, many fled to the mountains in search of safety and freedom.

It was in the Caribbean where clerics first protested against abuse of the natives. In 1511 Antonio de Montesinos shocked a congregation in Santo Domingo by denouncing maltreatment of the Indian population. Soon afterward Bartolomé de las Casas began his fervent campaign to protect the Indians from adventurers and conquerors. In response to these pleas, the crown ultimately agreed to regulate the treatment of the native population. But to protect the American natives, Las Casas also made a fateful suggestion: that Spain import African slaves as a source of necessary labor.

Thus began the tragic history of forced migration from the western coast of Africa. Of the 10 to 15 million people sent to the New World as slaves, approximately 2 million found their way to the Caribbean— where they would work on sugar plantations, alter the racial composition of the area, and, ultimately, help to establish foundations for the Industrial Revolution in nineteenth-century Europe.

In no other region of the Americas was the destruction of the native population as complete as in the Caribbean. Some Indians managed to escape to the mountains, as in Cuba, but in most places geography was an impediment. The islands were so small that there was nowhere to hide. As in New France and New England, the native population fell victim to virtual elimination.

Aboard the ships came the way of life, the language, the creeds, and political institutions of contemporary Europe. One of the vessels brought some sugarcane cuttings from the Canary Islands (as others had brought domestic animals and plants), and this altered the course of history. Sugarcane grew bountifully, but as prospects for sugar pro-

duction rose, the need for labor became all the more apparent. The demand for slaves seemed almost infinite: as one seventeenth-century witness observed, there was need "for at least eighty to one hundred Negroes, working all the time, and even one-hundred and twenty or more." Sugarcane and Africans came together to disrupt socioeconomic organization and to transform the Caribbean into the new frontier of sixteenth-century Europe.

The discovery of precious minerals in Mexico and Peru promptly distracted Spain's attention from the Caribbean, which became little more than a stopping place on the way to the now-prosperous mainland. Hispaniola, Cuba, and Puerto Rico served as supply stations and military garrisons for the increasing number of ships loaded with gold and silver from the Mexican and Peruvian mines. This income from the New World created the myth of a wealthy Spain, but the long-term reality was that this windfall weakened the mother country—and created temptations for European rivals.

Though the Spanish crown proclaimed authority over the entire Caribbean, it was unable to sustain a commercial and political monopoly. The area was simply too large, royal forces and settlements too thinly dispersed, and the economic stakes too high. By the late sixteenth century and throughout the seventeenth century, the Caribbean Sea was an open and inviting target for privateers and buccaneers, who raided coastal settlements and pursued the royal fleets. Spain's European rivals, especially England, encouraged and sometimes outfitted these pirates; Francis Drake, John Hawkins, and Henry Morgan all became knights of the English realm.

European powers established settlements as well. The English seized Jamaica in 1655. The French took the western half of Hispaniola in 1679. Having occupied northeastern Brazil from 1630 to 1654, the Dutch then moved onto a number of islands off the coast of Venezuela. Little by little, Spain ceded or accepted de facto loss of some of its colonial claims. Caribbean holdings became pawns in European wars, handed back and forth between winners and losers like the proceeds in a poker game.

Meanwhile the demand for sugar was steadily increasing in Europe. Soon the cultivation of sugar not only dominated trade from the Caribbean (controlled largely by the English, Dutch, and French) but profoundly affected the agricultural and racial composition of the islands. In time the islands became overwhelmingly populated by people of African descent—a situation that persists to this day. Only in the larger Spanish islands of Puerto Rico and Cuba, where the emphasis on sugar came relatively late, did those of African origin fail to become a majority.

Another consequence of sugarcane cultivation was the transformation of once-diversified systems of production into single-product econ-

omies, emphasizing sugar for export. Most of their consumption needs had to be imported, from other islands, the mainland, or Spain itself. Only on the smaller islands, such as Grenada, were other products (in this case coffee) more important than sugar. Since most of the original population had died and Spanish settlers did not like to work with their hands, the demand for slaves continued through the eighteenth century.

Of course, the slave trade became highly profitable, and sugar from the New World created a demand for European products that later helped to stimulate the Industrial Revolution. Sugar became the king of the Caribbean until, like other primary products, it faced stiff competition from other parts of the world (including the U.S. South). By the 1850s Cuba alone could produce almost all the sugar needed for export to Europe. The invention of the modern mill, or *ingenio*, with its use of steam and mechanization, decreased the need for slave labor and ultimately set the stage for the abolition of slavery (though planters in Cuba and Puerto Rico remained recalcitrant to the last). By the end of the nineteenth century sugar no longer ruled supreme. With its demise the imperial strategists of great-power Europe turned away from the Caribbean to focus on more lucrative areas of expansion, such as India and Southeast Asia.

The political legacies of conquest and colonization would be complex. With so many European masters, the islands of the Caribbean would have diverse legal and cultural traditions. They would achieve independence at different times and through various means. As we shall see, Haiti would break away from France by 1804; some of the British possessions would not acquire sovereignty until the 1960s and 1970s. Some have been struggling with the challenges of nationhood for nearly 200 years; others are even now following the same generation of leaders who first brought them independence.

Apart from Cuba and Haiti, none of the possessions of the Caribbean engaged in a protracted war for liberation. Because of their small size and geography, they have never developed military establishments. As colonies, they could rely on the military might of their mother countries; as islands, they were at least partly protected from hostile invasions by the sea. With the exception of Haiti and the Dominican Republic, independent nations of the Caribbean have not endured the kind of military domination so apparent in nineteenth- and twentieth-century mainland Latin America.

Finally, the smallness and the poverty of Caribbean islands would keep their governments from becoming strong; unlike the powerful state machineries evident in contemporary Chile or Mexico, they would be chronically weak. With scant resources and modest populations, they would actually constitute "mini-states." Such vulnerability would, in turn, create imposing challenges for this promising but convoluted part of the world.

Overview: Economic Growth and Social Change

The preeminence of sugar in the Caribbean and the requirements for labor led to a continuing threat of worker scarcity. Attempting to delineate a suitable demographic policy for the New World, the Spanish crown tried a number of alternatives. Spain sent convicts and white slaves; allowed for free emigration to the colonies; and enslaved the native Indians, though this led to their decimation. It was the African slave trade that kept sugar in its place of privilege throughout the colonial era.

All other crops became secondary activities. Economic diversification, such as it was, stemmed in part from geographic realities. In the more mountainous islands, such as Guadeloupe, sugar cultivation was difficult, so bananas and other products were raised. Tobacco appeared on small farms among the rolling hills of Cuba and on some other islands. But the rush to profit from the production of sugarcane caused the diversion of most of the flatlands into vast plantations, a tendency that reached its most extravagant point in Barbados, whose residents had to import almost their entire food supply.

The loosely organized society of the sixteenth century, dominated by whites and small-household production units, gave way to a strictly organized and hierarchical society of masters and slaves by the seventeenth century. Production was firmly controlled by the mother countries. With the exception of England, each European country formed its own trading company: in addition to the *casa de contratación* of Spain, there were the Dutch West Indies Company and the French *Compagnie des Isles D'Amerique*.

Perhaps the primary social consequence of these transformations was the creation of a rigid system of racial stratification. Virtually everywhere a three-tiered pyramid existed: whites at the top, browns in the middle, and blacks at the bottom. As whites eventually withdrew and Indians disappeared, the African heritage became dominant.

European demand for sugar permitted many of the settlers to make large fortunes, which they used to build great manorial houses and to purchase acceptance into the political and social life of the mother country. As in Mexico and Peru, the colonists never felt at ease on the islands; most longed to return home, and, in fact, some went back to positions of power and prominence. If there appeared a plantation aristocracy in some parts of the Caribbean, it was not a deeply rooted one.

The emphasis on sugar not only destroyed the once-diversified local economies but also consolidated the dependence on a single market. Most trade would take place with the mother country alone. In this way the agricultural production of the New World came into contact with the emerging manufacturing centers of Europe. By the twentieth century, most Caribbean trade would be with the United States.

Exploitation led to occasional rebellion by slaves, some of whom took to the mountains and created runaway "maroon" communities. (The term apparently comes from the Spanish word for a runaway steer, *cimarrón*, later used to mean fugitive in general.) From there they would raid plantations for goods, arms, and even women. White planters responded to this threat with a combination of persuasion, negotiation, and paramilitary force.

As sugar production declined and the population grew, countries in the Caribbean began making efforts to diversify their economic bases. Coconuts, spices, bananas, citrus fruits, and pineapples have acquired an increased role in agricultural production. (Sugar remains a major export for Jamaica and the Dominican Republic.) In countries such as Jamaica and Trinidad and Tobago, two mineral commodities—bauxite and petroleum—have gained prominence. Other islands now have viable industries in textiles, manufacturing, financial services, assembly plants, and tourism. For some these activities represent the principal source of foreign exchange.

However, these industries do not provide much employment, and, outside of postwar Puerto Rico, manufacturing has not played a major role in the area. The scarcity of job opportunities has led to a massive outmigration—to other islands, to the United States, and to parts of Europe. Indeed, some have said that the Caribbean exports not only its products but also its people. As a result, the islands of the Caribbean have never formed an industrial working class. Here, in contrast to such nations as Argentina and Brazil, there is hardly any proletariat.

The Caribbean mini-states remain largely rural as well. There are no major cities; Santo Domingo, by far the largest, had around 800,000 inhabitants in the early 1980s. Most people live in the countryside. About three-quarters of the population of Haiti and over half of the population of Jamaica still reside in communities with less than 2000 inhabitants. The Caribbean has neither the problems nor the amenities of modern city life.

As a result of all these developments, societies in the Caribbean are relatively "classless." With blacks as the predominant cultural element and with still-underdeveloped economies, these tend to be homogeneous societies. There are diverse social layers, perhaps most evident in Haiti and the Dominican Republic, but the kind of ethnic aristocracy that characterized the colonial era no longer exists. This fact is expressed in a bitter joke: "All Caribbeans have equal access to their fair share of poverty."

In confronting the challenges of economic and social development, the Caribbean has witnessed two dominant experiences or models: those of Puerto Rico and Cuba. Under the name of "Operation Bootstrap," the Puerto Rican model began in the early 1960s as an example for not only the Caribbean but for all of Latin America: its pillars were close cooperation with the United States and reliance on foreign invest-

ment. Although there were some impressive statistical achievements, the model revealed inherent weaknesses as unemployment swelled, and growing numbers of Puerto Ricans migrated to New York and other parts of the United States.

The Cuban model, described in Chapter Eight, entailed the construction of a socialist "command" economy. Despite its social achievements, the political and economic costs have discouraged acceptance as a revolutionary example by other countries in the area. Instead, there have been efforts to find a third alternative—one attempt was made on Jamaica and another on the tiny island of Grenada. Both of these would fail.

Social and economic development in the Caribbean has remained under international influence ever since the sixteenth century. The idea of a common organization to unify the region and to consolidate developmental gains took form in the creation of the Caribbean Community and Common Market (CARICOM). Long subject to internal ideological and political fragmentation, CARICOM has shown signs of revitalization in the 1990s. In an additional display of solidarity, island nations (including Cuba) joined with neighbors in Middle America and northern South America in the mid-1990s to form the Association of Caribbean States.

Haiti: Slave Republic, Voodoo Dictatorship

What is now Haiti, on the island of Hispaniola, was once one of the most prosperous overseas possessions of France; today it is one of the poorest countries in the world. With a population of about 6.7 million, Haiti has a per capita income of approximately $370.

The island's original inhabitants were almost entirely replaced by African slaves imported to work on sugar estates. During the French Revolution Haiti's residents were granted full citizenship, a move that white estate owners resented. Resulting conflicts led to a wave of rebellions. This time the slaves wanted not only personal freedom but national independence as well.

Under the leadership of Toussaint L'Ouverture, the blacks of Haiti revolted in 1791 and in 1804 declared national sovereignty. This was to be the second free nation in the Americas and the first independent black country in the world. Although Toussaint led the rebellion he was persuaded to go to France, where he eventually died in an obscure dungeon. It was one of his lieutenants, Jean Jacques Dessalines, who proclaimed the country to be free from colonial rule.

The Wars of Independence broke up and destroyed the large sugar estates. Land was at first worked collectively under a system called the *corvée*, but the highly individualistic tendencies of the postwar period led to distribution of parcels to freeholders. Thus the legacy of large oligarchic landowners, so prevalent elsewhere in Latin America, did

not take root in independent Haiti. Instead, a large number of small holdings replaced the sugar estates, and production decreased drastically. Modern sugar technology has not been suited to small farms and as a result Haiti did not reproduce the economic innovations of turn-of-the-century Cuba.

The country has experienced repeated civil war and foreign intervention. Independence gave power to the blacks, who now form about 90 percent of the population, a fact that light-skinned mulattoes have resented all along. Indeed, the mulattoes constitute a prosperous minority, still clinging to an ideal of French civilization and speaking French on a regular basis. The majority black population, by contrast, speaks a native language, Creole, and finds spiritual inspiration in *vodum*, an eclectic blend of Dahomian religions and Catholicism. Ever since the colonial era a kind of caste system has divided the mulattoes from the blacks, and conflict between the two elements has been a persisting theme in Haitian history.

Internal tension produced increasing instability. From 1804 to 1867 Haiti had only ten chief executives. From 1867 to 1915 there were sixteen presidents, with an average term of only three years. And from 1911 to 1915 Haiti faced one of its most chaotic periods, during which time six presidents met violent deaths.

Confronting World War I and accustomed to "dollar diplomacy," the United States occupied Haiti in 1915 and stayed there until 1934. The invading troops were charged with the general administration of the country. One of their first tasks was to abolish the army and to replace it by a national police force. The financial administration of the country was entrusted to a cadre of technicians and bureaucrats, who ensured payments of all foreign debt obligations (especially those owed to the United States). Some new public works were initiated and old ones were repaired, but the majority of the population regarded the invaders with smoldering resentment.

One reason for this feeling was dismay over the loss of sovereignty, as the United States took over the management of the country and even the administration of the customshouses. (In fact the financial commission would not leave Haiti until 1941, years after the departure of the military garrisons.) Another reason was the marked preference of U.S. officials for the mulattoes, whom they brought to power in a variety of ways—including the superficial election and reelection of Sténio Vincent as president during the 1930s.

In time the black population, backed by the Haitian Guard (as the police force was known), ousted another mulatto president and installed Dumarsais Estimé in 1946. He replaced mulatto officials with blacks and undertook a series of reforms designed to benefit both urban workers and agricultural producers. Estimé discharged the country's debt to the United States and signed an agreement with the Export-Import Bank for the development of the Artibonite Valley. In

Racist depiction of Haiti as a naive, inept black child formed and reflected U.S. attitudes about military intervention. Occasioned by the political and economic crisis of 1915, this cartoon expressed the helplessness of Haiti—through the caption "I'm in for something now!"—and Uncle Sam's determination to take charge. (Hanny, *St. Joseph News-Press*, 1915. Courtesy of the St. Joseph News-Press/Gazette.)

1950 he tried to have the constitution amended so he could remain in power, and for this he was deposed by the army and sent into exile.

Control passed to Colonel Paul E. Magloire, a black leader who was influential within the army and popular among the nation's masses. At his inauguration he promised to safeguard the rights guaranteed by the constitution, to continue irrigation projects and other public works, and to promote improved education. In the international arena Magloire sought good relations with the United States, while the increase in export prices brought on by the Korean War helped to stimulate economic growth. Resented by a group of ambitious rivals, he was overthrown in a coup in 1956.

After months of uncertainty there emerged the figure of François Duvalier, who had himself elected president in September 1957. So began one of the most backward, unfortunate, and tyrannical periods in the history of this land.

Soon after seizing power Duvalier set out to bend the nation to his will. The army, the police, and the security forces became accountable to him alone. He created a special police force which came to be known as the *Tontons Macoutes,* or "bogey men," the most dreaded repressive force in the country. Through sheer terror he rid himself of his opponents and maneuevered elections to become president for life *(président a vie).*

A proponent of *noirisme,* a movement that looked to Africa for inspiration, Duvalier expelled mulattoes from the national bureaucracy. He assumed near-total control of the state, and it is said that he himself appointed members of the security police. He gained influence over the masses by cannily associating himself with the figure of Baron Samedi, the earthly keeper of the *vodum* tombs. He created a sort of latter-day court, whose favorites gained riches through the dispensation of state favors. To institutionalize a system of kickbacks, Duvalier even set up an umbrella organization, the Movement for National Renovation, which collected contributions from business and high government employees for the ostensible purpose of building public facilities. Needless to say, the money was never used for such ends.

Until his death in 1971 Duvalier took the side of the United States in most international arenas, including the United Nations and the Organization of American States. On occasion pro-U.S. votes would lead to increased aid or loans for his corrupt regime. During his tenure Haiti became more and more isolated, a kind of international pariah, notwithstanding his personal desire to establish stronger ties with Africa.

What Duvalier really wanted was the opportunity to rule Haiti according to his own design. Ill-starred invasions by revolutionary forces took place from time to time but always ended in failure. Their principal shortcoming stemmed from their inability to cultivate support among the peasantry, which "Papa Doc" continued to dominate through a combination of fear and mystification. Each attack from abroad led to more brutal repression and to propagation of the official slogan: *Dieu, Duvalier, et le frapeu, un et indivisible*—God, Duvalier, and flag, one and indivisible.

As death neared, Duvalier persuaded the National Assembly to lower the minimum age for president from forty to eighteen and proceeded to install his son as his successor and *président a vie.* Why he selected his son instead of his daughter Marie-Denise, with whom he had a closer relationship, remains a source of mystery and speculation.

Young Jean-Claude Duvalier, or "Baby Doc" as he was sometimes known, inherited a bitterly impoverished country. Though he took some steps to promote development and may have been less brutal than his father, he retained a parasitical group of favorites—a "kleptocracy" of sorts. Government became the means of self-enrichment. His

marriage to a wealthy mulatto woman led to some dissatisfaction among his father's supporters. Popular discontent and internecine struggles continued to mount and finally led to his demise in February 1986, when he boarded a U.S. Air Force plane and departed for France.

He left behind an anguished nation. Between 1980 and 1986 the economy had shrunk by 10 percent. Three out of four adults could not read, one out of every five children died before the age of five. The nation's per capita income was barely over $300. It was about half that level in the countryside. Thirty years of rule by the Duvaliers had made Haiti the poorest country in the Western Hemisphere.

Political recovery was tentative. For decades the opposition had been suppressed, labor unions controlled, and the media corrupted. As Baby Doc left the country, there were cries for liberty and calls for *dechoukaj*, an "uprooting" of the Duvalier regime: tombs and statues fell, policemen felt popular wrath, and collaborators with the dictatorship fled from office. A transitional government was formed under Lieutenant General Henri Namphy, a seemingly apolitical officer—who had, however, served as army chief-of-staff under Duvalier. Elections scheduled for late 1987 resulted in a bloodbath, as paramilitary forces assaulted voters and opposition candidates. A subsequent ballot resulted in the controversial election of Leslie Manigat, a well-known social scientist, but Namphy threw him out of office in the spring of 1988. He himself was soon replaced by General Prosper Avril, an ambitious young military officer who revived the *Tontons Macoutes* and imposed a wave of repression. To many observers it appeared that Haiti was suffering from "Duvalier without Duvalier."

Authentic change began in 1990. Early in the year protest demonstrations and a general strike persuaded General Avril to leave the country. Under a woman interim president, Ertha Pascal-Trouillot, open elections took place in December 1990. Emerging with two-thirds of the vote was Jean-Bertrand Aristide, a Roman Catholic priest who espoused liberation theology and advocated far-reaching political and social change. In January 1991 disgruntled "Duvalierists" attempted a military coup to prevent the "Communist" Aristide from taking office: the effort failed but left 74 dead and 150 injured. Since his party did not win even a plurality in the legislature, Aristide had to construct an effective ruling coalition from a position of weakness. Late in 1991, unruly elements within the military ousted him from office. The United States and other nations promptly condemned the coup, and the OAS slapped an embargo on trade with Haiti, but diplomatic negotiations for a peaceful solution to the crisis dragged on for years.

As Haitians sought to escape the oppression imposed by the new military regime of General Raoul Cédras, it was the prospect of a large-scale flood of immigrants that gave shape to U.S. policy. The Coast

Guard started picking up thousands of Haitians attempting to reach U.S. shores on homemade rafts, and took them to an encampment at the U.S. naval station at Guantánamo (in Cuba). In May 1992 President George Bush ordered the Coast Guard to return all Haitian rafters to their homeland without any screening for political amnesty. Democratic presidential candidate Bill Clinton denounced the Bush policy as "a callous response to a terrible human tragedy," but then consented to its continuation after the November 1992 election. A UN-sponsored negotiation nearly brought a settlement to the impasse in 1993, but collapsed in the face of defiance from the Cédras regime. By early 1994 leaders of the African-American community mounted sharp criticism of Washington's inaction, and Clinton reversed himself by announcing that U.S. authorities would process rafters at sea and grant asylum to victims of political repression. News of the change led to yet another wave of rafters. Despite public skepticism, Clinton began to contemplate the use of military force. In mid-September Clinton denounced the Cédras government as "the most violent regime in our hemisphere," and stressed the dangers of inaction: "As long as Cédras rules, Haitians will continue to seek sanctuary in our nation. . . . Three hundred thousand more Haitians, 5 percent of their entire population, are in hiding in their own country. If we don't act, they could be the next wave of refugees at our door. We will continue to face the threat of a mass exodus of refugees and its constant threat to stability in our region and control of our borders."

As tension mounted, Clinton dispatched a high-level delegation under former president Jimmy Carter for a last-ditch effort at negotiation. At the final minute, as U.S. troops were already en route for an invasion of Haiti, Carter reached an agreement with the Cédras government. Clinton canceled the invasion but sent instead an occupation force; in less than a week there were more than 15,000 American troops on the ground. Aristide returned to office in mid-October and the U.S. occupation gave way to an international peacekeeping force in early 1995. It remained to be seen whether Haiti could, at last, develop and consolidate a meaningful democracy.

The Dominican Republic: Unfinished Experiment

The history of the Dominican Republic shares some dramatic features with Haiti. Both nations together constitute the island of Hispaniola, and they have been intertwined ever since the Spaniards first arrived. It is in the Dominican Republic that the failure of Spanish invaders to establish a strong settlement is most notable; here the notorious *encomienda* system got its start.

Decisions abroad have largely determined the fate of this region, most notably the Treaty of Ryswijk in 1695, when the entire island was

ceded to France without the slightest consultation with the colonists. During the movement for Haitian independence at the end of the eighteenth century, revolutionary forces took control of the Dominican Republic, an act that still arouses animosity between the two countries. Spanish colonists eventually regained control and then precipitated a long period of *caudillo* wars.

The strategic position of Hispaniola made the island important to the United States, committed by the early nineteenth century to keeping European powers from intervening in the hemisphere. Anarchy and chaos have at various times prompted the United States to intervene. From 1916 to 1922 U.S. marines occupied the Dominican Republic (as well as neighboring Haiti). As in Haiti, a National Guard was created to fight guerrilla bands. Among the most brilliant disciples of the American occupational force was Rafael Leonidas Trujillo, an ambitious soldier who would eventually become one of the most ruthless dictators in the hemisphere.

Thanks to the economic stimulus of World War I, which boosted prices for sugar and cocoa, economic conditions improved in the Dominican Republic during the American occupation. As in Haiti, the U.S. troops strengthened the country's infrastructure, upgrading the educational system and imposing control on public finances. Critics nonetheless began to complain about the "dumping" of inferior U.S.-made products on the local market and about the general disdain the invaders displayed for local citizens.

An agreement between the United States and Dominican leaders in 1922 led to the formation of a provisional government. Two years later elections gave power to Horacio Vázquez, a politician of long standing. He completed a peaceful term in office. Foreign investors came to the island and bought small businesses, woodlands, plantations, and estates. The National Guard maintained law and order. In 1929 Vázquez made the error that has plagued so many of the leaders in Latin America's history: he tried to revise the constitution so he could run for office again.

A rebellion erupted, and Trujillo presented himself as a candidate in the 1930 elections. Wielding his power base (the National Guard), he made it clear that he would win at any cost and claimed victory with 95 percent of the vote. He quickly began banishing political opponents from the scene. The future belonged to Trujillo, and he would rule the nation without mercy until his death in 1961.

As with so many dictators, Trujillo exploited the country's resources in order to amass his own personal wealth. During the 1950s the average annual growth rate was 8 percent, an impressive performance by any standard, but the benefits failed to reach the general population. Much of the nation's income was appropriated and stashed in foreign bank accounts. The masses remained as poor as their neighbors in

Haiti. Contradictions between Trujillo and his coterie of admirers swelled as the economy prospered: the more he took for himself, the more discontented his collaborators became. In 1961 his former friends and cronies, not his enemies, staged a coup against Trujillo and masterminded his assassination.

In 1962 free and fair elections led to the triumph of Juan Bosch, a former journalist and social reformer who sought to confiscate and redistribute Trujillo's landownings as part of a program of agrarian reform. But his efforts at improving the lot of the masses aroused discomfort among the traditional elites, who saw his innovations as dangerously akin to similar actions taken in Castro's Cuba. A military coup ousted Bosch in 1963. A countermovement then sought to reinstate him as president. The resulting conflict led to a civil war between the armed forces and the pro-Bosch "constitutionalists," mainly workers and students. As the struggle intensified, the United States grew fearful of "another Cuba" and took over the country in April 1965. The invading force consisted of 22,000 marines, a contingent whose size amazed even American civilian officials on the scene.

Why the United States deployed such massive force remains a subject of debate. The constitutionalists received endorsement from the country's tiny Communist Party, but there is no evidence to suggest that this was a radically left-wing movement as a whole. Instead, the insurrection was an attempt to oust Donald Reid Cabral (the head of the ruling junta and ex-automobile salesman) and to restore Bosch to his rightful place in power.

In early 1965 Bosch's Revolutionary Dominican Party (PRD) and its main rival on the moderate left, the Revolutionary Social Christian Party (PRSC), had reached an agreement to collaborate in efforts to restore constitutional government. Young military officers acting in concert with PRD leaders seized the army chief-of-staff and declared themselves in opposition to the junta. Soon they were joined by intellectuals and professionals, and the constitutionalists began passing out arms to thousands of civilians (estimates range from 2500 to 10,000). It was in this context, as the insurrection swelled, that the United States reached its decision to support the anti-Bosch forces.

To justify its action, the United States tried to engage the participation of other countries from Latin America through the Organization of American States. Favorable responses came only from Paraguay and Brazil, both under right-wing military rulers. The U.S. attempt to form an "inter-American peacekeeping force" not only failed to legitimize the intervention, but it also discredited the OAS as a whole and contributed to the subsequent demise of that institution.

The U.S. intervention led to the formation of an interim government and, eventually, to the holding of elections in June 1966. Victory went to Joaquín Balaguer, an ex-Trujillo official and favorite of the United

States. With full blessing from Washington, the Balaguer government implemented a number of important developmental programs. Housing was built, land was distributed, education was strengthened and improved. Austerity programs reduced severe problems with the balance of payments, and, to assist with these and other challenges, aid from the United States climbed to more than $132 million for 1968. Agricultural production rebounded, and foreign investment responded. As a result of all these factors, economic growth was substantial.

The Dominican armed forces underwent moderate reform, and its most recalcitrant elements were dispatched abroad, often on fictitious diplomatic missions. But the perpetuation of inequality and deprivation led to increased political polarization, most notably shown by the leftward tilt of ex-president Bosch and his PRD.

The tentative and gradual transition toward democracy nonetheless continued. Free elections survived minor threats in 1970 and in 1978, when the armed forces threatened to annul the results, but on both occasions the outcome was eventually allowed to stand. Balaguer's opponents won the elections of 1978 and 1982.

A deteriorating trade imbalance and mounting external debt led to a serious internal crisis in the mid-1980s, just as the country was preparing for the election of 1986. Polarization intensified, but in a heated campaign victory went once again to Joaquín Balaguer. In 1990, at the age of eighty-three, Balaguer won another term and declared his intent to stay in politics for the remainder of his life. In 1994 he triumphed yet again, to the displeasure of the United States (which resented his tacit support of the Cédras regime in neighboring Haiti). Upon taking office Balaguer agreed to reduce his term to only two years and to promise not to run in 1996. This effort at conciliation may have been too little and too late. In a society divided by color and class, he had appealed to racial prejudice by suggesting that his dark-skinned opponent, Francisco Gómez Peña, wanted to reunite the Dominican Republic with Haiti.

From the mid-1980s onward the Dominican Republic sought in some ways to imitate and modify the Puerto Rican model of development by incorporating elements from contemporary Asia. Taking advantage of low-cost labor, the idea was to transform the country into a kind of Singapore. Despite some spurts of economic growth, however, poverty continued: by the mid-1990s about one-quarter of the adult population was unemployed, and the infant mortality rate was one of the highest in the hemisphere. Occasional riots revealed the potential for violence. "We are divided," as a newspaper editor observed, "and anything could bring this division to civil war. The United States should be careful, because it could have a bigger problem than Haiti if this country explodes."

Jamaica: Runaways and Revolutionary Socialism

Jamaica, "the land of the rivers," has one of the most fascinating histories of the Caribbean. African slaves on the sugar plantations rebelled soon after arrival and fled to the mountains. There they created viable and self-sustaining fugitive maroon communities. The economic base of this alternative society was small peasant agriculture. The maroon legacy has since played a major role in the cultural traditions of Jamaica. It is said, for instance, that the legendary "Nanny" of the maroons was the mother of all free Jamaicans.

Particularly after 1870, Jamaica began to turn away from monocultural dependence on sugar and to develop other agricultural products. The island's varied climate permitted the raising of cacao, sea-island cotton, and bananas—which became a special favorite of Jamaican planters, small and large landholders alike. As a result banana production did not lead to the displacement of smallholders, as sugar had done centuries before. Only population growth proved a challenge to the smallholder, as productivity could not keep pace with the increasing needs of a larger population.

As in the rest of the Caribbean, modern Jamaica took shape during and after the 1930s. Strikes and demonstrations broke out against white minority rule, and such leaders as Marcus Garvey emerged to articulate the aspirations of the masses. Garvey founded the "Back to Africa" movement, which had strong repercussions in the United States, and he became a prominent figure in the Harlem Renaissance of that era. His organization had the purpose of promoting fraternity and unity among blacks in many lands, improving conditions in black communities, helping to "civilize" African tribes, and, ultimately, founding one or more independent black nations, preferably in Africa. Sometimes jailed for his activities in the United States, Garvey returned to Jamaica where he was elected several times to the city council of Kingston. He remained a prominent figure in U.S. black artistic and intellectual circles, which included such luminaries as Langston Hughes, Duke Ellington, Lena Horne, James Baldwin, and Richard Wright.

Jamaica gained independence from Britain in 1962. The struggle for sovereignty required and produced political leaders of renown. Two of the most important among them were Norman Manley and Alexander Bustamante. Working from opposite ends of the political spectrum, they and their followers helped forge one of the most dynamic political systems in the region.

Together with Trinidad and Tobago, Jamaica constitutes one of the "English-speaking giants," largely because of vast resources of bauxite and oil. In addition, two-thirds of the island is covered by a blanket of tertiary limestone. This has alleviated the dependence on agriculture, which is more varied in Jamaica than in most other parts of the Carib-

bean (the country still produces sugar, of course, used partly to make high-quality rum). In principle, this combination of strategic materials and diversified agriculture would appear to provide the basis for continuing economic development.

In practice, however, Jamaica has suffered greatly from the oscillations of the world economy. After the OPEC-induced oil shock of 1973, the country lacked the foreign exchange needed to cover its increasing oil consumption and, like so many other nations, began to borrow heavily abroad. By 1978 the foreign debt reached the staggering figure of $1.4 billion and by 1980 it had climbed to $1.9 billion.

Between 1972 and 1980 the government of prime minister Michael Manley (Norman's son) tried to establish a welfare state within a democratic context, one that would differ significantly from both the discredited Puerto Rican model and the unproven Cuban model. As Manley conceived it, "democratic socialism" would not be a transitional stage toward communism but an end in itself. It would respond to and respect the cultural particularities of Jamaican society. The state would control the "commanding heights" of the economy in order to direct and ensure development, but the private sector would play a central role. As the minister of finance would describe the concept:

> In our mixed economy the primary emphasis, the primary impetus for economic growth, comes through the private sector. They carry the substantial burden of economic growth. For this to take place, the private sector needs a number of things, in particular it needs capital and of course it needs skill. Insofar as capital is concerned, the government has at its disposal various institutions for assisting in the process of capital formation and equally important, in the distribution of such capital as it is accumulated to those areas of production which are likely to be most beneficial for economic growth.

Democratic socialism would also be a multi-party competitive system that would strive to prohibit all forms of exploitation. Moreover, the assertion of national self-determination would lead to a reduction of dependency on external forces and markets.

Reaction to the ideals of democratic socialism was mixed. The chamber of commerce vigorously denounced the state's proposed "incursions" into the private sector, and the business community as a whole was unenthusiastic. But there was support among the workers and the masses, as shown by the electoral strength of the People's National Party (PNP) standing behind Manley and his programs. At least for a while, the strategy appeared to work: the economic conditions of the laboring class improved between 1972 and 1975, but then started to deteriorate again. By April 1976 the unemployment rate was up to 20.5 percent.

Manley and the United States were on a collision course. As though his domestic policies were not enough, Manley established diplomatic

and commercial relations with East Europe and other socialist countries. He was instrumental in the formation of the bauxite producers' association. He took trips to Cuba and to Africa, where he denounced racism and imperialism and proclaimed allegiance to the nonaligned movement of the Third World.

Opposition mounted abroad (especially in the United States) and at home. Economic conditions weakened the base of the PNP's support. The PNP made mistakes, of course, but might have been able to rectify its errors with a modest amount of economic support from international organizations, particularly the World Bank and the International Monetary Fund. A technical error in the calculation of the foreign-exchange potential of the economy set the stage for a well-orchestrated campaign accusing the government of mismanagement and lack of leadership. Tension mounted as elections approached in 1980. In a country of barely 2 million citizens, nearly 1 million were legally registered, and of these, 87 percent went to the polls. The winner was Edward Seaga, of the moderate Jamaica Labor Party (JLP), who received 58 percent of the tally; Manley got 41 percent.

With a palpable expression of relief, the international community offered to Seaga the help it had denied Manley. The IMF agreed to liberal terms for refinancing the Jamaican debt. Mexico and Venezuela assisted with the supply of oil (as they were also doing for Central America). U.S. aid increased, and David Rockefeller chaired a special committee to mobilize private investment for Jamaica. U.S. President Ronald Reagan singled out Seaga and his country for praise in launching his ultimately stillborn Carribbean Basin Initiative.

In spite (or because) of this aid, Jamaica's foreign debt escalated to $4.5 billion by the end of the decade, making Jamaica one of the largest per capita debtors in the developing world. And while economic growth continued, Seaga allowed a sharp deterioration in social services. During the 1980s per capita income declined by nearly 6 percent, sliding to less than $1300 by 1989, a figure even lower than the level of 1970.

In the 1989 elections, Michael Manley's PNP returned to power with a decisive 57-to-43-percent margin over Seaga's JLP. As in the 1970s, Manley received especially strong support from manual workers and the poor. This time, however, Manley presented himself to the international community as a political moderate with extensive contacts in Europe and the United States. He distanced himself from Fidel Castro, sought a rapprochement with the United States, courted foreign capital, entered into negotiations with the IMF on a debt-payment strategy, and announced his intention to privatize state-owned industry. Ironically, it was Seaga, his conservative opponent, who denounced Manley's sale of government shares in the Jamaican telecommunications industry as "a national disgrace."

In 1992 Manley retired as prime minister and turned over power to

his long-time political heir, P. J. Patterson, who accelerated the move toward free-market economic policies. In March 1993 elections the PNP triumphed with a 55 percent majority over a weakened and divided JLP. By 1995 inflation was down to single digits, but serious problems remained. One was the challenge from drug trafficking and the violence and corruption that it spawned. Another was unhealthy reliance on only two industries, mining and tourism, for economic growth. A steady stream of Jamaicans continued to migrate to the United States. Amid these circumstances, uncertainty prevailed. As P. J. Patterson observed, "The greatest danger is that we might become forgotten. We have to avoid the danger of becoming marginalized."

Puerto Rico: From Settler Colony to Capitalist Showcase

Puerto Rico became part of the United States as a result of the Spanish-American War. In July 1898, in retaliation for the sinking of the U.S. vessel *Maine,* American troops landed on the bay of Guánico, initiating the country's first act of colonial expansion. Puerto Rico thus became the pawn in a war between Cuban patriots and Spanish garrisons. It had not anticipated occupation.

Quite the contrary. Spain had already agreed to grant Puerto Rico autonomy the year before, and preparations were under way to devise some sort of "home rule" for the island. The U.S. invasion changed all of this. Suddenly, Puerto Rico became a crucial factor in U.S. global strategy—not only because of its potential for investment and commerce but also because of its geopolitical role in consolidating U.S. naval power.

The idealistic desire to expel Spain from the hemisphere no doubt played an important part in the U.S. decision to move into Puerto Rico. After all, Spain appeared to represent the most reactionary elements of European society. A colonial monarchy, Spain stood for everything the United States claimed to oppose, and it represented a continuing violation of the Monroe Doctrine. Popular acceptance of the "black legend" about Spanish atrocities in the Americas further contributed to this conviction.

But there remains a basic question: Why did the United States take Puerto Rico as a colony while helping Cuba achieve independence? The difference may well reside in the histories of the two islands. There was a longstanding, armed insurrectionary movement against Spain in Cuba, an island which would have been much more difficult to invade. Puerto Rico, however, was on the way to a negotiated settlement, and therefore could present less resistance to outside forces. While in the course of these negotiations, Puerto Rico became caught in a complex struggle between major powers and Cuba's insurgents.

Turn-of-the-century Puerto Rico bore clear signs of Spanish domination. During the colonial period the island had served as an important

military garrison and commercial center, a role that intensified as the slave trade reached its peak in the 1700s. Sugar production became the predominant agricultural enterprise. There were also small farmers, *jíbaros*, rugged individualists who cultivated staple crops and helped maintain a diversified economy. Because of this, the slave population always remained a minority. The destruction of the peasant economy would not come about because of colonial emphasis on sugar, as happened elsewhere in the Caribbean, but because of technological innovation in the twentieth century.

Ever since the arrival of the marines, Puerto Rico has had a peculiar relationship with the United States. After 1898 residents of the island had no clear status at all. In 1917 they were granted citizenship in the United States. In 1947, nearly a half-century after the invasion, Puerto Rico was permitted to attempt self-government. In 1952 the island was granted "commonwealth" status within the United States. This remains an ambiguous situation: Puerto Rico is neither a colony nor a state but something in between.

To develop the island and to provide an inspiration for Latin America, the United States collaborated with dynamic governor Luis Muñoz Marín to undertake "Operation Bootstrap" during the 1950s and 1960s. Under this plan the U.S. federal government would encourage investments in Puerto Rico through a series of tax holidays and other allowances. This would stimulate investment in basic infrastructure and in the improvement of the labor force. The result was to attract relatively small industries, mostly of the labor-intensive type.

Operation Bootstrap wrought tremendous changes in the social and economic life of Puerto Rico. Sugar estates and small farms were replaced by factories; as industrialization thrived, citizens joined the ranks of the laboring class. But the overseas investments did not provide enough jobs to absorb the growth in the working-age population, and the result was massive unemployment.

One consequence of this trend was to accelerate the flow of migrants to the U.S. mainland, where 40 percent of Puerto Ricans now reside. Fully one-half of the migrant population—that is, 20 percent of the total—now lives in New York City. In a sense there are now two Puerto Ricos: one on the island and one on the mainland. There is movement and communication back and forth, of course, but social tensions and cultural differences separate the two communities. As though in demonstration of this fact, Puerto Rican residents of New York are sometimes known as "*neo-ricans.*"

Political life on the island is active and orderly. The chief executive is the governor, who is elected every four years. The dominant issue has been the island's relationship with the United States. In a 1967 plebiscite on this question, 60 percent favored the continuation and improvement of the commonwealth status, and 38 percent came out in favor of statehood. Those who favored complete independence chose

to boycott the plebiscite, but this faction has been vocal and visible (in 1947, in fact, a pro-independence group made an attempt on the life of U.S. President Harry S Truman).

The pro-statehood forces, represented by the New Progressive Party (PNP), won gubernatorial elections in 1968, 1976, and 1980. Under the leadership of Carlos Romero Barceló, this group subscribed to the belief that full statehood would provide working-class Puerto Ricans with increased access to federal welfare programs, stimulate economic growth, and remove the stigma of "second-class citizenship" associated with commonwealth status. Popular support for this movement came especially from urban areas. Romero Barceló was himself mayor of San Juan before serving as governor.

The pro-commonwealth party, or Popular Democratic Party (PDP), won the elections of 1972, 1984, and 1988. Its most prominent leader was Rafael Hernández Colón, who believed that some measure of autonomy is necessary to preserve the island's historic and cultural identity. Within the commonwealth relationship Hernández Colón also called for a greater degree of meaningful autonomy. As governor Hernández Colón actively promoted worldwide economic relations for the island and played a leadership role in the development of the "twin plant" concept—dividing the production process into separate parts, with initial phases to be done in some other area of the Caribbean and final assembly in Puerto Rico. To a large extent, this strategy represented a response to changes in the implementation of the U.S. revenue code, which reduced the tax-free havens for U.S. companies in Puerto Rico. The twin-plant idea would allow such firms to extend their operations into other parts of the Caribbean and still retain their privileged tax status.

By the 1980s pro-independence forces were divided between three parties: the Puerto Rican Independence Party (PIP), led by Senator Rubén Barrios; the Puerto Rican Socialist Party, under leftist Carlos Gallisá; and the Popular Boricua Army (PAB), also known as the *macheteros,* and their mainland collaborators in the Armed Forces for National Liberation (FALN). These latter groups called for armed resistance against colonial domination and sponsored occasional acts of organized violence.

Concern steadily mounted over economic issues and, largely as a result of a U.S. recession, Puerto Rico faced a downturn in the early 1990s. Amid this atmosphere the 1992 gubernatorial election went to the PNP's Pedro Rosselló, who vowed to press for statehood. His first act in office was to sign a bill giving English equal status with Spanish as an official language. And in November 1993, fulfilling a campaign promise, Rosselló held a new plebiscite on the island's status. To the surprise of many observers, the pro-commonwealth position won with 48.4 percent of the vote; statehood obtained 46.2 percent; the pro-independence stance got only 4.4 percent. This outcome appeared to

settle the issue. For the foreseeable future, Puerto Rico would continue its curious relationship with the United States.

Lesser Antilles: Struggle of the Micro-States

The Lesser Antilles are a series of small islands to the northeast of Venezuela. Some have gained independence within the last three decades. Others remain under colonial tutelage or in commonwealth status. As in Puerto Rico, the question of national independence continues to persist.

Such is the case with the Turks and Caicos Islands, the British Virgin Islands, the Cayman Islands, and Monserrat. All are tied to the British crown. The economies of most of these islands rely heavily upon agriculture and tourism, with modest inroads by manufacturing of late. Most infrastructure is related to tourism, especially the hotel industry and transportation (including airports). Export products include leather goods, plastic bags, textiles, fiberglass, and electronic components. Fishing is also an important activity and a common source of tension between the islands as stocks become depleted.

The French Caribbean consists of the islands of Martinique and Guadeloupe. Both were settled by the French in the 1630s, and, along with Haiti and other possessions, they formed the core of the mother country's lucrative sugar and coffee interests in the Americas. ("We have lost Canada," a French minister of state would acknowledge after the catastrophic outcome of the Seven Years' War, "but we have retained Martinique!")

In 1946 the residents of Martinique and Guadeloupe voted to unite with France as "overseas departments" *(departements d'outre-mer),* under which arrangement they received the full rights of French citizenship. Local administration is in the hands of a democratically elected council. Pro-independence forces have been mostly on the political fringe, though a significant movement has appeared in Guadeloupe under the leadership of a French-trained physician, Dr. Claude Makouke. As of the moment, most people seem to prefer the stability of French authority to the uncertainties of independence.

Under the socialist government of François Mitterand in France, a political reform gave modest advantages to leftist groups. In Guadeloupe the left captured twenty-four of forty-two seats in parliament, thus becoming a majority; not so in Martinique, where the Progressive Party won only ten out of forty-eight seats. This reveals both a difference between the two islands and the overall moderation of the political scene.

Both Guadeloupe and Martinique rely heavily on agriculture, tourism, and direct aid from France. The educational and health-care systems are of high quality. The public sector employs 33 to 35 percent of the labor force. Unemployment nonetheless remains high, around

28–30 percent, and many residents migrate to France in search of jobs. The future of these islands will continue to depend on Paris, though the United States has recently begun to heighten its presence in the area.

The Netherlands Antilles are St. Martin, St. Eustatius, and the so-called ABC group—Aruba, Bonaire, and Curaçao—all of which form an administrative federation with headquarters in the city of Willemstad (Curaçao). The official head of state is the queen of Holland, represented in the islands by a governor and a prime minister. The islands are self-governing, though Holland retains responsibility for external and military affairs. The federation has nonetheless begun to pull apart. Aruba has long sought national sovereignty, and will attain this goal in 1996.

The islands are heavily dependent on tourism, offshore banking, and oil refining. In the late 1980s the Exxon Corporation announced its decision to leave the islands, however, given the decline in demand for fuel in the northeastern area of the United States. Changes in U.S. laws threatened to challenge the offshore banking industry, which could lead to revenue losses of 25 to 40 percent for the local government. And tourism from Venezuela was hurt by the devaluation in 1983 of that country's currency, the bolivar, an event that made vacations in Aruba and elsewhere much more expensive for Venezuelan travelers. Economic and political prospects for the Netherlands Antilles thus had an air of uncertainty.

The largest single nation in the Lesser Antilles is Trinidad and Tobago, first colonized by Spain and then taken over by the British in 1797. The islands were administered as a British crown colony until 1962, and they gained full independence as a republic in 1976. A diversified economy (including oil production) at one time provided the islands with a per capita income of approximately $7000. This was by far the highest in the region, though unemployment and inflation were constant problems. By the mid-1980s per capita income had declined to less than $3000.

The main spokesman for the independence movement in Trinidad and Tobago, Eric Williams, became one of the most famous politicians in the Caribbean. Under his leadership the People's National Movement (PNM) won every one of the seats in the local council in 1971, and he became prime minister on a crest of popularity. Conflict with organized labor prompted challenges to his authority, however, and the persistence of chronic unemployment helped encourage the formation of new parties on the left. Williams died in 1981, still a revered figure, and he was succeeded by his minister of agriculture.

Mounting economic difficulties in the mid-1980s finally led to a loss of control by the PNM and to the victory of the National Alliance for Reconstruction under the leadership of A. N. Robinson. Plans were under way to increase tourism and foreign investment. Even so, Trini-

dad and Tobago remains determined to safeguard the political and economic independence achieved with so much effort.

The most dramatic event in the political life of the Lesser Antilles occurred on the tiny island of Grenada in October 1983: military invasion by the United States. Precipitating this action was a series of internal events which culminated in the assassination of several government members and the prime minister himself, Maurice Bishop. The public justification for this intervention offered by the Reagan administration was the need to protect U.S. citizens on the island, including students in a local medical school. This explanation was reinforced by a request for decisive action by conservative governments in neighboring countries, including the Seaga team in Jamaica. But the real reason, according to many observers, was the leftward tilt of Grenadian policy in domestic and especially international affairs.

The story begins with the formation of the People's Revolutionary Government (PRG), a populist movement that developed in reaction to the corrupt and irresponsible leadership of Eric Gairy after the achievement of self-government under British rule. Under the charismatic leadership of the popular Maurice Bishop, the PRG won power in 1979 and undertook a number of reforms. New institutions replaced the obsolete and decrepit state apparatus; a labor code established the legality of unions and led to a massive rise in union membership; organizations everywhere encouraged daily participation in public and local affairs. In an effort to diversify the economy, largely dependent on the export of a single product (nutmeg), Bishop and his advisers sought to strengthen the country's infrastructure—improving roads and cultivation techniques—and to explore new methods of marketing and packaging. The PRG also expanded the role of the state in economic affairs, quadrupling public-sector investments in comparison to the Gairy regime. Apparently as a result of this stimulation, real wages grew, employment increased, and production went up by 5.5 percent in 1982.

Grenada sought to develop tourism as well, and this became a major bone of contention. A key requirement would be the construction of a new airport, one that could handle jetliner traffic from Europe and the United States. But as the project proceeded, the Reagan administration began to depict the airport as a military threat, claiming that it would be used by Soviet or anti-U.S. forces. The charge was never substantiated but became a dominant theme in regional controversy over the subject.

It was in foreign affairs that the PRG made its boldest moves, aligning itself with Cuba and declaring solidarity with revolutionary movements throughout the Third World. This entailed the cultivation of fairly close ties with the Soviet Union and with East Europe. The U.S. government reacted negatively to these developments and began issuing ominous pronouncements about the economic and military sig-

nificance of the Caribbean sea lanes. Once again, the strategic location of the area would make it a pawn in the global arena.

The Reagan administration attempted—without success—to dissuade West European allies from extending aid and support to Grenada. On one occasion the United States offered to provide funding for the new airport on the condition that the United States could lease it for ninety-nine years. Bishop angrily responded that his country would be "nobody's backyard" and promptly gave the contract to a British firm.

Internecine struggles within the government of Grenada mounted in the fall of 1983. In early October a group of hard-line radicals, under Bernard and Phyllis Coard, seized power from Bishop and led him to a brutal execution. Chaos ensued and a power vacuum was created. The Reagan administration saw its opportunity and decided to invade. On October 25 a contingent of U.S. troops landed on Grenada, overwhelmed modest resistance, and quickly subdued the island.

The eventual challenge was the restoration of political legitimacy. Though Great Britain retained a semblance of formal authority, as Grenada was a member of the Commonwealth, the United States exercised de facto power. A new election finally took place in December 1984, just over a year after the invasion, and victory went to a moderate element known as the National Party. Subsequent elections were desultory affairs and produced no significant statesmen.

The country endured a slow and painful transition toward a market economy throughout the 1980s and early 1990s. Export earnings declined because of crop diseases and falling world prices (for nutmeg, bananas, and cocoa). Unemployment rose to 40 percent, and the number of people below the poverty line actually doubled. In early 1995 Nicholas Brathwaite finally stepped down as prime minister and subsequent elections brought to office Keith Mitchell, a gregarious populist of the center-left New National Party. As people of this tiny island looked forward to economic change, it appeared that Grenada remained in a kind of political limbo: it was neither independent nor a colony.

By the 1990s small countries of the Caribbean, often known as "micro-states" or "mini-states" because of their modest size and scant resources, faced daunting challenges of governance. States found it increasingly difficult to impose economic and social policy and to establish effective authority. An illustration of this problem came in July 1990, when 113 members of a fringe Muslim group attempted a coup d'etat in Trinidad and Tobago, holding the prime minister and much of the cabinet hostage for several days. Though ultimately defeated, the plotters managed their effort with only 130 weapons, most of them purchased in Florida gunshops. Perhaps more insidious, in the long run, was the arrival of drug traffickers who sought to use Caribbean

islands as a trans-shipment corridor for marijuana, cocaine, and heroin. "It would not be difficult for drug people to take over a political party, target key constituencies and gain power," said the prime minister of St. Lucia. "Then you would have the entire paraphernalia of government serving local or international drug interests." To this degree, the sovereignty of these island-states became a concern for the entire hemisphere.

TEN

CENTRAL AMERICA
Colonialism, Dictatorship, and Revolution

Central America has received scant attention from scholars in the United States. This neglect is partly due to the relative paucity of archives, libraries, and research centers in the nations of the isthmus. It is partly due to the smallness of the individual countries, which makes them appear less significant than Argentina, Brazil, or Mexico. And it is also due to the common assumption that the countries of Central America are backward: the least developed area in a developing world. Dominated by dictators, the "banana republics" of the isthmus were viewed as sleepy relics of the past, the last places where popular revolt would strike.

Upheavals in Nicaragua and El Salvador have sharply challenged this image. If we discard modernization theory and instead follow an approach which places Central America within the context of the capitalist world economy, the history of the region begins to acquire new meaning. As we shall see, Central America came to develop classic plantation economies, and this fact had a decisive effect on social-class relations and political outcomes. The isthmus thus provides yet another variation on the interplay between socioeconomic change and political transition.

Central American history furthermore offers an opportunity to examine and comprehend the policies of the United States. Through trade, investment, invasion, and diplomacy, the United States obtained extraordinary influence over trends and events in the region. The use (and abuse) of this power not only yields insight into the behavior of the United States. It also enriches our understanding of the ways that Latin Americans have interpreted the motives and actions of the "colossus of the north."

Colonial Background

Nature endowed Central America with singular beauty. And from Guatemala to Panama, the isthmus exhibits many contrasts: a spectacu-

lar mountain range, studded by volcanoes of 10,000 feet or more; some
arid zones; and verdant jungles along the coasts. Much of the soil is
fertile and the year-round climate is temperate, warm in the mornings
and cool in the afternoons. There are lakes in the mountainous areas
but no major navigable rivers (with the possible exception of the Río
San Juan in Nicaragua). Nor does either coast have adequate natural
deep-sea harbors. Nature can bring calamity, too, through torrential
rains, hurricanes, and violent earthquakes.

During precolonial times Central America was a meeting ground for
Meso-American, South American, and Caribbean cultures. In contrast
to Mexico and Peru, it was not the site of a centralized Indian empire.
Indigenous peoples lived in stable, autonomous communities and engaged in trade with one another. After 500 B.C. a relatively advanced
civilization appeared in the highlands of Guatemala and El Salvador,
and it was greatly influenced by Olmec culture from the Veracruz-
Tabasco coast of Mexico. Nahuatl settlements later followed, and classic
Mayan culture appeared in the lowlands of northern Guatemala. The
period from A.D. 600 to 900 marked the apex of the Old Maya Empire,
as it was formerly called, though it did not constitute a highly organized political unit.

Spaniards first came to the area in 1501. Vasco Núñez de Balboa
sighted the Pacific Ocean in 1513 and established his power in what is
now Panama. In the 1520s, already under pressure from Spanish
crown authorities, Hernán Cortés went as far south as Honduras. Pedro de Alvarado launched an expedition from Mexico City around this
same time, and, like other conquerors, he was able to take advantage
of hostilities between two Indian groups, the Quichés and the Cakchi-
quels (who became his allies). As also happened elsewhere, the conquerors soon fought among themselves. Alvarado's column reached a
stalemate in its encounter with the forces of Pedro Arias de Ávila, better known as Pedrarias Dávila, who founded Panama City in 1524.

In the mid-sixteenth century the Spanish crown established the Kingdom of Guatemala as part of the viceroyalty of New Spain. The kingdom included what later became Costa Rica, Nicaragua, El Salvador,
Honduras, Guatemala, Belize (British Honduras), and the Mexican
state of Chiapas. Its capital was the highland city of Antigua, Guatemala, then referred to as Santiago de Guatemala. Panama fell under
the jurisdiction of Peru.

The diversity of Indian cultures meant that Spaniards penetrated
Central America in stages, not all at once, and each conquest required
the establishment of a new government. The result was decentralization. Municipalities assumed day-to-day authority, and town councils
(*ayuntamientos*) became the most important governing bodies. Nomi-
nally under the control of distant viceroys, Spanish residents of the
isthmus functioned under separate royal orders for all intents and purposes.

MAP 4 Central America

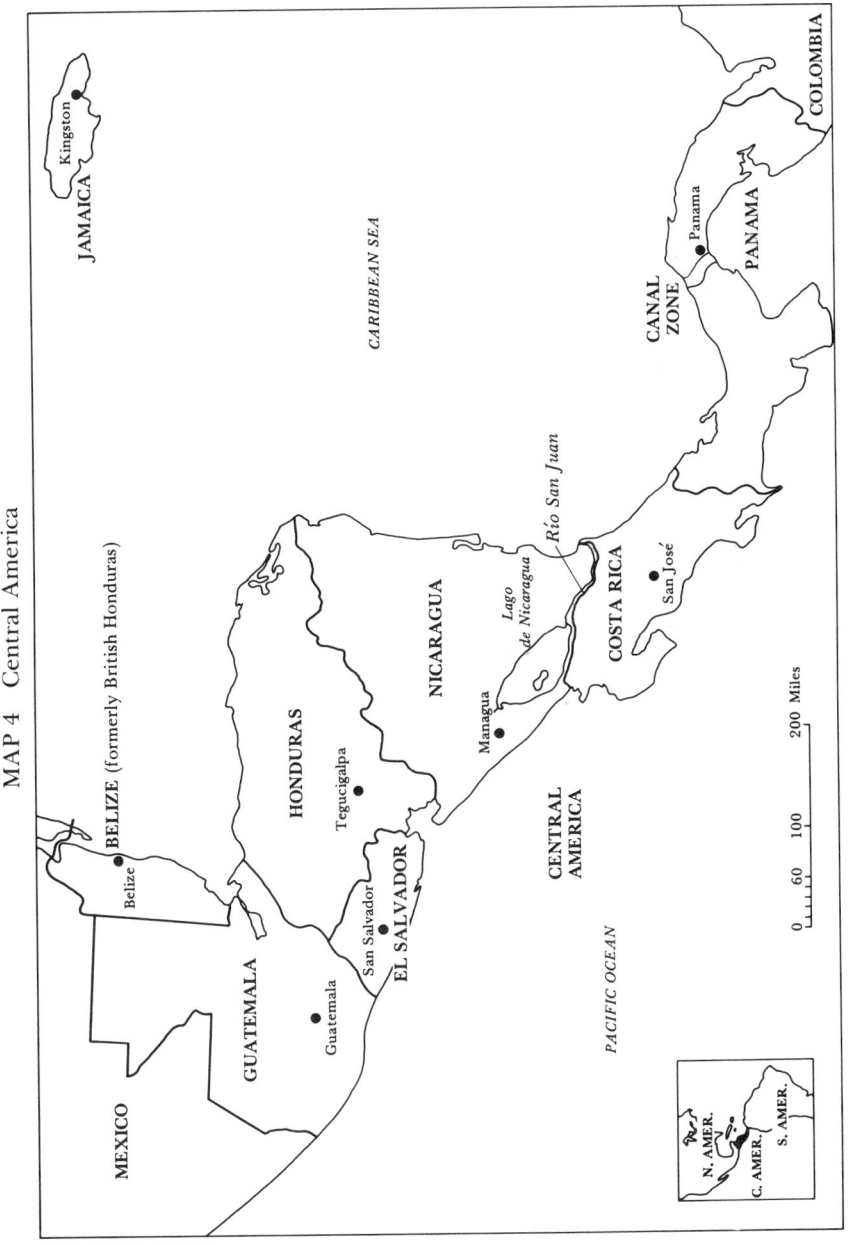

The church followed closely on the heels of conquest. Secular and regular clergy, especially Franciscans and Dominicans, took an active part in missionary efforts. By the late seventeenth century there were 759 churches throughout the area, which would acquire an archbishop in 1745. Early on, the church became a powerful source of authority.

Economic activity was modest. Mining was from the beginning a small-scale operation. The first major export was cacao, though Venezuela soon preempted this market. Indigo then took over as the leading export, and there was a bustling contraband trade in tobacco. In the 1660s the English established a foothold at the mouth of the Belize River (later British Honduras), which they used as a base for commerce in dyewood and mahogany and for buccaneering raids. But for the most part Central America was not a source of great wealth, and it received correspondingly little attention from the Spanish crown.

The social structure was controlled, at the top, by a two-part elite. One element consisted of Spanish-born bureaucrats (*peninsulares*) whose political base was the imperial court (*audiencia*) in Guatemala; the other consisted of locally born landholders (*criollos*) whose strength resided in town councils. At the bottom was the labor force, comprised of Indians and African slaves. There also emerged a stratum of mixed-bloods, known as *ladinos* in Central America, who worked as wage laborers or small farmers in the countryside and as artisans, merchants, and peddlers in the towns. Near the end of the colonial era approximately 4 percent of the region's population was white (either Spanish or creole), about 65 percent was Indian, and 31 percent was *ladino* (including those of black descent).

Independence: The Struggle for Unification

During the eighteenth century the Bourbon monarchy attempted to reassert royal control of Spanish America, a move that everywhere reduced the political autonomy of the landed creole class. In Central America a continuing decline in cacao production and a precipitous drop in the indigo trade between the 1790s and the 1810s led to further discontent within the creole ranks. These factors heightened longstanding differences between the imperial bureaucracy and the local aristocracy, between the capital and the provinces.

As independence movements flourished elsewhere, Captain-General José de Bustamante managed to maintain Spanish control of Central America, mainly by forging an alliance with *ladinos* and Indians against the upstart creoles. He was forced out in 1818, however, and Spain's adoption of a liberal constitution in 1820 sent repercussions throughout the area. In mid-1821 Agustín de Iturbide's declaration of the *Plan de Iguala* in Mexico forced the issue. Partly fearing "liberation" by Mexican troops, the socially conservative landowners decided to break with now-radical Spain, and in January 1822 proclaimed annexation of the

isthmus to royalist Mexico. The following year Iturbide's abdication led to complete independence. Chiapas remained with Mexico. The other states, from Costa Rica to Guatemala (excluding Panama), became the United Provinces of Central America.

Despite discord and disagreement, Central America managed to separate itself from Spain—and from Mexico—in a relatively peaceful fashion. The peoples of the isthmus did not suffer nearly the same level of physical destruction that occurred elsewhere in the 1810s. And as a result, the colonial social order survived almost intact.

The isthmus thus achieved full independence as a politically unified republic. The ideal of unification had long inspired local patriots and would continue to be a highly valued goal in years to come. But it also proved hard to sustain.

As happened elsewhere in Latin America, the Central American political elite divided into two factions: Liberals and Conservatives. The Liberals advocated continuation of reforms started by the Bourbon monarchy. They called for increased restrictions on clerical power, for the abolition of slavery, for the elimination of burdensome taxes, and for the promotion of economic development. They drew their support from emerging professional classes, white and *ladino*, and from upper-middle sectors excluded from the circles of the landed creole aristocracy. They espoused the idea of unification, too, and had considerable strength in the outlying provinces.

Conservatives stood for order, moderation, and stability. They upheld Hispanic institutions, especially the church, and they expressed suspicion of progressive reform. Led by creole landowners, they first advocated free trade, then reverted to a protectionist stance when they felt the impact of British commercial competition.

Violence erupted in the 1820's and Liberals appeared to have the upper hand. The constitution of 1824 bore resemblances to both the U.S. constitution and the Spanish constitution of 1812. In 1829 Liberal forces under Francisco Morazán of Honduras defeated a Conservative army, and in the wake of victory Liberals began a campaign to eliminate Conservatives from positions of power. Before long the tide would turn.

Rafael Carrera and Conservative Supremacy

The year 1837 brought shocks to Central America. In the mountain regions of Guatemala there began a massive rebellion, a peasant revolt that challenged the Liberal state. Village priests exhorted their poverty-stricken parishioners to join the uprising, proclaiming that a cholera epidemic—which started late the year before—was a sign of heavenly wrath. Indians flocked to the cause. Race war spread from Guatemala to the other provinces.

Leader of this movement was José Rafael Carrera, a *ladino* swineherd

with no formal education. In mid-1837 he defined the goals of the revolt as: (1) reinstatement of traditional judicial procedures, (2) restoration of religious orders and ecclesiastical privilege, (3) amnesty for all those exiled in 1829, and (4) obedience to Carrera himself. His forces soon controlled Guatemala, and in 1840 he defeated Morazán.

The triumphant Carrera emerged as the dominant figure in Central American political life, a position he held until his death in 1865. Starting in 1839, the Guatemalan legislature proceeded to dismantle the Liberal program. Merchant guilds were reestablished, the archbishop was asked to return, education was turned over to the church. Roman Catholicism became the official state religion, and priests regained protection of the ecclesiastical *fuero*. Carrera maintained his grip on the military, and in 1851 he assumed the Guatemalan presidency as well. It was an era of Conservative ascendancy.

The Carrera revolt ushered in some lasting changes. A few Indians, but more especially *ladinos,* began to play active roles in political life. The white elite regained its social supremacy but lost its monopoly over the state. And under Carrera, the government abandoned the goal of trying to assimilate Indian masses. It adopted instead a policy aimed at protecting the Indians, much as the Spanish crown had done, and this policy helped contribute to the segregation that has persisted to this day.

The Carrera era also brought an end to the Central American confederation. In 1838 the congress declared each of the states to be "sovereign, free, and independent political bodies." The ideal of unification had come to be identified with the Liberal period of 1823–37 and was seen as a failure, so Carrera discarded the dream. At the same time, he sought to impose like-minded Conservatives in the (increasingly sovereign) states. In Nicaragua this impulse eventually culminated in one of the more bizarre episodes in the history of inter-American relations— the William Walker affair.

Geographical and economic considerations had long stimulated interest in the idea of an interoceanic route through Central America. Having failed to discover a system of lakes and rivers connecting the Pacific Ocean and the Caribbean Sea, planners and visionaries pondered the possibility of an isthmian canal. The British and French gave early expressions of curiosity. The United States followed suit after taking Oregon and California, especially after the gold rush of 1849.

Because of its extensive lakes and the San Juan River, Nicaragua seemed a natural site for the canal project, and in late 1849 Cornelius Vanderbilt and his associates secured a concession from a Liberal government. Intrigue rapidly thickened. Costa Rica claimed jurisdiction over the proposed terminus at the eastern end of the route for the canal. Hoping to block their U.S. rivals, the British supported Costa Rica. By 1853 Conservatives had gained power in Nicaragua and, with-

out conceding territorial rights, they chose to take sides with the British.

Frustrated Liberals turned to the United States for help. What they got was William Walker, the glib and intellectually gifted son of an austere, frontier-fundamentalist family from Tennessee. As a young man Walker studied medicine in the United States and Europe, then took up law in New Orleans. Under a contract with the Liberals, Walker hired a small army and invaded Nicaragua in 1855. He seized one of Vanderbilt's passenger vessels, won a quick victory, named himself head of the armed forces, and settled in as the country's authoritative ruler.

The U.S. government took a permissive view of these developments, openly tolerating intervention by a North American citizen in the affairs of another state. Walker staffed his forces with veterans from the 1846–48 war with Mexico, accepted support from Vanderbilt's business competitors, and invited migrants from the U.S. South—who brought slavery along. Opposition mounted from the British and from Conservatives in other states, however, and Walker was driven from power in 1857. He tried to return and met his death in 1860.

Thus ended the "National War," an event with long-lasting implications. It discredited both Liberals and the United States, and helps explain why Conservatives stayed in power much longer in Nicaragua than in other parts of nineteenth-century Central America.

Liberal Theory and "Republican Dictatorships"

After Rafael Carrera died in 1865, the Liberals began a resurgence. It had begun in the late 1850s, when Gerardo Barrios became president of El Salvador, and it soon picked up in other countries. In 1870 Tomás Guardia, an army officer, assumed the presidency of Costa Rica. In 1873 Justo Rufino Barrios took power in Guatemala, as did Marco Aurelio Soto in Honduras in 1876. The legacy of the Walker expedition delayed the Liberal comeback in Nicaragua until 1893, when José Santos Zelaya became chief executive.

Like other elites in late nineteenth-century Latin America, the Central American Liberals believed in the notions of progress and economic development. They sought to integrate their countries with the rest of the world, to acquire the trappings of civilization, and to promote material improvement. In outlook they shared the views of Argentina's Generation of 1880 and, more particularly, of the Mexican *científicos*. In politics they would closely follow the example of Porfirio Díaz.

Notwithstanding their liberal commitments, these leaders set up what came to be known as "republican dictatorships." They centralized authority, rigged elections, controlled institutions, and kept themselves in

power for extended periods of time. They drew domestic support from the landed aristocracy and from some middle-sector elements. They forged close alliances with foreign interests—British, German, and North American. They modernized their military establishments and police forces, which they freely used to intimidate and suppress the opposition.

This pattern produced some social alterations. Where Conservative-Liberal distinctions were clearest (Guatemala and Costa Rica), they led to the near-total eclipse of power of old Conservative families. Where partisan lines were blurred (Honduras, El Salvador), some dynasties managed to hang on. Nicaragua proved to be an exception, as Conservative families had managed to consolidate their position by the time Zelaya came to power. Liberal ascendancy in general opened opportunities to middle-sector professionals and *ladinos* and, as we shall see, it fostered policies that led to the formation of new elites.

Moreover, it stripped the church of power and prestige. Its economic role was diminished and its legal privileges were abolished. As one historian would later write, "The major role the clergy had played in rural Central America became minor. This was one of the most important changes ever to take place in Central America." The demise of the church left an institutional vacuum in Central American society. It would eventually be filled, at least in part, by a new kind of Roman Catholic Church.

Overview: Economic Growth and Social Change

The economic program of the resurgent Liberals stressed the promotion of exports, especially of raw materials, in exchange for imports of manufactured goods. This entailed a reliance on agricultural production—and, in particular, on the cultivation of coffee and bananas.

Colonial Central America grew modest amounts of coffee. Costa Rica began serious production in the 1830s, shipping exports first to Chile and later to Europe. Guatemala promptly followed suit, and by 1870 coffee was the country's leading export, a position it has held ever since. El Salvador, Nicaragua, and Honduras joined the coffee trade in the 1870s and 1880s. Central American coffee exports have not risen to enormous volumes—never accounting for more than 15 percent of the world supply—but they have always been of high quality.

Coffee had important social consequences. Since it was grown in the cool highlands, along the mountain slopes, it did not everywhere require large-scale usurpation of land from the peasants. There were substantial takeovers in Guatemala and El Salvador, though perhaps less dramatic than occurred in Porfirian Mexico. In Honduras, Nicaragua, and Costa Rica, most peasants lived in the lowlands, however, so dislocations were less common. Also many coffee plantations were modest in size, and they were usually owned by Central Americans.

Foreign investors came to play an important part in coffee production in late nineteenth-century Nicaragua, and Germans acquired substantial amounts of coffee-growing land in Guatemala. But in general, coffee production remained in Central American hands.

Though Liberal leaders sought to encourage immigration (more than Mexico's *científicos*), Central America never received the kind of massive, working-class influx that went to Argentina, Brazil, and the United States. Labor for coffee cultivation instead came from the mostly Indian and *mestizo* peasants. In time they fell into two groups: *colonos*, who lived on the plantation and leased small plots of land for subsistence cultivation; and *jornaleros*, day laborers who worked for wages while living at home and retaining control of some land. In either case they retained close contact with the earth and retained outlooks of traditional peasants, rather than forging class consciousness as a rural proletariat.

The banana trade would eventually become emblematic of Central American culture, but it had a small-scale start. In 1870 a New England sea captain named Lorenzo Baker began shipments from Jamaica to the east coast of the United States, and in 1885 he joined with Andrew Preston to form the Boston Fruit Company. In the meantime Costa Rica had engaged Henry Meiggs, the dynamic railway entrepreneur, to lay track along the Caribbean coast in exchange for grants of land. Meiggs turned the contract over to two of his nephews, Minor Cooper Keith and Henry Meiggs Keith. In 1878 Minor Keith began shipping bananas to New Orleans, and soon established the Tropical Trading and Transport Company.

In 1899 the two companies merged, with Preston as president and Minor Keith as vice president, to form a singular enterprise: the United Fruit Company (UFCO). Here began a remarkable chapter in the history of U.S. investment, penetration, and control in Central America.

UFCO, or *la frutera* (the fruitery), as Central Americans called it, established a virtual monopoly on the production and distribution of bananas. Through government concessions and other means, the company acquired vast tracts of land in the hot, humid, sparsely settled Caribbean lowlands. Through the Meiggs connection it dominated transportation networks and owned a major corporation, International Railways of Central America. It built docks and port facilities. In 1913 UFCO created the Tropical Radio and Telegraph Company. *La frutera* possessed a large number of ships, widely known as the "great white fleet," and it had enormous influence on marketing in the United States. UFCO tolerated and even encouraged small-scale competition, but it was never seriously challenged in the decades after World War I.

The banana trade created enclave economies par excellence. UFCO supervisors and managers came from the United States, most notably from the South, and black workers were imported from Jamaica and

the West Indies. One result was to alter the racial composition of the eastern lowland population. Another was to create harshly enforced racial divisions within *la frutera* itself.

The industry became a giant foreign corporation. Some banana lands remained in local hands, but UFCO possessed control of technology, loans, and access to the U.S. market. Because of natural threats from hurricanes and plant disease, UFCO also sought to keep substantial amounts of land in reserve. These could usually be obtained only by government concession, a fact which required the company to enter local politics. The picture is clear: UFCO provided relatively scant stimulus for Central America's economic development, but became directly involved in local matters of state.

Coffee and bananas dominated economics after the turn of the century. As shown in Figure 10-1, the two products accounted for more than 70 percent of Central America's exports in 1913 and 1938 and nearly as much in 1960. The proportion declined by the early 1970s, as cotton and such other goods as meat, sugar, shrimp, refined petroleum (from refineries in Panama), and light manufactures gained in importance, but the traditional products still carried great weight.

One implication of this fact is obvious: the Central American economy became thoroughly dependent on the export of two commercial crops, coffee and bananas. The economic fortunes of the isthmus now depended almost entirely on the vagaries of the international market. When coffee or banana prices were down, earnings were down, and there was little room for flexible response—since coffee and banana plantations could not be easily or quickly converted to producing basic foodstuffs (assuming that the owners wanted to do so, which was hardly the case). It is worth noting, too, that coffee consistently formed a larger share of exports than bananas, and UFCO could not control the coffee market. In strict economic terms only Costa Rica, Honduras, and Panama were "banana republics." Guatemala, El Salvador, and Nicaragua were mainly coffee countries.

The coffee-banana strategy moreover led to heavy reliance on trade with a single partner: the United States. In the late nineteenth and early twentieth centuries Central America had a flourishing trade with Europe—Germany, in fact, was the biggest coffee customer. But after World War I the United States asserted its supremacy. From the 1920s through the 1950s, as Table 10-1 demonstrates, the United States purchased 60–90 percent of the region's exports and provided a similar share of imports. The North American predominance in international transactions faded to 30–40 percent in the 1970s for most countries, which were doing more trade than before with each other and with West Europe (and fell to nearly zero in the 1980s for Nicaragua, largely as the result of a U.S.-imposed boycott). One way or another, the United States still had considerable commercial leverage over the nations of the isthmus.

FIGURE 10-1   Principal Central American Exports, 1913–1980

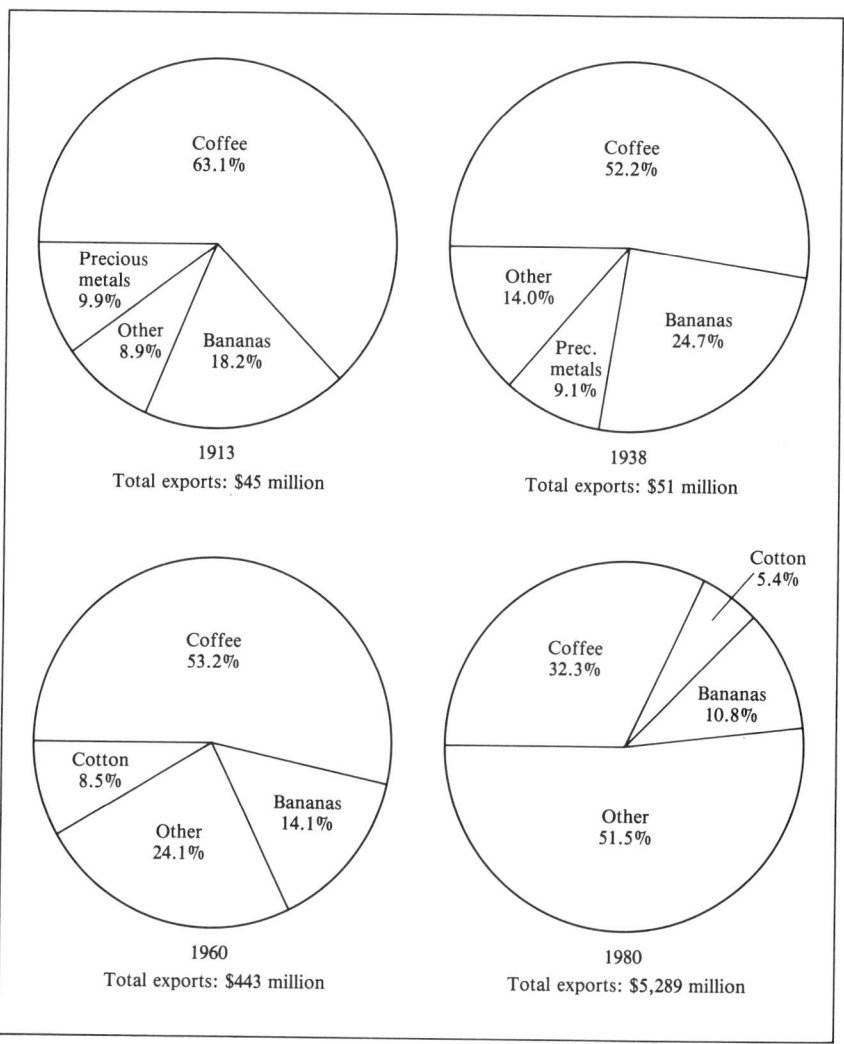

Source: Ralph Lee Woodward, Jr., *Central America: A Nation Divided*, 2nd ed. (New York: Oxford University Press, 1985), pp. 277, 366. Copyright © 1985 by Oxford University Press, Inc. Reprinted by permission.

The stress on agricultural exports and the persistence of peasant economies combined to discourage industrialization in Central America. The small scale of national markets presented another major obstacle to industrial growth.

In recognition of these difficulties, leaders of the isthmus decided to create the Central American Common Market (CACM) in 1960. The idea was to stimulate industrial development through a twofold strat-

TABLE 10-1   Central American Trade with the United States, 1920–88
(as % of total)

	1920		1950		1988	
	Exports	*Imports*	*Exports*	*Imports*	*Exports*	*Imports*
Costa Rica	71	52	70	67	44	39
El Salvador	56	79	86	67	39	42
Guatemala	67	61	88	79	40	43
Honduras	87	85	77	74	49	57
Nicaragua	78	73	54	72	0.4	1.3
Panama	93	73	80	69	50	19

Sources: James W. Wilkie, *Statistics and National Policy*, Supplement 3, *Statistical Abstract of Latin America* (Los Angeles: UCLA Latin American Center, 1974), Table XV-3; James W. Wilkie and Peter Reich, eds., *Statistical Abstract of Latin America*, 28 (Los Angeles: UCLA Latin American Center, 1990), Table 2602.

egy: promoting free trade among member countries and creating common tariffs to protect infant enterprises. Costa Rica, El Salvador, Guatemala, Honduras, and Nicaragua formed the CACM membership; Panama did not belong because of its special status with the canal, but expressed from the start a continuing interest in association with the group.

The Common Market met with instant success. Commerce among the member countries multiplied, growing from 7.5 percent of all exports in 1960 to 26.9 percent by 1970. Manufacturing flourished—in such areas as electrical equipment, prepared foods, pulp and paper products, and fertilizers—as the isthmus embarked on the path of import-substitution industrialization. But CACM made little headway in the agricultural sector, where protectionist policies remained the rule, and it failed to meet the challenge of unemployment (which stood at 9.4 percent in 1970).

CACM also suffered from political disputes. Honduras had begun to complain about the distribution of benefits in the mid-1960s and withdrew from the market after an armed clash with El Salvador in 1969 (described below). Honduras subsequently negotiated bilateral agreements with the other countries—except El Salvador—but CACM lost a good deal of precious momentum.

Notwithstanding these efforts the regional economy has remained primarily agricultural, and its society has continued to be mostly rural. Around 1900 less than 10 percent of the population lived in cities. By 1970 the figure ranged between 20 and 40 percent (compared to 66 percent for Argentina, for example, and 61 percent for Chile). Even the biggest cities of Central America have been small by international standards. In 1970 Guatemala City, by far the largest, had well under a million inhabitants (731,000), and the other capitals had populations

ranging from 200,000 to just over 400,000. Urbanization came late to Central America.

This delay has, in turn, produced a major social fact: Central America has never had a substantial urban working class. There are some workers in the cities, of course, and there have been sporadic efforts at unionization since the 1920s. But the de-emphasis on manufacturing and the smallness of the cities have not given rise to the scale of working-class movements that appeared in Argentina, Chile, Brazil, or Mexico. (In Nicaragua, for example, the unionized share of the economically active population in 1973 came to a paltry 2 percent.) A principal collective actor has been largely absent from the scene: just as Argentina has not had a classic peasantry, so Central America has lacked an urban proletariat.

Furthermore, the historical de-emphasis on manufacturing meant that the agricultural sector would never be challenged by an industrial sector. To be sure, CACM helped give shape and strength to a fledgling business group, but it did not lead to an outright assault upon the social order. Consequently, there would be little incentive to form the sort of multiclass populist alliance that often emerges from sectoral conflict (as in Argentina, where Perón joined together industrial workers and entrepreneurs in a common attack upon the rural aristocracy). In the absence of an industrial threat, landlords and peasants, mostly Indian in Guatemala, faced each other in the Central American countryside. When conflict occurred, it would accordingly tend to follow class lines. Control of land would be the overriding issue.

The most active groups in Central America's cities have generally consisted of middle-sector merchants and professionals—lawyers, journalists, intellectuals, and students. They have not displayed much middle-class consciousness, but they have spawned some reformist political movements and produced a considerable number of civilian political leaders. As time passed their role in state and society steadily increased.

Despite this growth and development, the masses of Central America have remained poor. Per capita incomes are low (with a median around $1100 in 1992) and wealth is tightly concentrated. As indicators of social welfare, national literacy rates reveal that in 1970 only 45–60 percent of adults could read and write in El Salvador, Guatemala, Honduras, and Nicaragua (Costa Rica, always the exception, was near to 90 percent). The proportion of young children (ages seven to thirteen) who went to school was 70 percent or more for all countries but Guatemala—the largest country, where it was 50 percent. But this was more than offset by the low proportions of teenagers attending secondary schools: less than 20 percent in El Salvador, Guatemala, Honduras, and Nicaragua, only 33 percent in Costa Rica. If Central America's children

were getting any education at all, their exposure to school was generally brief.

In time, this pattern of development and deprivation would exert tremendous pressure on the region's political systems. Understanding this process requires a brief examination of each country's recent history.

Panama: A Nation and a Zone

Panama did not become an independent republic until after the turn of the century. During the 1800s it was a province of Colombia. Because of poor communications and distance from Bogotá it had become semiautonomous. The energetic president of Colombia, Rafael Núñez, tried to assert central control during his tenure in office (1885–94), and partially succeeded. After his death the country fell into disarray, and Liberals and Conservatives began a frightful struggle that culminated in Colombia's "War of the Thousand Days" (1899–1903). The conflict ended with the Liberals in defeat, the economy in paralysis, the government near bankruptcy.

In view of these conditions Panama might have sought independence on its own, as it had long been chafing under rule from Bogotá. Ultimately, however, Panamanian sovereignty would not arise from an indigenous popular movement. It would grow out of big-power diplomacy and international intrigue.

At issue was an interoceanic canal, a time-honored vision for Central America. Plans went back as far as the seventeenth century. In 1878 the government of Colombia authorized a French group under Ferdinand de Lesseps, builder of the Suez Canal, to dig a route through Panama. U.S. engineers tended to favor Nicaragua, and a North American firm received a contract to begin digging in that country. The race was on. Then came the financial Panic of 1893, when both groups ran out of money and quit.

Popular opinion in the United States favored bold action in Latin America. The works of Rudyard Kipling spread deep convictions about the "white man's burden." Popularized versions of Social Darwinism helped convince North Americans they were among history's "fittest." The acquisition of Cuba and a Pacific empire through the Spanish-American War prompted interest in far-flung possessions. And in such magisterial books as *The Influence of Sea Power upon History* (1890), the historian-publicist Alfred Thayer Mahan forcefully argued that naval power was the key to international influence, a doctrine requiring a two-ocean navy for the United States. After Theodore Roosevelt became president in 1901, it was clear that Washington would make a move. What kind?

Colombian politics led to resolution of the issue. As the War of the Thousand Days was nearing its end in 1903, Washington dispatched

troops to quell disorder in Panama. This resulted in a diplomatic crisis which eventually produced the Hay-Herrán Treaty, an agreement that authorized the United States to build a canal in Panama. The U.S. Congress eagerly approved the document—but the Colombian legislature, unwilling to compromise national sovereignty, refused to go along.

The next step was insurrection. The de Lesseps chief engineer, Philippe Bunau-Varilla, was continuing to push for a canal, and now he seized his chance. With Roosevelt's full knowledge, Bunau-Varilla started laying plans for a separatist rebellion in Panama. As the uprising began, U.S. ships prevented Colombian troops from crossing the isthmus to Panama City. The revolt was a success.

Within days Washington extended recognition to the newly sovereign government of Panama, and received Bunau-Varilla (still a French citizen) as its official representative. U.S. Secretary of State John Hay and Bunau-Varilla hastily signed a treaty giving the United States control of a ten-mile-wide canal zone "in perpetuity . . . as if it were sovereign." A pliant Panamanian legislature soon approved the document. Bunau-Varilla and the administration lobbyists then turned their attention to the U.S. Senate, where pro-Nicaragua sentiment was still fairly strong. On the morning of the decisive vote Bunau-Varilla placed on each senator's desk a Nicaraguan postage stamp depicting a volcanic eruption, and the silent message took hold. The Senate approved the measure by a sixty-six to fourteen margin, and the die was cast.

Panama thus acquired nationhood through big-stick diplomacy. Opened in 1914, the canal immediately became a major international waterway, and the government began receiving steady annuities. The Canal Zone became a de facto U.S. colony, an area of legal privilege and country-club prosperity that stood in sharp and conspicuous contrast to local society. Outside the Zone, Panama developed the characteristics that typified Central America as a whole: dependence on agricultural exports (especially bananas), reliance on the U.S. market, and domestic control by a tightly knit landed oligarchy. Notwithstanding the Hay-Bunau-Varilla Treaty, the situation could not last forever.

In 1952 Colonel José Antonio Remón became president and began renegotiation of the treaty of 1903. Three years later his efforts resulted in an agreement that increased the annuity payable to Panama, curtailed economic privileges for U.S. citizens, and sought to equalize wage rates for North Americans and Panamanians. But the question of sovereignty was left untouched. It came up in 1956, after Egypt's seizure of the Suez Canal. When President Ricardo Arias bitterly protested Panama's exclusion from a conference on the Suez crisis, Secretary of State John Foster Dulles frostily replied that the United States had "rights of sovereignty over the Panama Canal . . . to the entire exclusion of the Republic of Panama of any such sovereign rights, power, or authority."

Tensions and consultations continued. In January 1964 American

students at Balboa High School (in the Canal Zone) raised a U.S. flag alone, without the accompaniment of a Panamanian banner, and large-scale rioting ensued. Ellsworth Bunker, U.S. ambassador to the Organization of American States, charged that assaults on the Canal Zone came from mobs "infiltrated and led by extremists, including persons trained in Communist countries." Panamanian President Roberto F. Chiari broke off relations with Washington.

Tempers cooled and discussions resumed. In 1968 Dr. Arnulfo Arias won a disputed election, a campaign that he himself called "one of the most shameful in the history of the country." After holding office for merely eleven days he was overthrown by the National Guard, which set up a junta under Brigadier General Omar Torrijos Herrera. This marked a clear assertion by the Guard of hegemony in politics. It led to the emergence of Torrijos as the nation's strong man. And it yielded continuity in leadership, as Torrijos patiently pursued negotiations with the Nixon, Ford, and Carter administrations in the United States.

The United States finally accepted a treaty that provided for complete Panamanian sovereignty over the canal by 1999. Ronald Reagan (long before assuming the presidency) and other U.S. conservatives vigorously denounced the agreement as a sellout, but Democratic President Jimmy Carter eventually obtained Senate approval. Intellectuals and statesmen throughout the hemisphere applauded the move. However briefly, U.S.-Latin American relations took a positive turn.

But Panama continued to have troubles of its own. Torrijos died in an air accident in 1981. On July 30, 1982, the civilian president Aristides Royo abruptly resigned, turning the office over to the vice president. Royo cited poor health but most observers believed that he had been dismissed by the National Guard. The new strong man was General Manuel Antonio Noriega, reputed to be deeply involved in the corruption plaguing Panama. (He had also been a part-time agent for the CIA.) Panamanian nationalism flared when that government refused to renew the agreement with the United States under which the "School of the Americas"—a U.S. financed and directed training program for Latin American military—had operated in Panama.

Anti-American feeling surged again in 1988, as the U.S. government imposed an economic boycott in an effort to oust Noriega from power. In the following year Noriega annulled elections apparently won by Guillermo Endara, leader of an oppositionist "civic crusade" against the dictatorship, and the United States tightened the screws. In December 1989 the Bush administration dispatched more than 20,000 U.S. troops to crush the Noriega regime. The invasion force met stiff but sporadic resistance, then overwhelmed the Panamanian defenses and captured Noriega himself—and took him to Miami, where he would stand trial for alleged complicity in drug trafficking. U.S. officials reported that only twenty-three American servicemen had lost their lives, but there

would be continuing controversy over the death toll for Panamanians (estimates ranged from several hundred to several thousand). Economic damages from the invasion may have been as high as $2 billion.

Many Panamanians greeted the U.S. troops with enthusiasm but eventually became disenchanted. The United States was slow to provide economic assistance in the wake of its military operation. As a result of the sanctions and then the invasion, the gross national product shrank by 22 percent between 1988 and early 1991. Unemployment was above 20 percent, and 40 percent of the population lived below the poverty line. Overweight and uninspiring, Endara proved to be an ineffective leader: according to opinion polls, popular support for his ruling coalition declined from 73 percent in mid-1989 to 17 percent in March 1991. An opposition party accused the hapless president of links with money-laundering schemes, the very charge that the United States had used to justify its invasion in the first place.

The Panamanian economy began picking up in the early 1990s, as growth rates averaged more than 7 percent, and Endara finally managed to normalize relations with external creditors. High levels of poverty and unemployment continued, however, and leaders of the traditional elite fell into partisan bickering. Presidential elections in May 1994 added an ironic twist to the situation by granting victory to Ernesto Pérez Balladares, a former Noriega crony who topped a crowded field of candidates (including salsa star Rubén Blades) by appealing to widespread frustration. During the campaign Pérez Balladares distanced both himself and his party from Noriega, invoking instead the legacy of Omar Torrijos, but the basic fact remained: Noriega loyalists were back in power. Years after the U.S. invasion, politics in Panama was much the same.

As the year 1999 began to loom on the horizon, attention returned to the Panama Canal. While there was little doubt about Panama's ability to manage the waterway itself, there arose considerable concern about its capacity to maintain support facilities and properties at an operating cost of $500 million per year. Squatters were entering the Canal Zone at the rate of 300 per day, and buildings were falling into disrepair. One opinion poll showed 75 percent of Panamanians favored the continuation of a U.S. military presence in the area for both economic and political reasons. Governance of the Panama Canal remained an issue for Panama, the United States, and the hemisphere at large.

Costa Rica: Fragile Democracy

Costa Rica has long been unique. Despite its name ("rich coast"), it was of minimal economic importance to Spain, and as the southernmost area in the kingdom of Guatemala it was relatively remote from the

rest of Central America. Sparsely populated from the outset, it never developed a large-scale black or Indian subservient class. Nor did it have a wealthy landed oligarchy.

Coffee cultivation began on modest, family-sized farms in the 1830s. The flourishing commerce gave rise to a substantial and prosperous agrarian middle sector—and to a merchant class in the cities—without creating a landless peasantry. United Fruit established banana plantations on the east coast in the late nineteenth and early twentieth centuries, and bananas soon became the country's leading export.

For economic and demographic reasons Costa Rica emerged as a racially and socially homogeneous society. By 1925 about 80 percent of the population was white, 4 percent was black (mostly workers on banana plantations), 14 percent was *mestizo,* and less than 1 percent was Indian. Middle-class culture prevailed, and racial conflict was largely absent.

Social consensus led to broad acceptance of constitutional politics. Early twentieth-century governments fostered welfare programs (so Costa Rica, like Uruguay, inevitably came to be compared to Switzerland). Conservatives exchanged power with Liberals. There was not much to fight about, and democratic traditions began to take root.

Then worldwide depression in the 1930s bred social discontent. The National Republican Party came forward as an alternative to communism. The Liberal-Conservative distinction faded, and with leftist support, National Republicans won the presidential elections of 1936, 1940, and 1944. A progressive social security system and labor code were put in place.

Two leading factions then emerged. One was the vehemently anticommunist National Union Party, led by Otilio Ulate Blanco. The other was the left-of-center (but anticommunist) Social Democratic Party, organized by former Conservative José ("Pepe") Figueres Ferrer. In the 1948 elections both movements joined in a coalition against the National Republicans. Violence flared, disputes erupted, and Figueres assumed authority. Acting with vigor and decisiveness, he dissolved the army, levied new taxes, called a constituent assembly, and—as was occurring at this time elsewhere in Latin America—outlawed the communist Popular Vanguard. The dust settled and Ulate Blanco took office in 1949.

Figueres won the presidency in 1952 and normalcy returned. He stimulated agricultural exports and negotiated a new contract with United Fruit, under which the Costa Rican share of profits increased from 10 to 30 percent. With support from Washington he withstood an uprising in 1955. The election of 1958 went smoothly. Said Figueres after the loss by his party's candidate: "I consider our defeat as a contribution, in a way, to democracy in Latin America. It is not customary for a party in power to lose an election."

Subsequent events would bear out his claim. Voter participation in

Costa Rica has generally been over 80 percent, one of the highest rates in the world (compared to 55–60 percent in the United States). And moderation has prevailed: less than 10 percent of the vote has gone to extremist candidates of left or right. In contrast to so much of Central America, Costa Rica has a strong and viable political center.

Economic straits, as always, put the system to a stringent test. Under the dubious administration of Rodrigo Carazo (1978–82) Costa Rica ran up a foreign debt of $4 billion U.S. dollars, enormous for a country its size. The growth rate declined from 8.9 percent in 1977 to −2.4 percent in 1981, during which year the local currency (the *colón*) was devalued by more than 400 percent. Unemployment climbed to 10 percent and appeared to be still rising.

Costa Rican democracy managed to survive the terrible financial pressures of the post-1982 years, as the nation's two dominant political parties continued their tradition of alternating back and forth in power. But the country had to pay a heavy price for being a neighbor of Nicaragua. As the Sandinista-Contra war deepened, Nicaraguan refugees and Contra elements ensconced themselves in Costa Rica, another nation put at risk by the Cold War in Central America.

Oscar Arias Sánchez, elected to the presidency in 1986, chose to confront these problems directly. With skillful diplomacy and dogged determination, he persuaded chief executives from other Central American countries to come together in negotiations. The result of this process was the so-called Esquipulas accords, named after the town where the first meeting took place, that called on the war-torn nations of the region to (1) initiate a cease-fire, (2) engage in dialogue with opposition movements, (3) prevent the use of their territory for aggression against other states, and (4) cease and prohibit aid to irregular forces or insurrectionary movements—these last two points directed especially at Nicaragua and the United States. The August 1987 agreement also called for free elections and democratization of all nations in the region. It was an ambitious plan, one that seemed too good to be true, but it had the unequivocal merit of being a Central American solution to Central American problems. In fact it helped bring a measure of peace to the region, and it earned for Arias a Nobel Prize.

Subsequent elections revealed the continuing importance of eminent political families. Victory in 1990 went to Rafael Angel Calderón, son of one former president, and victory in 1994 went to José María Figueres, son of another former president. In the wake of the debt crisis of the 1980s, Figueres sought to combine economic reform with a progressive stand on social and economic issues. Such multinational enterprises as Motorola and Coca Cola continued to invest in Costa Rica, and economic growth was for the most part strong. By 1993 open unemployment had fallen to just over 4 percent, one of the lowest rates in the Americas. In comparison to neighboring countries of Central America, Costa Rica remained an exception.

Nicaragua: From Dynasty to Revolution

For much of its history Nicaragua has been a pawn of outside powers, especially the United States. During the nineteenth century it received unceasing attention from avaricious adventurers, many of whom sought to build a canal, and it endured the brief but ignominious presence of William Walker. The pattern would continue into the twentieth century.

The British conceded the Caribbean basin to the U.S. sphere of influence in the 1890s, and Washington eagerly seized the opportunity. The United States occupied Cuba, "took" Panama, and established a protectorate in the Dominican Republic. To justify these and future actions, Theodore Roosevelt proclaimed in 1904:

> Any country whose people conduct themselves well can count upon our hearty friendship. If a nation shows that it knows how to act with reasonable efficiency and decency in social and political matters, if it keeps order and pays its obligations, it need fear no interference from the United States. Chronic wrong-doing, or an impotence which results in a general loosening of the ties of society, may in America, as elsewhere, ultimately require intervention by some civilized nation, and in the Western Hemisphere the adherence of the United States to the Monroe Doctrine may force the United States, however reluctantly, in flagrant cases of such wrong-doing or impotence, to the exercise of an international police power.

Known as the "Roosevelt Corollary" to the Monroe Doctrine, the rationale took on clear meaning. To prevent intervention by Europe, the United States could intervene in Latin America at will.

Nicaragua would soon find out what the dictum meant in practice. Washington had developed a strong dislike for José Santos Zelaya, the Liberal dictator who had staunchly resisted foreign control in negotiations over a canal route. In 1909 Zelaya ordered the execution of two North American adventurers. Secretary of State Philander C. Knox denounced Zelaya as "a blot on the history of his country" and expelled Nicaragua's ambassador from the United States. Subsequent U.S. support for an anti-Zelaya revolt helped to force the president to resign.

Financial chaos ensued. European creditors began demanding payment on their debts. In desperation the new president, Conservative Adolfo Díaz, asked the United States to send military aid to protect North American economic interests from the threat of civil war within Nicaragua and to "extend its protection to all the inhabitants of the republic." Citing the Roosevelt statement, President William Howard Taft dispatched the marines. A plan for fiscal recovery obtained a guarantee from a New York banking conglomerate, which received control of the national bank and the railway system as security on its investment. Politically and economically, Nicaragua became a full-fledged protectorate of the United States.

Most Americans accepted the claim that the U.S. protectorate in Nicaragua was justified by the "Roosevelt corollary" to the Monroe Doctrine; the original caption for this cartoon was "cutting a switch for a bad boy." (McKee Barclay, *Baltimore Sun*, 1910. Courtesy of The Baltimore Sun.)

This condition lasted till 1933. In the mid-1920s a dispute arose over presidential succession. The United States imposed the trusty Adolfo Díaz and agreed to supervise upcoming elections. As a result of this compromise, a Liberal named Juan Bautista Sacasa won the presidency in 1932 and called for withdrawal of U.S. troops. The New York bankers had already recovered their investment and Franklin Delano Roosevelt was about to proclaim the Good Neighbor policy. In 1933 the marines left Nicaragua.

But one Liberal activist, Augusto César Sandino, refused to abide by the bargain of the late 1920s. A fervent patriot, a nationalist and social moderate, Sandino had waged a guerrilla campaign against U.S. intervention and Nicaraguan collaborationists. He gained a widespread popular following, but the United States worried about the presence of leftists among his supporters. U.S. marines joined in the campaign against him but he consistently eluded his pursuers. Even after U.S. forces left, Sandino continued the fight. He saw the U.S.-trained Na-

tional Guard as the major threat to an independent Nicaragua. The Guard and the Sandinistas were now competing fiercely for influence over the new government. With Sacasa firmly installed in the presidency, Sandino agreed to a meeting to try and reach a peace agreement. After leaving the presidential palace, Sandino and two supporting generals were seized by Nicaraguan National Guard officers and promptly executed. A genuine national hero, Sandino now became a martyr as well.

Political power resided not in the electoral system but in the National Guard, a domestic police force created during the U.S. occupation. At its head was General Anastasio ("Tacho") Somoza García, an ambitious and ruthless tyrant who had given the order to execute Sandino. He eventually unseated Sacasa and took over the presidency in 1937. Thus began the Somoza regime.

A wily politician, Somoza drew support from several sources: the National Guard, which he constantly nurtured and protected; the landed elite, with whose members he entered into numerous partnerships; and the United States, whose political power structure he knew so well how to manipulate.

Amassing an enormous fortune for himself and his family, Somoza promoted Nicaragua's economic growth and cultivated U.S. aid. He was struck by an assassin in 1956 and rushed to a hospital in the American-controlled Canal Zone. Ever grateful for Somoza's rabid anticommunism, President Eisenhower sent his personal surgeon to try to save the dictator's life. Somoza nonetheless succumbed.

The family enterprise endured. The elder son, Luis Somoza Debayle, took over and was elected to the presidency by an 89 percent majority in 1957. A trusted family associate, René Schick, assumed office in 1963. Four years later Luis Somoza died of a heart attack. Power then passed to Anastasio Somoza Debayle, a West Point graduate and, like his father, head of the National Guard. Self-seeking, ruthless, corrupt, Somoza clamped an iron rule on the country, but offended thoughtful Nicaraguans by his excesses. It was rumored, for example, that he exacted large-scale economic profits from the reconstruction of Managua after a devastating earthquake in 1972.

The complete absence of representative institutions meant that opposition to Somoza could take only one form: armed resistance. In the 1960s a guerrilla movement emerged, and it eventually consisted of three major groups. One, known as Prolonged Popular War (GPP), formed in the early 1960s among rural peasants of the mountainous north. Second were the *Proletarios,* who splintered from the GPP in 1973 to carry the movement to workers and especially intellectuals in the cities. Third were the *Terceristas,* a politically moderate and non-Marxist group led by Edén Pastora, renowned from his military exploits as the legendary "Commander Zero." Taking their name from Augusto César Sandino, the movements all combined to form the San-

dinista National Liberation Front (the name of the core group since 1961).

After a year and a half of struggle the Somoza regime suddenly collapsed in 1979, just as Batista had given way in Cuba two decades before. Safely exiled in Miami, Somoza would lay the blame on forces beyond his control. "The Nicaraguan people have not thrown me out. I was thrown out by an international conspiracy that today has a majority of communists and that today desires Nicaragua to be a communist country." Somoza later met his death at the hands of assassins in Paraguay.

Once in power, the Sandinistas proclaimed two broad policy goals. One called for implementation of an "independent and nonaligned" foreign policy, which meant no further submission to the United States. The other envisioned the creation of a "mixed economy" in order to achieve socioeconomic justice.

The revolution got off to a promising start. President Carter invited Nicaragua's leaders to the White House, sent $8 million in emergency relief to Managua, and secured an additional authorization from Congress for a $75 million aid package. Their domestic task was eased by the magnitude of the Somoza family fortune, which included about 20 percent of the country's cultivable land. It was therefore possible for the revolutionary government to nationalize these holdings and to initiate an agrarian reform without having to face the diehard opposition of an entrenched landed aristocracy.

The Nicaraguan revolutionaries quickly attacked the same social problems the Cubans had become famous for liquidating, such as illiteracy, inadequate preventive medicine, and insufficient vocational and higher education. Nicaragua welcomed approximately 2500 Cubans (the count was carefully monitored by the CIA and State Department)—doctors, nurses, school teachers, sanitary engineers—to help the revolutionary government raise basic living standards. Cuban military, police, and intelligence personnel also arrived to help consolidate the regime against what the Sandinistas (and Cubans) were convinced would be counterrevolutionary attacks from within and without.

The Nicaraguans also solicited help from the United States, which responded in 1980 with its modest program of $75 million. Far more important financial help came from West Europe—especially West Germany, France, and Spain. The Soviets, on the other hand, offered no hard-currency credits. The Nicaraguans therefore appeared to have a better chance to avoid complete trade and financial dependence on one ideological bloc than did Cuba in 1959–61.

The Nicaraguan euphoria did not last long, however. In the United States the Republican Party electoral platform of 1980 formally deplored "the Marxist Sandinista takeover of Nicaragua," and the Reagan administration thereafter began a persistent campaign to undermine the Sandinista government. The United States launched a trade em-

bargo against Nicaragua, thereby pushing the Sandinistas into greater
dependence on Cuba and the Soviet Union. Outsiders disagreed vigorously over the nature of the Nicaraguan regime. The Sandinistas
clearly controlled the police, the military, and virtually all executive officials. They had closed down almost all the opposition media. On the
other hand, most of the land and the service sector remained under
private ownership, opposition parties functioned, and foreign multinationals (such as Esso) continued to operate. At least Nicaragua was
more open than Cuba. That was no coincidence, since Fidel had urged
the Sandinistas not to repeat his mistake of breaking completely with
the capitalist camp.

The Sandinistas' steady movement toward the Cuban model was accelerated by attacks from Contras, an exile army funded by the United
States and commanded in part by former Somoza army officers, along
with some anti-Somocistas disillusioned with Sandinista rule. Although
the Contras could harass the Sandinista regime, they could not take
and hold major targets within Nicaragua. They did, however, force the
government to spend half of its budget on defense and to alienate its
citizens with wartime measures. Partly as a result of these factors the
economy went into a serious tailspin. Output declined by 4 percent in
1987 and 8 percent in 1988, when inflation reached the spectacular
level of 33,000 percent!

It was in this context that elections took place in February 1990. With
Daniel Ortega as their candidate, the Sandinistas confidently anticipated victory. Their opponent was Violeta Barrios de Chamorro, the
widow of a distinguished anti-Somocista and leader of a fragmented
opposition coalition (UNO from its Spanish initials). Most pollsters announced that Ortega's lead was widening as the election approached.
Then came the stunning results: UNO captured 54.7 percent of the
vote, against 40.8 percent for the Sandinistas. At the urging of Jimmy
Carter (present as an international observer), Ortega made a gracious
concession speech. To near-universal surprise the opposition prepared
to take power.

Chamorro proclaimed an end to the fighting and, at her inauguration, announced an "unconditional amnesty" for political crimes and
an end to the draft. Nonetheless she was unable to broaden her political base. With strong representation in the national assembly, Sandinistas could prevent constitutional reform; they also continued to control
the armed forces and other key institutions, including labor unions.
Chamorro also had to deal with discontented former Contras who were
insistent on their rights to land. Antonio Lacayo, her son-in-law and
leading adviser, inspired distrust in many circles. Ravaged by war and
devastation, Nicaragua was proving difficult to govern.

Instability and uncertainty prevailed throughout the early 1990s. Assisted by over $860 million in direct foreign aid and more than $200
million in debt write-offs, Chamorro's economic team managed to

bring down inflation, but overall growth remained sluggish. Unemployment rose from 12 percent in 1990 to 22 percent in 1993 (with under-employment affecting another 28 percent). Now known as *recontras*, former Contras engaged in occasional skirmishes with demobilized Sandinistas, known as *recampas*, but the two sides accepted a peace agreement in April 1994. (At times they also joined together in common cause, as in efforts to seek appropriate benefits for ex-soldiers in general.) Sporadic clashes nonetheless continued, as the national government proved unable to maintain law and order in the countryside.

Political developments began to focus on upcoming elections in 1996. A series of constitutional reforms in February 1995 reduced the presidential term from six to five years, placed a ban on immediate reelection, and—in an attempt to thwart the ambitious Lacayo, as well as to prevent dynastic rule—prohibited the president from being succeeded by a close family relative. Both the FSLN and UNO coalitions started to unravel, and in this unsettled context the Somoza clan (represented by three brothers) started to make a political comeback. Dispirited by years of military conflict and economic distress, many Nicaraguans began to see the Somoza family as the lesser of evils. Expressing sharp criticism of the Sandinistas, one middle-aged worker reflected: "Somoza was ten times better than those thieves. Yes, he stole. But our foreign debt was $1 billion after fifty years of the Somozas, and a decade later it was $12 billion and we had nothing to show for it. So, I ask you: who was the bigger thief?"

Honduras: Rule by Military

Honduras has undergone the least transition of all the Central American republics. Liberal-Conservative party rivalries have persisted to the mid-twentieth century, popular agitation has been minimal, and power has rested in the hands of a triangular alliance: landowners, foreign investors (mainly United Fruit), and the military. Because of its economic and political weakness, Honduras has been especially vulnerable to outside influence. It remains, in many respects, a stereotypical "banana republic."

With vivid clarity, Honduran history reveals a fundamental fact of Central American political life: the emergence of the military as an autonomous caste and as a supreme arbiter in national affairs. In Honduras, as elsewhere in the isthmus, a career in the armed forces (or national guard) offered chances for upward mobility to middle-sector *ladinos*. Land was already controlled by the aristocracy, the universities were restrictive, there was hardly any industrial development: an ambitious young person of middling origin had almost no other alternative. As a result recruits and cadets took immense pride in the honor and dignity of the military as an institution, and officers tended to look down on politicians and civilians. To this extent the armed forces stood

apart from civil society—but their consent (if not support) was essential to the survival of any political coalition.

The leading figure in early twentieth-century Honduras was Tiburcio Carías Andino, whose conservative-oriented National Party was prevented from taking the presidency in 1923. After some dispute Carías' candidate was permitted to govern till 1929, when the Liberals recaptured office. In 1932 Carías himself won the presidency, and he held the position till 1948.

In 1957 a group of young military officers supervised the election of Dr. Ramón Villeda Morales, a progressive Liberal who became an outspoken supporter of John Fitzgerald Kennedy and the Alliance for Progress. But senior commanders resented his reformist tendencies, and in 1963 they dismissed him in favor of armed forces leader Oswaldo López Arellano, who ruled till 1975 (when he was toppled, of course, by a military coup).

Officers retained control until 1981. Partly as a result of international pressure, especially from the United States, fairly open elections took place in that year. The candidate of the Liberal Party, Roberto Suazo Córdova, won 54.1 percent of the vote; the military decided to accept the result. Until further notice, at least, Suazo had a chance to govern, and in 1985 peaceful elections led to the triumph of another Liberal Party candidate, José Azcona Hoyo.

Strife with neighboring countries has played an important part in Honduran history. During the 1960s, for instance, tensions with El Salvador mounted steadily. There had been longstanding, although minor, boundary disputes. The principal source of the trouble was economic: El Salvador was densely populated, with 390 persons per square mile, while Honduras had only 55 per square mile. *Salvadoreños* consequently went to look for jobs and land in Honduras, and the Hondurans understandably resented their presence. A 1963 law prohibited companies from employing more than 10 percent foreigners (that is, Salvadorans), and a 1968 decree prevented Salvadorans from gaining title to Honduran land.

Violence erupted in both capitals in 1969, when national soccer teams from the two countries played an elimination round in the World Cup competition. Antagonism reached such a point that the third and deciding game had to be played in Mexico. El Salvador won the contest, and frustrated Hondurans attacked Salvadoran residents in Tegucigalpa and other cities. El Salvador severed diplomatic relations and demanded reparations. So did Honduras. The case was going to the OAS when El Salvador claimed that Honduran planes had conducted bombing raids.

As the world looked on in astonishment, the so-called soccer war began in earnest and lasted two weeks. In one sense, it cast light on the political function of organized sport in Latin American society. In an-

other, it revealed the latent hostility and rivalry among the republics of Central America.

Geography made it inevitable that Honduras would be drawn into the U.S.-sponsored Contra war against the Sandinistas. The United States rapidly transformed Honduras into a launching pad for Contra attacks against neighboring Nicaragua. The land was soon dotted with airfields, supply dumps, and base camps for Contra troops. Thousands of regular U.S. military and National Guard units rotated duty in Honduras, and the economy was inundated by the influx of hundreds of millions of U.S. dollars. All these activities reinforced the power of the Honduran military.

Civilian government survived, at least in name, as Rafael Leonardo Callejas ascended to the presidency in 1990 in a smooth transfer of office. Elections in 1993 went to Carlos Roberto Reina, of the Liberal Party, who struggled to deal with economic decline. His biggest political challenge came from the armed forces, which resisted the president's efforts to crack down on military collusion with international drug traffickers. And to protest investigations into alleged abuse of human rights, the army sent tanks into the streets of Tegucigalpa in August 1995. The display of power spoke for itself.

El Salvador: From Stability to Insurgence

Oligarchic control eventually took hold in nineteenth-century El Salvador. In 1863 Rafael Carrera launched an invasion from Guatemala and imposed a Conservative of his own liking, but Liberals countered with a successful revolt in 1871. Legal decrees in the 1880s prohibited the collective ownership of land by Indian communities and thus paved the way for the usurpation and consolidation of land by a tiny aristocracy— *las catorce,* the notorious "fourteen" families (which have meanwhile expanded in number and size). Coffee became the leading export crop, commerce flourished, and from 1907 to 1931 political power rested in the hands of a single family, the patriarchical Meléndez clan.

Peasants did not accept this passively. Angered by the loss of land, they staged four separate revolts between 1870 and 1900. The movements were crushed but they carried a message: like the *zapatistas* of Mexico, the peasants of El Salvador were willing to fight for their rights.

The ruling coalition of coffee-growing oligarchs, foreign investors, military officers—and church prelates—prevailed throughout the 1920s. The crash of 1929 had severe repercussions in El Salvador, since independent small farmers and plantation laborers suffered greatly from the drop in coffee prices. In 1931 a U.S. military attaché, Major A. R. Harris, filed this report:

There appears to be nothing between . . . high-priced cars and the oxcart with its barefooted attendant. There is practically no middle class between the very rich and the very poor. . . . Roughly 90 percent of the wealth in the country is held by about one-half of one percent of the population. Thirty or forty families own nearly everything in the country. They live in regal splendor [while] the rest of the country has practically nothing. . . . A socialistic or communistic revolution in El Salvador may be delayed for several years, ten or even twenty, but when it comes it will be a bloody one.

It would not take that long.

On May Day 1930 a popular throng of 80,000 held a demonstration in downtown San Salvador against deteriorating wages and living conditions. The next year an idealistic landowner and admirer of the British Labour Party, Arturo Araujo, won the presidential election with the support of students, peasants, and workers. Somewhat naively, he announced that the Salvadoran Communist Party would be permitted to take part in municipal elections in December 1931. Exasperated by this prospect, the armed forces dismissed him from office on December 2 and imposed a right-wing general, Maximiliano Hernández Martínez.

The peasants broke out in rebellion. In late January 1932, as a chain of volcanoes erupted in Guatemala and northwest El Salvador, bands of Indians armed with machetes made their way out of the ravines and tangled hillsides down into the towns of the area. Led by Agustín Farabundo Martí, a dedicated communist who had fought alongside Sandino in Nicaragua, the peasants murdered some landlords and plunged the country into a state of revolt.

Hernández Martínez responded with ferocity. Military units moved on the rebels and the conflict took on the appearance of a racial war, as Indians—or anyone resembling Indians—suffered from the government attack. In the tiny country of 1.4 million inhabitants, between 10,000 and 20,000 Salvadorans lost their lives.

The events of 1932 sent several messages. Peasants learned to distrust city-bred revolutionaries who might lead them to destruction. Indians began to seek safety in casting off indigenous habits and clothes. On the political level, leftists concluded that they could still cultivate followings in rural areas, especially in the absence of a reformist alternative. The right drew a stark lesson of its own: the way to deal with popular agitation was by repression.

A proto-fascist sympathizer, among the first to recognize the 1936 Franco regime in Spain, Hernández Martínez stayed on till 1944. Military officers seized power with the consent and blessing of *las catorce*. Major Oscar Osorio headed a moderate dictatorship in 1950–56. In 1960 his handpicked successor, Colonel José María Lemus, was overthrown by a civilian-military group with slightly leftist leanings under Colonel César Yanes Urías. Just one year later Yanes Urías was ousted by rightists under Lt. Col. Julio A. Rivera, whose Party of National Conciliation (PCN) took control of the state. This alliance of civilian

conservatives and military officers would reign supreme till the late 1970s.

A reformist challenge finally came from José Napoleón Duarte, who founded the Christian Democratic Party (PDC). As mayor of San Salvador (1964–70), the dynamic and articulate Duarte built up a strong following among the intellectuals, professionals, and other urban middle-sector groups. The PDC bore a commitment to peaceful reform through electoral means. Though Duarte may have won the presidential election of 1972, the recalcitrant military turned power over to one of its own, Colonel Arturo Armando Molina. Duarte himself was imprisoned, tortured, and exiled—but he did not take to the hills.

Conditions in the meantime worsened for the peasants. Coffee exports were thriving but the poor were suffering. About 80 percent of the people lived in the countryside, and by 1975 about 40 percent of the peasants had no land at all—compared to only 12 percent in 1960. Increasingly unable to gain access to the soil, the *campesinos* of El Salvador were getting ready to rebel.

The reform-oriented option gradually disappeared during the 1970s. The first attempt took the electoral road in 1972, and that was defeated by the military. The next step involved the formation of "popular organizations," apolitical groups that sought nonviolent routes to change. Sometimes organized by exiles like Duarte, they found support and stimulus from a revitalized institution: the Roman Catholic Church.

Indeed, the reawakening of the church has been one of the most decisive processes in contemporary El Salvador. The trend goes back to two events: the Second Ecumenical Council of the early 1960s (Vatican II), and the conference of Latin American bishops at Medellín, Colombia in 1968. The Medellín conference, particularly, denounced capitalism and communism as equal affronts to human dignity and placed the blame for hunger and misery on the rich and powerful. To redress these inequalities the bishops called for more education, increased social awareness, and the creation of *comunidades de base*, Christian communities of twelve to fifteen people each.

These developments had a profound impact on the ecclesiastical hierarchy in El Salvador, then under Archbishop Oscar Arnulfo Romero. As one priest firmly explained:

> it is not communism to make the right to organize and defend oneself known to the peasants. They have a right to defend their interests and to promote the political order of their choice, to defend their rights effectively, to denounce abuse by authorities or agents of the powerful. It is simply carrying out the Gospel mandate, a duty the church must not refuse or avoid.

As repression mounted, the church eventually acknowledged, in Romero's own words, "the case for insurrection . . . when all recourses

to peaceful means have been exhausted." No one was immune to violence: in 1980 the archbishop himself was shot dead in the cathedral of San Salvador.

Nonetheless, the realignment of the Salvadoran church had far-reaching implications. Other partners still belonged to the elite coalition: *las catorce* families, the military, and the foreign sector. But the church—at least, an influential segment of the church—had defected from the alliance and thrown its support to the masses. This altered the structure of power.

But secular and religious grass-roots organizations met with continued repression in the mid-seventies. The armed forces carefully controlled the elections of 1977, which turned the presidency over to General Carlos Humberto Romero. One of his most significant acts was to promote a "law to defend and guarantee public order."

The next phase in the deterioration of the political system began in October 1979, when a group of junior officers ousted Romero and set up a new government. At first the outlook was promising. The junta sought support from the "popular organizations." Given its commitment to human rights, the Carter administration greeted the regime with pleasure and relief.

Then things took a turn for the worse. Government repression persisted and killings continued at the rate of 1000 per month. The periodic estimates came from church and human rights groups who tried to monitor the slaughter. The cabinet resigned in protest, but the minister of defense—General José Guillermo García—hung on to his government post. The liberal wing of the Christian Democratic Party defected from the coalition. Now appearing undeniably conservative, the beleaguered Duarte took over as head of the government and announced a plan for land reform.

By this time the opposition moved underground. One group, the Democratic Revolutionary Front, was led by Enrique Álvarez, minister of agriculture in the first post-October 1979 junta. Another key element was the Farabundo Martí National Liberation Front (FMLN)—named for the leader of the 1932 uprising. In November 1980 Álvarez and five top associates were killed by government forces, an act that eliminated an entire cadre of reformist politicians. Guillermo Ungo stepped in to replace Álvarez, but the cause of moderate reform had suffered a devastating blow. Increasingly, guerrilla opposition to the regime took on a radical tinge.

The next month four North American women, three nuns and a lay worker, met a brutal death. The Carter administration protested vigorously and Duarte promised an investigation. In early 1981 the Reagan administration, more concerned with anticommunism than with social change or human rights, softened the U.S. demands. By mid-1982 a few low-ranking members of the National Guard were im-

plicated in the crime, but there would be no serious prosecution. Thus the regime survived international furor.

Yet Washington pressed for elections in early 1982. The goal was to elect a constituent assembly that would in turn select an interim president, and the campaign had ominous signs. The right was led by a fiery ex-major, Roberto D'Aubuissón, whom former U.S. Ambassador Robert White had once called a "pathological killer." Known as "Major Bob" by his admirers, D'Aubuissón summarized his platform with a campaign slogan: "Another '32," meaning that it was time for El Salvador to repeat the slaughter of 1932. The center, or center-right, was represented by Duarte and his semiconservative Christian Democrats. For its part, the Democratic Revolutionary Front—and the left as a whole—decided to boycott the elections. Ungo and his spokesmen argued that in the atmosphere of violence, leftist candidates were likely to be killed; voters would be intimidated; and the army would rig the results anyway. Consequently the election became a contest between the political right and the tattered remnants of the center.

Voter participation was remarkably high, if official statistics can be believed, and the results gave power to the right. Duarte's Christian Democrats won 35.3 percent of the votes, or twenty-four out of sixty seats in the constituent assembly. D'Aubuissón's party, the National Republican Alliance (ARENA), won 25.7 percent, or nineteen seats, but managed to form a working coalition with other right-wing groups and take control of the assembly.

Hopes were high in Washington that Duarte, a Notre Dame graduate and a favorite of U.S. policymakers, would realize the reformist programs designed to undercut support for the Marxist-Leninist guerrillas. In fact, Duarte was less effective in San Salvador than in Washington. FMLN fighters were highly disciplined and deeply entrenched in zones they had controlled for years. Duarte's government did redistribute significant chunks of farmland, but he could not displace the oligarchy that had made El Salvador's gap between the rich and poor among the worst in the Third World.

U.S. public opinion became a major factor in El Salvador. As of early 1983 the United States was supplying $205 million in economic aid and $26 million in military assistance, with higher requests pending in Congress. Few observers doubted that without this aid the regime in El Salvador would collapse. Growing opposition to the U.S. aid came from congressional liberals and religious groups, especially the Catholic Church, still incensed over the 1980 killing of the four American Catholic women. The intensity of U.S. opposition feeling could be seen in the bumper stickers that read "El Salvador is Vietnam in Spanish."

The battle continued in the Salvadoran countryside. FMLN guerrillas made periodic raids. Aided by U.S. military "trainers" (not called "advisers," to prevent association with Vietnam), government forces

conducted sweeping search-and-destroy missions. Villagers and peasants grew fearful of both sides. A decade of continuous fighting appeared to result in a stalemate. It had also led to the loss of 75,000 lives.

Presidential elections in March 1989 led to a decisive triumph for ARENA and for Alfredo Cristiani with 53 percent of the vote. Many observers believed that Cristiani, an athletic playboy without political experience, would be merely a puppet for D'Aubuissón and right-wing forces. In November 1989 six Jesuit priests were brutally murdered, apparently by a military-sponsored death squad. Cristiani solemnly declared that his government would capture and prosecute the assassins, but little was accomplished. Once again, a rightist regime was paying scant attention to human rights.

In keeping with the Esquipulas accords, Cristiani agreed to negotiate with the FMLN under the supervision of the United Nations. The talks were stalled by the murder of a top rebel negotiator, apparently at the hand of right-wing hard-liners, and then dragged on as each side accused the other of violating cease-fire arrangements. In March 1991 ARENA lost its legislative majority as two splinter parties—both proclaiming the need for national "conciliation"—together took more than 20 percent of the vote. The people of El Salvador seemed to be eager for a peaceful settlement.

In late 1991 UN-sponsored negotiations intensified, and in January 1992 the government and the FMLN signed a historic agreement for peace and reform. Under the accord, the FMLN agreed to lay down arms in exchange for wide-ranging reforms in political and military structures, including a reduction in the role and size of the armed forces and a purge of flagrant human-rights abusers. Under the supervision of a UN peacekeeping mission, the police force also underwent reform. By December 1992 the FMLN disarmed its guerrilla forces and became a legal political party, and in early 1994 the FMLN established itself as the country's second-largest political force—in elections that were won by Armando Calderón Sol of ARENA, which regained control of the national legislature as well. As Calderón Sol took office, he confronted the continuing challenge of economic reconstruction after a generation of civil war. It would not be an easy task.

Guatemala: Reaction and Repression

Guatemala has a long history of strong-man rule. After Rafael Carrera died in 1865, Justo Rufino Barrios established a twelve-year dictatorship (1873–85) and Manuel Estrada Cabrera followed with a twenty-two year, iron-fisted regime (1898–1920), the longest uninterrupted one-man rule in Central America. In 1931 General Jorge Ubico came to power and immediately launched a campaign to crush the fledgling Communist Party. Instead of relying on coffee planters alone, Ubico built a tentative base among agrarian workers by abolishing debt slav-

U.S. policy in Central America prompted a great deal of public controversy, and debate in the 1980s. *Above*, cartoonist Tony Auth satirizes President Reagan's position and the hesitancy of the U.S. Congress in voicing opposition; *below*, Steve Benson dramatizes the left-wing threat to U.S. interests. (Reprinted with permission of Universal Press. All rights reserved.)

ery. The national police maintained law and order. As Ubico once said of his tactics: "I have no friends, only domesticated enemies."

A wave of strikes and protests led Ubico to resign in July 1944. He was replaced by a military triumvirate, and this in turn was ousted by a group of junior officers. Thus came the October Revolution of 1944, an event that signaled the beginning of a decade-long transformation.

In an open election the following year Guatemalans elected as president Juan José Arévalo Bermejo, an idealistic university professor who proclaimed a belief in "spiritual socialism." Arévalo oversaw the promulgation of a progressive new constitution in 1945, modeled in part on the Mexican charter of 1917, and he encouraged workers and peasants to organize. Industrial wages went up 80 percent between 1945 and 1950. Arévalo pushed education and other reforms as well. But the going was not easy: during his five-year term in office Arévalo weathered no less than twenty-two military revolts.

In 1950 Arévalo turned the presidency over to Colonel Jacobo Arbenz Guzmán, the minister of defense, who led a center-left coalition in the elections of that year. A central figure in the October Revolution of 1944, Arbenz developed profound social concerns—partly at the insistence of his wife, María Vilanova, a wealthy Salvadoran who resembled Argentina's Evita Perón in her ambition for her husband to win political power through greatly increased social welfare benefits. Arbenz accepted Communist support, both during and after the election, but he was a reformer at heart. At his inauguration he spelled out his hopes for the country's future:

> Our government proposes to begin the march toward the economic development of Guatemala, and proposes three fundamental objectives: to convert our country from a dependent nation with a semi-colonial economy to an economically independent country; to convert Guatemala from a backward country with a predominantly feudal economy into a modern capitalist state; and to make this transformation in a way that will raise the standard of living of the great mass of our people to the highest level.

To achieve these goals, Arbenz said, Guatemala would need to strengthen its own private sector, "in whose hands rests the fundamental economic activity of the country." Foreign capital would be needed too, "as long as it adjusts to local conditions, remains always subordinate to Guatemalan laws, cooperates with the economic development of the country, and strictly abstains from intervening in the nation's social and political life." Finally, the new president declared, Guatemala would embark on a program of agrarian reform.

Arbenz set quickly to work. He authorized construction of a public port on the Atlantic coast and the building of an east-west highway. He convinced the legislature to approve an income tax—a watered-down version of a mild proposal, to be sure, but the first in Guatemalan his-

tory. He pushed for expanded public works and the exploitation of energy resources, including petroleum.

The centerpiece of the Arbenz administration was agrarian reform. Enacted in June 1952, the bill empowered the government to expropriate only uncultivated portions of large plantations. All lands taken were to be paid for in twenty-five year bonds bearing a 3-percent interest rate, and the valuation of land was to be determined according to its taxable worth as of May 1952. During its eighteen months of operation the agrarian reform distributed 1.5 million acres to some 100,000 families. The expropriations included 1700 acres owned by Arbenz himself, who had become a landowner through the dowry of his wife.

Almost immediately, Arbenz and the agrarian reform ran into a serious obstacle: implacable opposition from the United Fruit Company and from the U.S. government. *La frutera* had obvious reasons for resisting the reform. The company held enormous tracts of land in Guatemala, 85 percent of which was unused—or, as the company maintained, it was being held in reserve against natural catastrophes. And in arranging tax payments, UFCO consistently undervalued its holdings. (On the basis of tax declarations, the Guatemalan government in 1953 offered UFCO $627,572 in bonds in compensation for a seized portion of property; on behalf of the company, the U.S. State Department countered with a demand for $15,854,849!)

Washington was deeply involved. Some of the ties were personal. Secretary of State John Foster Dulles and his brother, CIA Director Allen Dulles, for example, both came from a New York law firm with close links to United Fruit. The company's Washington lobbyist was Thomas Corcoran, a prominent lawyer who was on close terms with President Eisenhower's trusted aide and undersecretary of state, General Walter Bedell Smith, himself once interested in a management position with UFCO. More important than personal ties, however, was the anticommunist doctrine developed in Washington.

The early 1950s had seen a well-articulated rationale on the needs of U.S. national security in a Cold War era. The United States had no choice, so the reasoning went, but to fight back against the Soviet Union and its client powers, which were dedicated to the overthrow not only of capitalism but of all the Western democracies. The Third World (a term not yet used then) would be a favorite Soviet target, argued Cold War theorists, and would be subverted by communist parties or their fellow travelers. The most drastic challenges thus far had been in Europe (the Berlin Blockade, the Greek Civil War, the French and Italian elections) and in Asia (the Korean War, the fall of Nationalist China, and the Indo-Chinese civil war). Was Latin America to be immune?

U.S. policymakers had pushed a hard anticommunist line in relations with Latin America. The Rio Pact of 1947 had laid the groundwork

for collective action, or so the United States hoped, against communist advances in Latin America, whether from within or without. In early 1953 John Foster Dulles was clearly worried about Latin America, where, he said, conditions "are somewhat comparable to conditions as they were in China in the mid-thirties when the communist movement was getting started. . . . Well, if we don't look out, we will wake up some morning and read in the newspapers that there happened in South America the same kind of thing that happened in China in 1949." The test came in Guatemala.

UFCO publicists and the Dulles brothers accused the Arbenz regime of being "soft" on communism and branded it a threat to U.S. security and to the free world at large. They cultivated fears that defeat in Guatemala might lead to a Soviet takeover of the Panama Canal. They warned that if Guatemala fell, then the rest of Central America might go as well (the "domino theory"). But the principal issue was agrarian reform. Such writers as Daniel James of *The New Leader* warned that communists would use the program as a stepping-stone to gain control of Guatemala. Whatever his intentions, the United States insisted, Arbenz was just a "stooge" for the Russians.

In August 1953 the United States decided to act. John Foster Dulles led a campaign in the OAS to brand Guatemala as the agent of an extrahemispheric power (the Soviet Union) and therefore subject to OAS collective action under the Rio Treaty of 1947. When the Eisenhower administration pressed for this interpretation at a Caracas meeting of the OAS in early 1954, all it got was a declaration stating that communist domination of a member government would cause concern and should in theory lead to collective action—but with no specific mention of Guatemala.

The Arbenz government now saw that U.S. intervention was likely. The regime cracked down on domestic opposition and turned to East Europe for small arms, which were en route by May. Meanwhile the U.S. government was demanding, in increasingly blunt language, compensation for U.S. property in Guatemala, meaning, of course, United Fruit.

Having failed to get OAS sponsorship for intervention in Guatemala, the Eisenhower government opted for covert action. The State Department had mounted the diplomatic offensive; now it was the turn of Allen Dulles and the Central Intelligence Agency. The CIA organized an exile invasion under an obscure renegade Guatemalan colonel, Carlos Castillo Armas. A rebel column of a few hundred men was assembled across the border in neighboring Honduras. They were equipped and directed by the CIA, which set up and operated a rebel radio station and provided a few World War II fighter planes to strafe Guatemala City. Under attack by these planes, and convinced that a large army was approaching the capital, Arbenz lost his nerve and gave up. The Castillo Armas rebels rolled into the capital virtually unopposed.

The new government purged communists and radical nationalists, reversed the expropriation of United Fruit lands, and dutifully signed a mutual Defense Assistance Pact with the United States in 1955. The errant Central American republic had been brought back into line by a relatively cheap and efficient CIA operation.

The United States was strongly denounced by Latin American nationalists for its intervention in Guatemala, and to this day it is a symbol for Latin Americans of cynical U.S. action. As described twenty years later by a CIA officer who had been intimately involved in the overthrow of Arbenz: "Castillo Armas was a bad president, tolerating corruption throughout his government and kowtowing to the United Fruit Company more than to his own people. The United States could have prevented this with the vigorous exercise of diplomatic pressure on Castillo Armas to assure that he pursued social reform for the many rather than venal satisfaction for a few. Instead, Washington breathed a collective sigh of relief and turned to other problems." Even so, the fate of the Arbenz regime would serve as a warning to nationalist leaders who contemplated challenging U.S. corporations.

The 1954 coup marked a turning point in Guatemalan history. It virtually eliminated the forces of the political center (as represented by Arévalo and Arbenz). So the country had only a left and a right, and the right was in control. Coffee planters, other landowners, and foreign investors and their subsidiaries regained their power under the protection of neo-conservative military regimes. Since then individual rulers have come and gone, but this alignment has persisted. Castillo Armas was assassinated in 1957. General Miguel Ydígoras Fuentes, who had lost to Arbenz in 1950, ruled from 1958 to 1963. Then came Colonel Enrique Peralta Azurdia. From 1966 to 1970 the presidency was held by Julio César Méndez Montenegro, a talented civilian who traced his political lineage back to Arévalo, but the armed forces kept him tightly in check. He was followed by Colonel Carlos Arana Osorio (1970–74), by General Kjell Langerud García (1974–78), and by General Romeo Lucas García (1978–82). In March 1982 power was seized by Efraín Ríos Montt, a flamboya... retired officer and born-again evangelical Christian; in mid-1983 he was ousted by General Oscar Humberto Mejía Victores. The more the leaders changed, the more the system stayed the same.

One feature of this entire period, especially after the mid-1960s, was the frightful abuse of human rights. Paramilitary death squads, most notoriously *Mano Blanca* ("White Hand") and *Ojo por Ojo* ("Eye for an Eye"), carried on a murderous campaign against political dissenters. No fewer than 80,000 people were killed or "disappeared" between the 1960s and the 1990s. The government bore at least indirect responsibility for these killings, but worldwide protests did not bri   much respite.

By the mid-1980s the Guatemalan military judged their campaign

against the Marxist guerrillas successful enough to allow the election of a civilian president. The victor was Vinicio Cerezo, a Christian Democrat and a centrist, who would govern only at the pleasure of the military. In 1990 he was succeeded by Jorge Serrano Elías, a center-right candidate, whose triumph reflected popular disillusionment with traditional political parties. Unable to build a working majority in Congress, Serrano suddenly dissolved the legislature in 1993 and announced that he was assuming dictatorial powers, much as Alberto Fujimori had done in Peru the year before. As international and domestic condemnation of the *auto-golpe* came to a crescendo, however, support for Serrano collapsed and he quickly fled the country.

The remaining two years of the presidential term were filled by Ramiró de León Carpio, a former human-rights ombudsman who initially inspired hope for positive change. In early 1995 the United Nations oversaw negotiation of a peace accord between the government and guerrilla forces, as well as an agreement to respect the rights of indigenous peoples. Government forces continued to wage intermittent war with guerrilla groups, however, and de León Carpio found himself obliged to defend the Guatemalan military in a diplomatic uproar over the killings of an American citizen and the spouse of another. In the meantime Efraín Ríos Montt, whose regime in the early 1980s was noted for its cruelty, began to position himself (or his wife) as a candidate in presidential elections in late 1995. Under a patina of electoral democracy, the military forces continued to predominate in Guatemala.

By the mid-1990s Central America seemed far from realizing its dreams. The people of the isthmus were suffering poverty, they were enduring large-scale violence, and they were facing continuous outside pressure. Part of the solution might lie in unification, but even that goal appeared to be beyond reach. In one historian's phrase, Central America was still "a nation divided." But there were bright spots, too: the tiny country of Belize (1985 population 159,000), formerly British Honduras, achieved independence in late 1981; and there remained a hope that economic integration might create the conditions for long-term collaboration and unification. After decades of warfare in Central America, activists were striving to heal historical divisions and to achieve the potential that they believed to be rightfully theirs.

ELEVEN

LATIN AMERICA, THE UNITED STATES, AND THE WORLD

The history of Latin America has by no means evolved in a vacuum. Ever since the time of Columbus the region has been constantly subjected to external forces. In the sixteenth century Spain and Portugal conquered and settled the continent with varying degrees of violence. During the seventeenth and eighteenth centuries Latin America became a pawn and a prize in European politics, as ambitious peoples from enterprising nations—England, France, and Holland—sought and seized strongholds in Iberia's New World. The early decades after independence marked a temporary decline in outside influence, as the new nations of Latin America turned inward and the expansionist powers of Europe focused their attention on more lucrative possessions in Africa, India, and Asia. From the 1880s onward, for the last century or more, the Latin American economies have been deeply integrated into the capitalist global economy, a system dominated by countries at the industrialized center—West Europe, the United States, and later Japan. To this extent, Latin America has never escaped its dependency on the outside world. Despite persisting and valiant struggles to achieve autonomy, it has for the most part managed only to modify, not eliminate, the form, nature, and extent of its centuries-old dependency.

Even the name Latin America reflects an imperialistic legacy. The term was coined by the French in the 1860s, when they were supporting Maximilian in Mexico and laying the cultural groundwork for a political and economic offensive throughout Spanish and Portuguese America. They argued for a Latin essence—but embodied in French culture—which lay at the heart of Ibero-American civilization. This was not pure fancy, since French culture enjoyed enormous prestige among the New World elites.

Whatever the realities of the French claim to a homogeneous influence, Latin America began to exhibit a continental sentiment as early as the late eighteenth century. Indeed, the Spanish and Portuguese colonial system could hardly have failed to create a sense of common interest in the countries they spawned.

To understand modern Latin America's world position we must understand the nature of the Spanish and Portuguese colonial empires. Most important, the economic structures were mercantilist: designed to integrate the colonies totally into the mother country's economy. In Spanish America, that even meant no intracolonial trade—all commerce from each region had to be with Spain alone. The practical effect was to distort the economic development of the colonies. Had they been able to trade more openly, they might have had considerably different economic options from those they inherited at the time of independence. (Portuguese America, too, was allowed to trade only with the mother country.)

The results of the Iberian mercantilist policies were cultural and psychological as well as economic. Spanish and Portuguese America had been nurtured on the model of closed societies. They were constantly on guard against political and economic inroads from European rivals, as well as against religious heresies emanating from Protestant Britain and the Low Countries.

Significant non-Iberian influences made their mark on colonial Latin America. The 5.2 million slaves shipped from West Africa to South America, the Caribbean, Mexico, and Central America brought a tradition different from either the Amerindian or the European. Despite the efforts of their European or *mestizo* masters, the Afro-Americans of Latin America made lasting marks on society and culture. The process went deepest in Brazil and the Caribbean, but was also important in Colombia, Venezuela, Mexico, and Central America. Languages, food, sports, and music all show profound and continuing African influence. The slaves engaged in rebellion as well, fleeing and founding runaway communities in Brazil and the circum-Caribbean, especially in Guyana and Curaçao. In the seventeenth and eighteenth centuries Brazil had flourishing runaway settlements, which by the nineteenth century had been either wiped out or absorbed by the growth of surrounding societies.

The Spanish and Portuguese American possessions might have continued well into the nineteenth century as colonies if it had not been for war in Europe. It was Napoleon's invasion of the Iberian mainland that provoked the crisis of loyalty in the New World. The crisis deepened, the monarchies were impotent, and creoles took charge of their own affairs.

Once the French were defeated, England emerged as the preeminent military, economic, and political power. The English moved at least part way into the vacuum created in the Americas. Their principal objective was economic—to promote Britain's commercial interests, which had relied on contraband trade throughout the eighteenth century. The basis for this activity would be England's strong network of economic institutions: the best available in shipping, banking, insurance, and investment capital. In effect, the British were seeking to replace

the former Iberian colonial institutional infrastructure linking Latin America to the world economy. But they differed from Spain and Portugal in a key respect: they laid almost no territorial claims in Latin America. The English sought economic gain without the burden of direct political rule. It was an "informal imperialism," whereby Europe's chief investor and trader avoided the expensive link of territorial control—with its potential military entanglements.

What of the United States? It was by no means a hemispheric power in the early nineteenth century. On the contrary, the United States was unable even to prevent Washington (and the White House) from being devastated by the English in the War of 1812. Washington had significant contact in Mexico and the Caribbean, but even there it could not begin to rival English seapower.

Yet the United States had become at least a symbol of postcolonial success for Latin American creole elites. Most significantly, it had thrown off European control. Born in a revolution with a philosophical rationale from the Enlightenment, the United States demonstrated how a republic could emerge from European colonialism. The North American rebels had fought for the right of representation—something which Spanish and Portuguese Americans notably lacked. The federal structure of the United States was another feature that impressed the patriots among the creole elites of what were to become large countries such as Mexico, Brazil, and Argentina, where the reconciliation of regional autonomy with national power would become a longstanding issue of importance.

From Independence to Pan Americanism

The independent republics (plus the empire of Brazil) of nineteenth-century Latin America were an anomaly on the world scene. Along with Haiti and the United States, these Spanish- and Portuguese-speaking nations offered the first cases of successful popular rebellion against European colonial rule. Once independent, however, the new nations played but a minor role on the world scene in the 1800s. On the whole, they became absorbed in their own development.

That was not what Simón Bolívar, the great revolutionary of the Wars of Independence, had hoped. He had a vision of a united Spanish America emerging from the anticolonial wars. He dreamed that Spanish Americans could subordinate regional loyalties to unite into a single South American nation. But Bolívar's dream was denied. The rebellious colonies fractured along the lines of the old Spanish administrative units. Bolívar bitterly denounced his creole contemporaries for their provincialism. It was no use. Each new nation would have to find its own way in the world.

An early concern in postindependence Latin America was the nature of the relationship with the former mother countries. For Caribbean

islands such as Cuba, Puerto Rico, and Jamaica, colonial rule continued. In Brazil, the former colony had become a "co-kingdom," with a
monarch in Rio de Janeiro. The rest of Latin America had to adjust to
the new reality of dealing with Spain as merely another distant European
power. Yet that could never be the case—the ties of blood and of such
institutions as the church, higher education, and publishing still gave
Spain a special place in Spanish America. Spain was now so divided between liberals and conservatives, however, that it could no longer attempt to project a single influence over its former American colonies.

The United States attempted to assert its power with the "Monroe
Doctrine," promulgated by President James Monroe in 1823. Originally
aimed at czarist Russia's possible claims to the American Northwest, the
doctrine became better known for its challenge to an apparent design
of the European Holy Alliance to help Spain reconquer its former colonies. President Monroe firmly declared that "the American continents,
by the free and independent condition which they have assumed and
maintained, are henceforth not to be considered as subject for colonization by any European powers." Further strictures warned the Europeans against using indirect means to extend their political power in the
New World. As later put in a popular slogan, the basic message was
clear: "America for the Americans."

In fact the message evoked indifference or scorn in continental Europe, some worry in Britain, and considerable sympathy in Latin
America. Within the United States the Monroe Doctrine became highly
popular (the phrase dates only from the 1850s) and was a cornerstone
of U.S.-Latin American policy for the next century. Within Latin
America, however, the United States in the early nineteenth century
lacked the economic and military power, as well as the political will, to
make a significant impact.

France attempted to extend its influence in a brief 1838 military expedition to Mexico to enforce payment of damage claims against the
Mexican government by French citizens. A negotiated settlement allowed the French to withdraw. Along with the British, the French also
repeatedly blockaded the ports of the River Plate between 1838 and
1850. But it was Britain that exercised the predominant extrahemispheric influence during most of the nineteenth century. Britain had
by far the most powerful navy, capable of asserting itself throughout
Latin America, despite the fact that the territory was 8000 to 12,000
miles from British ports. Second, Britain had the capital, the merchants, the bankers, and the insurance and shipping brokers to facilitate trade between Latin America and Europe, the great market. And
Britain had an ideology—liberalism—which underlay its expansionism
and which Latin American creole elites quickly assimilated. It offered
a rationale for integrating Latin America into the world economy,
which, not coincidentally, the British controlled.

Between the 1820s and the 1850s, the British pressed their advan-

As World War I approached, the United States sought to assert its sphere of influence throughout the Western Hemisphere under the Monroe Doctrine. (Charles L. "Bart" Bartholomew [?], *Minneapolis Journal*, 1912. Reprinted with permission of the Minneapolis Star and Tribune Company.)

tage in Latin America. They quickly took over the financial and commercial services for the major countries. They rapidly arranged loans for governments and stock companies in Mexico, Brazil, Argentina, and Peru. But capitalist development of these South American republics proved more difficult than the British creditors had hoped. Most borrowers defaulted on their loans, and the London financial markets soon came to mark Latin America as a high risk for credit.

Mexico was the one place where an extra-Latin American power other than Britain made an impact before 1850. There the United States pursued its "manifest destiny" by seizing huge portions of Mexican territory. The United States also made threats in the Caribbean and Central America. By midcentury the Caribbean and Mexico, so close to the United States, had come to be seen as special cases in Latin America's relations with the world.

From 1850 to 1880 Latin America changed its posture toward the outside world. Liberalism, both political and economic, gained increasing sway. In Argentina, for example, the nationalist Rosas was toppled

by his Argentine enemies, thanks in part to British and French intervention. As the Argentine liberals came to power, they installed in their country such admired institutions of the English-speaking democracies as the U.S. constitution and the U.S. public school system. The liberals also believed in developing their country's economy by accepting foreign trade. That meant applying the principles of economic liberalism—minimal state interference with the market. In practice, that was likely to reinforce Britain's predominant economic position.

A similar application of liberalism was being attempted in imperial Brazil. Emperor Pedro II modeled his rule on the British monarchy, although he exercised much more discretionary political power than did Queen Victoria. The two parties in the Brazilian imperial parliament were modeled on Britain's two-party system; even congressional speeches were filled with references to British constitutional precedents.

The years between 1850 and 1880 also saw French influence grow in Latin America, especially in culture. French was the foreign language most widely spoken by the elite, reflecting traditional practice in Europe itself. French cultural prestige in Latin America lasted well into the twentieth century, remaining predominant in some countries until the end of World War II, well after France's decline as a world power. Why this resilience? Partly because France was spared the Latin American nationalist reaction directed at the British, partly because French cultural values coincided with the Latin American elite's self-image. However French and Latin elites might admire the economic prowess of the Anglo-Saxons, they scorned the "materialistic" values that had suffused Britain and the United States. The French intelligentsia had produced its own rationale in defense of their country against Britain— arguing that France had a more humane vision of society than Britain's dehumanizing Industrial Revolution, and the Latin American elites identified with France's self-proclaimed cultural superiority.

The Rise of U.S. Influence

Between 1880 and the outbreak of World War I, Britain lost supremacy in Latin America. Other European powers, especially France and Germany, increased their economic ties, competing with British investors and merchants. But the most important challenge came from the United States.

During these thirty-five years, U.S. influence spread southward. The deepest penetration was closest to home, in the Caribbean and Mexico. U.S. investors found the Mexico ruled by Porfirio Díaz to be attractive for investment and trade, and they poured capital into Cuba as well. Even before the Spanish-American War erupted in 1895, the United States had come to overshadow Spain in Cuba's economy.

U.S. entry into Cuba's revolt against Spain signaled a new phase in

North America's relations with Latin America. After the Mexican-American War (1846–48) the United States continued to occupy and settle portions of the present-day Southwest. When the United States entered Cuba in 1898 and decisively defeated the Spanish, it was more than a military victory. It was a symbolic struggle that impressed all of Latin America. The time was long past when Latin American nations were absorbed only in their own problems.

The 1898 war was an independence battle by Cuban patriots against Spain. Other Latin Americans immediately sympathized with the Cubans. Yet the Yankees suddenly seized control of the rebellion. This naturally demoralized many Cubans, who had desperately hoped to defeat the Spaniards on their own. Furthermore, it underlined a common Latin American worry: Were the Spaniards and their American descendants inherently (i.e., biologically) weaker than the North Americans? Was the United States racially "destined" to take over Latin America?

A similarly agonizing self-appraisal had gripped Mexican elites after their humiliating defeat by the United States in 1846–48. They explained their weakness on the battlefield by linking it to supposed defects in their character or, more ominously, their racial background. Now, in the 1890s, racism was at its height in the United States and Europe. The Jim Crow laws of the United States had institutionalized segregation, and the universities and churches overflowed with professors and preachers who calmly explained the scientific basis for believing in "inferior" and "superior" races.

These racist doctrines also penetrated Latin America. In country after country, leading intellectuals faithfully repeated the racist dogmas of the Europeans. Underlying such introspection was a lurking fear that some new conquistadores might take away their lands, manifested in the occasional nativist reaction against the British through the nineteenth century. The precipitating incidents were often petty, such as brawls by British sailors when ashore, but they called forth a deep hostility. The Latin American elites knew that both the British and the North Americans often viewed them with contempt.

It is therefore not surprising that the crushing defeat of the Spaniards by the United States in 1898 deeply troubled the Latin Americans. Having seized Puerto Rico and Cuba (the latter as a protectorate), it was widely asked, how much farther would the Yankees go? The pessimists in Mexico had long believed that the United States had designs on more of their land.

U.S. expansionism into the Caribbean showed that the Latin American fears were not entirely groundless. The U.S. search for a satisfactory transcontinental canal site revealed what the Yankees would do to achieve their economic and geopolitical interests. President Theodore Roosevelt, that arrogant embodiment of North American imperialism, rode roughshod over objections to his site for the canal in a northern-

most section of Colombian territory extending northward into Central America. Thus was born the new nation of Panama, and the canal was built between 1904 and 1914. The United States got its canal, but it took the bogus creation of a new republic.

Between 1880 and 1914 the United States also attempted to create a new hemispherewide alliance of nations. It began with the ambitious plans of the U.S. Secretary of State James Blaine. On his initiative the First Pan-American Conference was held in Washington in 1889. Between 1826 and 1864 there had been five international congresses, all attracting only a limited number of countries from Latin America. The 1889 meeting was the first to involve the United States as well as all the nations of Latin America. Ironically, this was when the United States was accelerating its imperialist offensive in Latin America.

Out of the 1889 conference came an authorization for a "Commercial Bureau of the American Republics," from which emerged the Pan-American Union and later the Organization of American States (OAS). Accomplishments were limited primarily to commercial issues.

The U.S. diplomatic, economic, and military offensive into Latin America after the 1880s prompted a strong reaction among Latin American intellectuals and students. The Spanish-American War aroused José Enrique Rodó, Uruguayan-born and one of the most celebrated of these militant critics of the United States. In 1900 he published *Ariel*, a slim essay in which he contrasted North America's excessive materialism to Latin America's superior cultural sensitivity. Of the United States he charged: "Her prosperity is as great as her inability to satisfy even a middling concept of human destiny. . . ."

Rodó's ideas had great influence in Spanish America, whose elites were well prepared to hear of their spiritual superiority. They were also spurred to think in terms of a Latin American identity arising from the cultural unity toward which Rodó was groping. The theme was picked up and spread widely by such writers as the Argentine Manuel Ugarte, who wrote newly elected President Woodrow Wilson in 1913 to demand that "the stars and stripes cease to be a symbol of oppression in the New World." Other famous intellectuals caught up in this pan–Latin American reaction were José Martí of Cuba (whose critique began in the 1870s), Rubén Darío of Nicaragua, and Rufino Blanco-Fombona of Venezuela. Their writings were published throughout the continent in what was one of Latin America's most culturally integrated eras. Many themes from this anti-*yanqui* movement have survived in inter-American relations down to our day.

European influence continued to be highly important in Latin America between 1880 and 1914, despite the growing U.S. role. One need only look at the size of investment and trade. Table 11-1 demonstrates the magnitude of North American investment, which came to embrace the Caribbean and South America as well as Mexico and Central America. Even so, as Table 11-2 indicates, Britain was far and away

TABLE 11-1   U.S. Investment in Latin America, 1897–1914
(millions of dollars at end of year)

	Caribbean Countries	Mexico and Central America	South America	Total
1897	4.5	221.4	37.9	304.3
1908	220.2	713.0	129.7	1,062.9
1914	329.0	946.7	365.7	1,641.4

Source: United Nations, Economic Commission for Latin America, *External Financing in Latin America* (New York: United Nations, 1965), p. 14.

the most important single source of capital on the eve of World War I; France and Germany were also prominent.

Another important sphere of European influence was military training and technology. From the time of their Wars of Independence Latin America elites had known that they lacked military skills and advanced weaponry. To defeat the Spanish garrisons, usually undermanned and cut off from resupply, the patriots often had to hire or appeal to foreign soldiers and sailors.

As the nineteenth century continued, other European military establishments offered examples to ambitious Latin American rulers. In both Chile (1885) and Argentina (1899), for example, German military missions were contracted for extended periods to introduce new weapons and teach their use. Equally important, they taught new methods of staff organization and command. In Brazil the rapidly growing state of São Paulo contracted in 1905 a French mission to train the state militia, which included cavalry. These European missions were expected, above all, to transmit the new professionalism that had transformed the military in Europe. War was no longer a matter for aristocrats; now it was a serious business which required thorough scientific and technical training.

TABLE 11-2   Long-Term Foreign Capital Investment in
Latin America, 1914 (millions of dollars)

Origin	Foreign Private Investment (and External Public Debt)	Percentage
Britain	5066	51.9
France	1013	10.4
Germany	367	3.8
United States	1487	15.2
Others	1821	18.6
Total	9754	100.0

Source: United Nations, Economic Commission for Latin America, *External Financing in Latin America* (New York: United Nations, 1965), pp. 16–17.

What did this transfer to military "professionalization" imply for Latin America? It had enormous consequences in politics, as the "professionalized" armies became more active, ironically, in the constitutional realms of Chile, Argentina, and Brazil. The political implications would become obvious only after World War I.

The Consolidation of U.S. Influence

The First World War, although of no direct concern to Latin America at the outset, fundamentally changed the region's relationship with the world. For one thing, it accelerated Britain's decline as the most important economic force in the hemisphere. Drained by the long and costly hostilities on the continent, England had to draw on its overseas investments to pay for the war. Furthermore, Britain had begun to experience a long-term decline in world economic competitiveness.

Second, the war highlighted the dynamic U.S. economy, based on a continent full of resources and now mature enough to become a net exporter of capital. The decisive U.S. intervention in the war proved it could now tip the balance of economic and military power: European powers could never again disregard the United States.

Latin America was essentially a bystander in the conflict. The United States pressed hard to use the war as an occasion to strengthen its political influence in Latin America, but with varying results. Only eight Latin American republics declared war on Germany: Brazil (the only major country), Cuba, Costa Rica, Guatemala, Haiti, Honduras, Nicaragua, and Panama. Another five broke diplomatic relations: Bolivia, the Dominican Republic, Ecuador, Peru, and Uruguay. Seven nations remained neutral: Argentina, Chile, Colombia, Mexico, Paraguay, El Salvador, and Venezuela.

By the conclusion of World War I, U.S. power and influence in Latin America were clearly on the rise. The United States now exercised virtual hegemony in the Caribbean basin, as could be seen in the military occupations of Nicaragua (1912–25 and 1926–33), Haiti (1915–34), the Dominican Republic (1916–24), and Cuba (1917–22). Even when the United States did not occupy these countries, it deeply influenced their development, wielding veto power over their domestic politics. Most of the elites in these countries took U.S. hegemony for granted; indeed, they would have been surprised to think the world could have been otherwise.

The two exceptions to this pattern were Mexico and Cuba. In Mexico, the Revolution of 1910 threatened to change the country's relationship with the United States. Washington was disturbed over the direction taken by the Revolution, and made repeated interventions, including the military landing at Veracruz in 1914 and the sending of General John J. Pershing's column in 1916 to pursue Pancho Villa's irregulars. Yet the Revolution succeeded in giving the Mexicans an

élan, a pride, a degree of mobilization that would eventually help check the traditional brand of U.S. intervention. It nonetheless remained true that the Mexican people could do little without taking due account of the continuous pressure of that powerful neighbor to the north.

Cuba was a different case. There the United States had established a protectorate after ending its military occupation in 1902. This U.S. presence generated a nationalist reaction, erupting most often among students at the University of Havana. Although easily suppressed, they represented an important dissenting opinion within the elite, and in the 1930s they seized center stage in Cuban politics.

The world depression hit Latin America hard. New capital ceased flowing into the region, and foreign investors found it hard to repatriate profits. Country after country defaulted (or declared a unilateral moratorium) on their debts.[1] The reason was obvious. The collapse of the world economy had reduced demand for the primary products on which Latin America depended for its foreign exchange earnings. Suddenly these countries had no way to earn the dollars, pounds, marks, or francs to repay their foreign creditors. So Latin Americans could not now expect to receive any net inflow of capital. Even more important, they would be short of foreign exchange to pay for the imports essential to domestic economic development, especially industrialization.

The 1930s were thus a time when Latin American countries had to look inward. Not coincidentally, it was a period of high nationalist feeling. With the options for foreign economic help so reduced, it was logical to concentrate on domestic resources. Argentina, Brazil, and Mexico, for example, all took steps to increase national control over the oil industry. Mexico went the farthest, when in 1938 President Cárdenas nationalized all foreign oil firms. Argentina had already created an autonomous government oil enterprise in the 1920s; in Brazil, by contrast, the state oil enterprise was given its definitive shape in 1953.

The 1930s also brought one new import to Latin America: fascist ideology. The clearest version came from Italy, where Mussolini's movement grew in response to the emerging revolutionary left—especially the syndicalists and the communists. With the rise of Nazism in Germany, European fascism gained an even more powerful exemplar. Especially important for Latin America was the fact that Spain and Portugal had both fallen prey to authoritarian regimes (Franco after 1936 in Spain and Salazar after 1928 in Portugal) which had corporatist, many would say fascist, overtones.

Europe's swing to the right gave ammunition and prestige to those antidemocratic and antiliberal groups in Latin America that had their own reasons for wanting to create authoritarian governments. None of the Latin American "fascist" movements were exact copies of European

---

1. With the exceptions of Argentina, Haiti, and the Dominican Republic.

cases. In Brazil, the Integralists were a primarily middle-class movement, preaching the need for order, with an essentially corporatist message. Argentina had several rightist paramilitary groups that resembled European fascist organizations, but the deepest influence was in the military. The secret "lodges" within the officer corps produced initial stimulus for the Peronist movement, which owed part of its inspiration to the Italian model. But Peronism never gained a doctrinal and organization form like that of Italian fascism. In Mexico there was a small rightist movement, *sinarquismo*, which showed European influence; it also drew primarily on right-wing Christian sentiment. In Chile, finally, there was a National Socialist Party, clearly an imitation of the Nazi Party, which provoked a furious response from the Chilean left.

None of these parties or movements came close to winning power in the 1930s. In Chile and Mexico they faded, as the civilian political system proved capable of containing and absorbing them. In Argentina fascist sentiments were overshadowed by the Peronist movement, once World War II entered its final phase. In Brazil the Integralists almost appeared to be on the threshold of power in 1936, but they were swept away by Vargas's *Estado Nôvo* in 1937.

As the 1930s continued, Latin America became a staging ground for geopolitical competition. The Germans and the Italians both looked to Latin America to increase their economic and political influence. The Italians were especially active in Brazil and Argentina, where large Italian communities offered a possible base of operations. The Mussolini government sought to keep the Latin American republics pro-Italian. That meant stimulating and reinforcing anti-British (and, by extension, anti-U.S.) opinion. The Italian government also contributed directly to Brazilian fascists, by using its embassy to pass large sums of money to the Integralists.

As the most powerful partner in the European Axis, Hitler's Germany was even better situated to exert pressure in Latin America. The Nazi regime used its fully owned subsidiary airline, Condor, and its diplomatic service to create a network of agents and contacts throughout South America. Hitler and his more aggressive advisers convinced themselves that Latin Americans of German descent were eager to join the Fatherland and create separatist territories, alert to Berlin's every direction.

But the Germanic population in Brazil proved to be extremely loyal to its New World nation. The effect of Nazi measures in Brazil was just the opposite of what the Germans had intended: the Vargas government adopted stringent new laws requiring that all school classes be taught in Portuguese, thereby eliminating the exclusively German language schools in the south of the country. The result was to accelerate the assimilation of the German descendants in Brazil.

The Axis powers were of course not the only outside nations engag-

ing in this geopolitical competition. For many years the U.S. military had regarded Latin America as a vulnerable flank. This provided the military rationale for expansion in the Caribbean: the need to protect vital sea lanes. The vast, virtually unpatrolled boundary with Mexico was another area difficult to defend. The Panama Canal gave the United States an additional danger zone. What was needed, some officers thought, was an arc of military security stretching into northern South America.

As Nazi Germany built its war machine, U.S. officers cast a more worried look at the Latin American flank. With Japanese power growing in the Pacific, the United States was faced with the likelihood of a two-ocean war. Latin America could now be menaced from both east and west. The United States therefore examined its hemispheric ties in a new light: How could Washington gain maximum support from the Latin Americans in case of war?

Part of the answer would depend on cultural relations. Latin America between the wars saw a sharp rise in cultural programs sponsored by foreign governments. Because of their enormous cultural prestige, the French had the favored position. French was still the most widely spoken foreign language, and Paris was the point of reference for artists and writers of Latin America. Latin American universities, primarily copied from Iberian models, came under increasing French influence. In Brazil, for example, when the *paulistas* wanted to found a new university—the University of São Paulo, in 1933—they sent a delegation to France. In response, Paris sent a mission of leading professors in the social sciences, including such luminaries as Claude Lévi-Strauss, Jacques Lambert, and Pierre Monbeig. The French government established a fellowship program for study in France; similar programs also emerged elsewhere in Latin America.

While European influence remained dominant among the elites, North American influence was making great inroads through mass media—the unchallenged vehicles of popular culture in our century. By the 1930s Hollywood films were the craze throughout Latin America. U.S. film stars such as Jean Harlow and Clark Gable were household names in even provincial towns, as Argentine novelist Manuel Puig's *Betrayed by Rita Hayworth* poignantly shows. U.S. music, especially jazz, had a similar effect. The dynamic, fluid, dazzling, and futuristic North American society proved fascinating, although engendering a split between elite and popular reactions to the United States which was to deepen and prove increasingly important over time.

As the war broke out in Europe, the U.S. government stepped up its recruiting of military and political allies in Latin America. Franklin Roosevelt's government sought: (1) military bases, especially in the Caribbean and on the Atlantic coast of South America; (2) guaranteed accessibility to vital raw materials, such as natural rubber and quartz; (3) willingness to join the United States in an alliance against the Axis,

should the United States join the fight; or (4) at the very least, a neutrality that would prohibit hostile extracontinental powers from creating footholds in their countries.

The United States largely achieved these aims. The inter-American system, emerging with painful slowness since the 1880s, was now harnessed to the security interests of the United States. The U.S. military got the use of valuable bases, especially in Brazil, and U.S. war industry got access to the vital raw materials it needed.

World War II greatly increased U.S. influence in Latin America. Germany and Italy, once important powers in the region, were discredited: Italy, by its blustering lurch for new territory in North Africa; Germany, by the horrendous suffering its war machine unleashed on Europe. Fascism was a bankrupt legacy by 1945. Only the anomalous survival of Franco in Spain and Salazar in Portugal kept alive the right-wing ideology, and neither country, ironically, had great political influence in Latin America.

The British and French, although victors in the war, were gravely weakened. Both had to liquidate overseas investments to pay for the war. Neither had the resources or the will in 1945 to compete for influence in distant Latin America. The net result was that 1945 saw U.S. influence at an all-time high in Latin America.

The Expression of U.S. Influence

The United States emerged from the Second World War with greatly increased prestige and authority. The war brought the U.S. economy out of the depression and into a massive industrial effort. Unlike Europe or Japan, however, the United States suffered no war damage at home. The economy was intact and prosperous. In 1945 the United States accounted for half the world's manufacturing output and about two-thirds of global exports. The United States had built up the most imposing arsenal the world had seen, capped by the atomic bomb—the "ultimate weapon," which only the United States possessed. The war also provided the United States with a network of alliances that offered a strong power base in postwar international politics.

With the war over, many Latin American politicians hoped the new-found U.S. interest in Latin America would pay off. They expected increased attention to their problems, especially the obstacles to economic growth. After all, Washington could now afford to look south, where it would be logical to consolidate its greatly increased influence.

But it was not to be. U.S. government policy virtually lost track of Latin America after 1945. Attention focused, instead, on rebuilding Europe and Japan. The Truman administration (1945–53) and the Congress, along with farsighted business leaders, realized that an economically sound Europe was essential for a prosperous United States. The Marshall Plan of 1947 appealed to the U.S. public on both human-

itarian and economic grounds and helped to direct attention primarily toward Europe.

What about Latin America? It simply did not seem significant to U.S. policymakers. The Latin American specialists in the State Department and the military services found themselves downgraded or transferred. The Truman administration apparently assumed it would continue to receive loyal backing from Latin America, almost as a matter of course.

This relative U.S. indifference was broken again by an outside threat. As U.S. relations with the Soviets began to cool, the Truman administration decided to mount a Cold War offensive in Latin America. It assumed two aspects. First was pressure to get Latin American governments to sever diplomatic relations with the Soviet Union. This was remarkably successful, as every country with the exception of Mexico, Argentina, and Uruguay followed suit. The second aspect was to press Latin American governments to outlaw the local communist parties. Although not highly publicized in the United States, the success of this campaign demonstrated how responsive the Latin American political elites still were to U.S. direction.

The Truman administration also decided to make more permanent the military alliance created during the war. A special 1945 meeting of hemispheric foreign ministers in Mexico City agreed on the need to redesign the Pan-American system. The first step was taken in 1947, when delegations approved a treaty (the "Rio Pact") defining an attack on any American state, from inside or outside the hemisphere, as an attack on all, requiring collective measures to counter the aggression.

The second step was taken in Bogotá, Colombia, in March 1948, when a new body was created: the Organization of American States (OAS). The structure included a legal charter creating a council to deal with day-to-day business, inter-American conferences every five years, and foreign ministers' consultative meetings to handle threats to the hemisphere. A bureaucratic infrastructure took shape as the General Secretariat and the Pan-American Union. OAS member states committed themselves to continental solidarity (which the United States wanted) and total nonintervention (which the Latin Americans wanted), along with the principles of democracy, economic cooperation, social justice, and human rights. In short, the United States and Latin America created the world's most highly articulated regional association. Not surprisingly, the member states expected very different things from it.

The OAS got an important test when war broke out in Korea in 1950. The United States had troops stationed in South Korea and was thus immediately drawn into the conflict. As North Korean troops streaked south, Washington convinced the UN Security Council to brand North Korea as the aggressor. The Truman government then turned to the OAS, asking it to define the North Korean attacks as aggression against the United States—thereby obligating the OAS

members, by the Rio Pact, to join the battle. What the United States wanted was additional legitimacy and military contributions from Latin America.

The other OAS members balked at this appeal, except for Colombia, which sent a battalion of infantry. In their speeches the Latin Americans showed much more concern for their own economic problems than for the military clash in distant Korea. The OAS produced a compromise: in return for a commitment to improve their military defenses and increase military cooperation, the Latin Americans got a U.S. promise that it would submit proposals for solving the hemisphere's economic problems.

To U.S. policymakers the Korean War demonstrated that their battle with communism was worldwide, not just in Europe. The "fall" of China in 1949 further dramatized the point. In 1951 the Truman administration and Congress accordingly decided to extend to Latin America the U.S. Military Security Program of 1949, originally aimed at Europe. From 1952 to 1954 the United States signed bilateral mutual defense assistance pacts with ten Latin American countries: Ecuador, Cuba, Colombia, Peru, Chile, Brazil, the Dominican Republic, Uruguay, Nicaragua, and Honduras. (Argentina and Mexico were conspicuous by their absence.) Under these agreements the United States was to exchange military equipment and services in return for Latin American promises to expand defense capacities, to send strategic materials to the United States, and to restrict trade with the Soviet bloc.

The implications of these new defense arrangements were far-reaching. Washington was tying Latin America's armed forces into the U.S. web—once possessing American equipment, they would depend on the United States for parts, replacements, and ammunition. Furthermore, by frequent contact with U.S. military, in training programs and joint exercises, the Latin American officers could be expected to identify closely with the United States. No less important, the United States was offering far more equipment than the recipient countries could have bought through the normal appropriations of their governments. The armed services thereby gained power in their societies without having to fight budget battles at home. These U.S. military links were a revival of the structure the Roosevelt administration had created on the eve of U.S. entry into World War II. Now, in the early 1950s, the United States was building on its wartime prestige to expand and consolidate its influence.

One other country earned great prestige during the war: the Soviet Union. The Soviets had lost far more of their population (20 million) and endured much more suffering than the United States. Many Latin Americans admired the endurance and fortitude of the Russian people, and some of that admiration was transferred to the Latin American communist parties, which enjoyed a flurry of popularity in 1945.

The Brazilian Communist Party, for example, won 10 percent of the vote in the presidential election of 1946. Later that year three members of the Chilean cabinet were communist, and there were communist members of the national congress in Cuba, Colombia, Peru, Ecuador, Brazil, Chile, Bolivia, Uruguay, and Costa Rica.

The Truman administration began pursuing an aggressively anti-Soviet line in Latin America as early as 1946. The chain of bilateral military pacts was only one of the instruments to mobilize the Latin Americans against the Soviets. Since military officers were normally a bulwark of anticommunist and anti-Soviet opinion, the United States hoped to tie this pro-U.S. element even tighter and strengthen its anticommunist resolve. Military pacts and training programs now gave the United States a monopoly on foreign links among Latin American military.

At first glance this penetration might look similar to the relationship the Soviets established with its East European statellites after 1945. The Soviets, however, achieved a far greater integration of their client nations' military structure into the Soviet one—with direct link-up of command structure, communication systems, and Soviet intervention in day-to-day operations. The Soviets were especially careful in screening candidates for higher command positions. They took further advantage from the fact that large numbers of their own troops were still stationed in many of the satellites. Although the Soviet control of its satellites' military was more pervasive and complete, it resembled the United States in Latin America in that both superpowers had little opposition.

The Truman administration also expressed an interest in technical and economic help for Latin America, as a partial replication of the Marshall Plan—the oft-cited prototype of American overseas aid. But the situations proved to be very different. The Marshall Plan was directed at nations which were physically devastated but still possessed the most important economic ingredient of all: skilled and experienced manpower. An investment in these European countries—Britain, France, Italy, West Germany—could and did have a quick payoff. These industrial economies revived and were poised to participate in the rapid growth of world trade that was to occur after 1950.

The economic problem in Latin America was more fundamental. There was relatively little industry even in the largest countries. There was an enormous shortage of skilled labor and technological know-how. Infrastructure (roads, railroads) was often lacking. Economists did not know how to stimulate rapid economic development in such areas as Latin America.

Whatever the labels, economists and planners soon found that simply supplying more capital, in the form of dollars or investment goods, was not the answer. Economic development was (and is) a complex process that requires adjustment at every societal level. The Truman adminis-

tration favored a sympathetic look at the problem. In his 1949 inaugural speech President Truman proposed (in his address it was "Point Four," a label which then stuck) a U.S. government-coordinated technical assistance program to aid developing countries. This was in response partly to Latin American complaints that the United States was ignoring their fundamental economic problems, while concentrating its attention on Europe.

Before Truman's Point Four program had a chance to go far, the Democratic Party was turned out of the White House by the 1952 electoral victory of General Dwight D. Eisenhower. The Republicans brought a new philosophy, especially in international economics. They took a strongly laissez-faire stance, and thought government's first obligation was to let the free market operate. President Eisenhower's secretary of the treasury, Cleveland businessman George Humphrey, soon made clear that this free-enterprise philosophy would rule out any large-scale economic aid overseas, even through low-interest government loans. An equally cold shoulder was turned to current proposals for commodity agreements, designed to stabilize world-market prices for such products as coffee and cacao—primary products of great importance to Latin America. Above all, said the Republicans, Latin Americans must not discourage private foreign investment, which in these years meant primarily U.S. investment. (See Figure 11-1 for data on the growth and location of postwar American overseas investment: Latin America attracted the largest regional share of U.S. investment in 1950, though it would be far surpassed by Canada and West Europe by the 1980s.)

This Republican orientation toward Latin America provoked a strong reaction among reformist and development-minded younger politicians and technocrats in Latin America. Not coincidentally, Latin America was now beginning to produce its own analysis of its economic problems. This diagnosis was to help Latin American elites define their relationship to the outside world. Ideological competition was turning intense once again. In the 1930s the confrontation had pitted fascism and corporatism, linked to growing Axis power in Europe, against economic and political liberalism, linked to the United States and Britain. Now, in the 1950s, it was radical nationalism and Marxism versus neo-liberalism, with a position in between that we here refer to as reformist developmentalism.

The Nationalist Impulse

Nationalism has deep roots in Latin America. In early form it contributed to the rebellion against Spain and Portugal in the early nineteenth century. During the course of the nineteenth century, however, few nationalists were prepared to resist foreign economic intrusions. The dictator José Gaspar Rodríguez de Francia of Paraguay (1814–40) was

FIGURE 11-1   U.S. Overseas Investment, 1950 and 1985

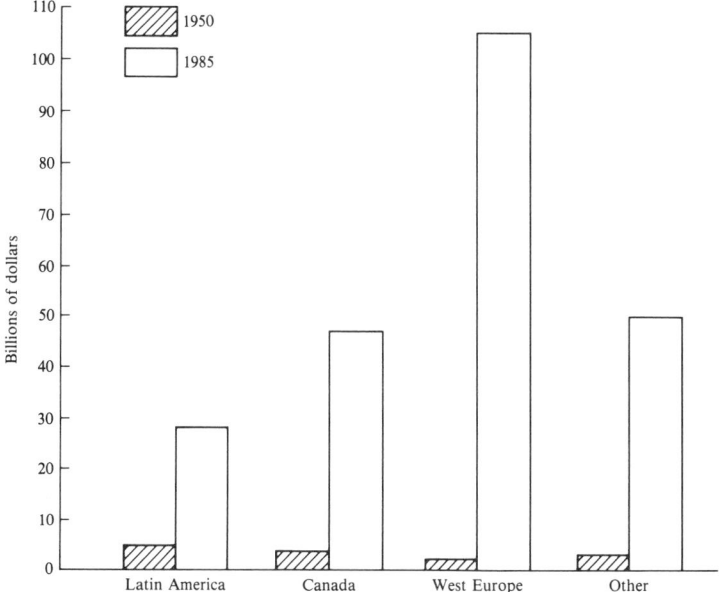

Sources: *New York Times*, 23 January 1967; and U.S. Department of Commerce, *U.S. Direct Investment Abroad* (Washington: U.S. Government Printing Office, 1990), Table 1397, p. 797.

one, as were his successors from the López family (1844–70). But Paraguay was a relatively insignificant country, and its defiance of British economic power made little difference to the rest of the continent. The occasional outbreaks of economic nationalism in Chile and Argentina before midcentury left little permanent effect. The nineteenth century saw the triumph of economic liberalism in Latin America. Its export-oriented strategies were based on the law of comparative advantage. A few nationalist voices cried out against this strategy, but without much practical impact.

It was not until the 1930s that the economic nationalists could hope to capture policymaking. One of their targets was oil exploration and production. Some nationalists, such as Alejandro Bunge in Argentina and Roberto Simonsen in Brazil, also argued vigorously for industrialization. They reasoned that their countries could never gain control of their economic future until they industrialized. In other words, they should break with a simple-minded application of the law of comparative advantage, which restricted them to exporting primary goods to pay for imported finished goods. They must create the productive capacity to survive slumps in the world economy, such as the 1929 crash.

The pro-industrialization arguments began to influence government policy in the 1930s in Brazil and Chile and, to a lesser extent, in Argen-

tina. In all three countries the state took the initiative by creating public enterprises (state oil and steel companies in Argentina and Brazil, CORFO in Chile, PEMEX in Mexico). This greatly increased the state role in the economy, which was anathema to believers in liberal economic doctrine.

After World War II, economic liberalism staged a comeback in Latin America. In part it was tied to the resurgence of political liberalism. It flourished also because the world demand for primary products was strong in 1946–47, boosting foreign exchange earnings of the Latin American economies. By the early 1950s, however, they had all run into trouble. The world demand for primary products became erratic, and their prices fluctuated dizzily. The pro-industrializationists reappeared, arguing that economic liberalism was no solution for Latin America's future.

In the 1950s the debate was joined by an articulate new Latin American voice: the Economic Commission for Latin American (ECLA), a United Nations regional agency created in 1948. It was to be a secretariat of technicians, primarily economists, who were to analyze systematically the economic problems of the Latin American region and its individual countries. Its executive secretary was Raúl Prebisch, a talented Argentine economist who had played a key role in the creation of the Argentine central bank in the 1930s. ECLA was located in Santiago, Chile, in a deliberate effort to obtain distance from the U.S.-dominated atmosphere of the OAS headquarters in Washington, D.C.

ECLA became an aggressive participant in the analysis of Latin America's relationship to the world economy. One of its major accomplishments was to train a generation of Latin American economists who learned, during their time with ECLA, to see their countries' problems in a continental perspective. They also got to know their counterparts in other nations and were able to compare notes on the problems and possibilities of economic policymaking. Essentially, ECLA helped to produce a Latin American mentality in economic analysis.

It would be misleading to imply that ECLA had a single message. It was careful to confine its publications to technical analyses of past economic trends, along with discussion of current policy options and projections of the effects of likely policy choices. Yet the choice of topics and the direction of the analysis showed the strong influence of Prebisch and his disciples.

The Prebisch-ECLA thesis, as it became known, was that the world economy since the 1880s had been working systematically to the disadvantage of the countries that relied on the export of primary products. It was argued that the prices of finished goods went up faster than prices for primary goods, so the developing countries found themselves in a steadily deteriorating position vis-à-vis the industrial countries. The way out of this dilemma? First, adopt international commodity agreements to protect primary-product exporters from huge market-

price fluctuations. Second, the larger developing countries should industrialize.

These conclusions were never clearly stated in ECLA documents, which had to remain technical studies produced by a branch of the UN bureaucracy. But they were the unmistakable implications, and they were widely recognized as such.

These arguments provoked fierce responses, both in Latin America and the United States. They irritated the Eisenhower administration (1953–61), which saw ECLA as a beehive of deluded statist thought, promoting policies likely to harm private enterprise. In fact, the Prebisch-ECLA analysis furnished ammunition for centrist politicians such as Vargas, Kubitschek, Frondizi, and Frei. Above all, ECLA gave reform-minded Latin Americans the confidence to shape their own development strategies, knowing they could call on an intellectual resource which was authentically Latin American.

In other words, ECLA offered a milestone in Latin America's search for self-knowledge. This was an important step, since Latin American analysts and politicians had long felt at a disadvantage when facing the economists, bankers, and businessmen of the industrial world.

The other important intellectual force gaining new strength in the 1950s was Marxism. To understand the force of Marxism in Latin America, one has to look closely at the history of the communist parties, since they were the primary channels of Marxist thought, at least into the 1960s. It is true that there were a few influential noncommunist party Marxist thinkers and politicians, of whom Peru's José Carlos Mariátegui was the most famous and influential. But they were the exceptions. It was the orthodox communist parties, affiliated with the Comintern in the 1920s and 1930s, which did most to spread the Marxist message. By the early 1950s they had lost the prestige that rubbed off from the Red Army in 1945. They were a negligible force in elections, except in Chile. They had some strength among student organizations and labor unions, but in general they suffered from identification with Stalinism.

Their Marxist analysis was a relatively crude form of economic determinism. Its application to individual countries was spelled out by the local parties, acting on Moscow's instructions. Few Latin Americans completely accepted the party line. When the communists moved to tap sentiments of economic nationalism, however, they could mobilize a wider following. Nationalist sentiment was usually directed against the United States, on which the communist parties concentrated their fire. The communists and many Marxists sought to develop a clearly antiimperialist (meaning, above all, anti-American) political stance. Nonetheless, Marxist thought was relatively isolated in Latin America in the 1940s and early 1950s, closely identified with a network of communist parties under Soviet discipline.

A crisis in the 1950s furnished strong evidence for the anti-

imperialist argument. As described in Chapter Ten, the United States strongly opposed the reformist program of Colonel Jacobo Arbenz in Guatemala. The CIA-organized exile invasion of Guatemala in 1954 imposed Carlos Castillo Armas as president, who promptly reversed the expropriation of United Fruit lands, and dutifully signed a mutual defense assistance pact with the United States in 1955. The Soviets, for their part, merely stood by and watched. A similarly managed coup succeeded in Iran in 1953, ousting Mossadegh and reinstating the Shah. With its power and wealth, the United States was flexing muscle on a global scale.

The State Department could count on a solid phalanx of Latin American votes in the UN. The dogma of free enterprise continued to dominate U.S. hemispheric policy, despite a chorus of Latin American voices calling for international commodity agreements, a multilateral development bank for the Americas, and expanded accessibility to technology. Democratic reformers in Latin America, such as José Figueres in Costa Rica, Rómulo Betancourt in Venezuela, and Eduardo Frei in Chile, urged the United States to encourage democracy and social reform. The United States had too often favored the dictators, argued Figueres, thereby nurturing the kind of extreme leftism that emerged in Guatemala. Time was growing short—"one minute to midnight," as Figueres liked to announce dramatically.

Such pleas fell on deaf ears in Eisenhower's Washington until Vice President Richard Nixon traveled to Latin America in 1958. The trip was largely ceremonial, to attend the inauguration of President Arturo Frondizi of Argentina, with stops in seven other South American countries. Nixon's well-publicized past as a virulent anticommunist made him a natural target for student leftists in Latin America. Despite repeated warnings from his security staff, Nixon insisted on making his public appearances. There were demonstrations everywhere, but they were especially ugly in Lima and Caracas. On one occasion Secret Service agents had their guns drawn against a mob attacking Nixon's car when the driver just succeeded in whisking the shaken vice president away. The events got worldwide publicity, and Washington worried over the possible causes of such anti-American feeling. The first important revisions in U.S.-Latin American policy began to appear in 1959, when the U.S. government dropped its opposition to the creation of an Inter-American Development Bank. There were also plans to expand U.S. bilateral aid. But events in the Caribbean soon overtook this modest change in official U.S. thinking.

The Revolutionary Option

During his guerrilla war against the dictator Batista, Fidel Castro had presented himself as a democratic reformer, fighting tyranny to restore

representative democracy. Once in Havana, however, Fidel moved steadily to the left. He knew that any threat to U.S. investment—or to U.S. political hegemony—in Cuba would bring U.S. intervention. Early in 1959, Fidel made overtures to the Soviets; by the end of the year Cuba was receiving economic aid from Moscow. By the end of another year Cuba had completed an almost total switch in trade relations— from overwhelming trade dependence on the United States to over- whelming trade dependence on the Soviet Union. Soviet military aid had also begun pouring into Cuba. The Soviets were obviously now willing to take a far bigger gamble than they ever contemplated in Gua- temala in 1954.

The U.S. reaction foreseen by Fidel now began. The attempted land- ing at the Bay of Pigs in early 1961 might have succeeded if President John F. Kennedy had ordered use of U.S. navy air cover, though he declined to do so. The dilemma was all the more acute because the Kennedy administration was working frantically to produce a new and more enlightened Latin American policy.

The United States was humiliated: first by the failure of the anti- Castro invaders, second by the clumsy cover used to conceal U.S. involvement. The defeat left Fidel stronger than ever. He had dramatic proof that the United States presented a constant threat to Cuba's secu- rity. He could now crack down against all domestic opposition.

The failed invasion also confirmed the entry of the other superpower into the hemisphere. The long-discussed U.S. fear—Soviet penetration of the Americas—now appeared to be fact. If the Soviets were willing to supply the Cubans, how many other guerrilla movements in Latin America might hope for the same support? That worry goaded the Kennedy policymakers to speed up the formulation of their new Latin American program.

The new policy, as announced in 1961, had two distinct aspects. The first was a major multilaterally sponsored economic and social develop- ment program, christened the "Alliance for Progress." It was to involve both economic growth and social reform, to be carried out by demo- cratic governments. Each Latin American nation would have to submit a detailed development plan to a council of nine "wise men," econo- mists and technocrats. The U.S. government promised to provide funding, up to $20 billion over ten years. It also promised to push multilateral authorities and private investors in the United States and Europe to increase their capital flows to Latin America. The United States launched this new policy in a blaze of publicity, including a gala White House dinner for the Latin American diplomatic corps.

Considerable groundwork for this new program had been laid by the Eisenhower administration. Yet the dramatic embrace of the cause of social reform, and the willingness to press for public funds, would have been difficult for a Republican administration. In effect, the U.S. gov-

ernment was now adopting the logic put forward by ECLA and by the democratic reformers, such as Figueres and Betancourt. To the surprise of many of its detractors, the U.S. government was now claiming the leadership of a peaceful social revolution in Latin America.

The second aspect of the Kennedy administration's new Latin American policy was less publicized. It was a stepped-up counterinsurgency program by which the U.S. government would help Latin American governments fight guerrilla movements. The Kennedy policymakers were not betting on economic development and social change alone. Even the most committed reformer, they reasoned, might run afoul to a well-organized guerrilla opponent. Conventional military and police could not do the job alone. New methods were needed, such as the British had perfected in their successful antiguerrilla campaign in Malaya. The Kennedy mystique was therefore now applied to the creation of a new, elite corps of counterinsurgency experts. They were given a distinctive green beret, and told that training counterpart forces in the Third World would be one of their main tasks.

As part of this effort to protect favored governments, the United States increased its military supply and training programs. There was a premium on modernizing the military, making them capable of the rapid and coordinated action needed to fight well-trained guerrillas. A new element was also added: the training and equipping of police forces to control urban riots and sabotage, and improvement of their techniques of interrogation and record retrieval. This activity was housed in the U.S. Agency for International Development (USAID) and bore the euphemistic label of "public safety."

The theory behind this two-pronged U.S. policy was that the United States had an interest in identifying and supporting strongly reformist leaders, giving economic aid to achieve growth, and at the same time offering the means to defeat armed domestic opponents who might have Soviet or, more likely, Cuban help. The United States was now betting on the anticommunist reformers to ride the historical tide in Latin America and to produce more prosperous, more egalitarian, and therefore stronger nations with interests compatible with those of the United States. Soviet penetration would be stopped, the U.S. sphere of influence would be preserved, and everyone in the hemisphere could be better off.

The Alliance for Progress was born to widespread applause. The glamor and eloquence of the new U.S. president, contrasting sharply with the pro-big business image of Eisenhower, excited Latin American imaginations. Liberals in the United States and Europe were thrilled, as were those long-time critics of U.S. policy who had excoriated Uncle Sam for always siding with the dictators.

Then Cuba hit the headlines again. The CIA, apparently with clearance from the White House (or at least Robert Kennedy), tried several plots to assassinate Castro. All failed. Meanwhile Fidel tightened his

grip on Cuba, and looked to spread the revolution elsewhere in Latin America.

The Soviets became emboldened by U.S. weakness at the Bay of Pigs and decided to act. During 1962 they flooded Cuba with military equipment, including missiles—hardly a defensive weapon, since the missiles were too close to be detected by the U.S. warning system. The United States got the OAS to approve a blockade against all ships carrying offensive weapons to Cuba. There followed a frightening face-off, and the Soviets finally agreed to remove the missiles, but only when the United States promised to lift the blockade and pledged not to invade Cuba in the future.

The "missile crisis" of October 1962 had great implications for Latin America's relations with the United States and the world. First, the Soviets had been forced, by threat of direct U.S. attack on their Cuban installations, to give up the strategic advantage they tried to win by stealth. In effect, the Soviets were ratifying anew the U.S. strategic military hegemony in Latin America outside of Cuba.

Second, the United States had promised to keep hands off of—at least by direct invasion—the "first socialist experiment in the Americas." Fidel now knew he could concentrate on building socialism at home and exporting revolution abroad. Nonetheless, he needed a large security force to ward off CIA and exile-sponsored raids, as well as to prevent Cubans from fleeing the island.

Castro was in a good position to extend support—including money, materials, and training—to revolutionaries elsewhere in Latin America. Cuba became a "platform" for the export of revolution. A prime target was Venezuela, where President Rómulo Betancourt was a prototype of the democratic reformer the Kennedy administration sought in Latin America. In November 1963 the Venezuelan government announced its discovery of arms sent to Venezuela from Cuba, and asked the OAS to act. The Venezuelans charged that the arms were intended for local guerrillas, precisely the kind of antidemocratic forces Washington had said would justify U.S. counterinsurgency aid. The OAS dutifully endorsed the Venezuelan charge and, in July 1964, asked member nations to sever all diplomatic ties with Cuba and to suspend all commercial and transportation links. Cuba was to be an outcast nation in the hemisphere. All OAS members complied, except Mexico, which had reason to want the U.S.-Cuban stalemate to continue, since it had partially replaced Cuba both as tourist attraction and as a supplier of sugar.

The pattern for Cuba's role in the hemisphere was now set. Backed up by an extraordinary Soviet subsidy—far greater per capita than the U.S. subsidy to any of its client regimes in Latin America—the Castro regime pressed ahead on two fronts. One was the difficult task of constructing socialism in Cuba, far more difficult than most revolutionaries had foreseen in 1960. The second was the promotion of revolution

abroad. It was best exemplified by Ché Guevara, who soon found that, like the achievement of socialism at home, the promotion of revolution abroad was easier said than done.

The Fidelistas were by no means the only source of the call to revolution in Latin America during the 1960s. Just as Soviet penetration via the Cuban revolution challenged U.S. hegemony in the hemisphere, so the Sino-Soviet split in world communism fragmented the revolutionary left. Back in the 1940s and 1950s, the Latin American left had been dominated by national communist parties, almost all under Moscow's direction. In the late 1950s, however, the split between Beijing and Moscow created new opportunities for Latin American revolutionaries. The Chinese communist Maoists gave inspiration to Latin Americans long dissatisfied with the bureaucratized orthodox communist parties. They yearned for revolution now, not in some distant era when "objective conditions" would be favorable. Revolutionaries of this bent arose in every major country in the 1960s. Almost invariably they met a bloody end.

Democratic Reformers and the Alliance for Progress

The framers of the Alliance for Progress banked on an improbable combination of favorable factors: elected governments promoting economic growth while achieving social reform. If Latin America had all the problems its analysts so often described, how could politicans suddenly produce the consensus necessary to carry out such ambitious programs? Why would the wealthy and privileged stand by? Could economic growth be achieved if governments mounted social reform threats to established producers?

Elsewhere we have seen the fate of democratic reformers in Chile, Brazil, Argentina, and Peru. In Chile, Frei (1964–70) fell far short of his goals in key areas such as land reform and redistribution. Power then passed to a more radical reformist, Salvador Allende (1970–73), under whom politics polarized dangerously and the economy spun out of control, due in part to U.S. economic warfare against Chile. Reformism—of which Allende was still a representative—had failed dramatically. Chile's repressive military regime after 1973 was part of what the Alliance for Progress was designed to avoid.

In Brazil, Jânio Quadros (1961) at first appeared to be the dynamic vote-getting reformist Brazil needed. But he resigned after only seven months in office, opening a three-year political crisis that culminated in a coup against his ineffectual successor, João Goulart (1961–64). The military-dominated governments after 1964 emphasized economic growth, with little effort toward social reform. The result was an increase in social inequality, combined with a steady move in the 1960s away from direct elections—a path the architects of the Alliance for Progress would have thought was an invitation for trouble.

In Argentina, the best hope for democratic reformism was Arturo Frondizi's government (1958–62), but it was soon jeopardized by a highly unpopular anti-inflation program and became a casualty of the long-standing Peronist-military confrontation. None of the succeeding governments in Argentina came close to the model envisioned in the Alliance for Progress.

Venezuela, one of the few countries to maintain a continuous civilian government, was a bright spot for the U.S.-promoted reformist policy. Yet it produced little significant social reform and its relevance was always compromised by its windfall of oil revenues.

In Peru, President Fernando Belaúnde Terry (1963–68) looked like a democratic reformer, and he bet on economic development, especially by opening up the Peruvian Amazon. But he ran afoul of nationalists in the military and also met severe economic difficulties. He was then deposed in a military coup led by General Juan Velasco Alvarado, whose military regime carried out a land reform more radical than any contemplated by Belaúnde. Velasco was in turn replaced by a more pro-private-sector military regime, which allowed elections again in 1980, and Belaúnde, long in exile in the United States, was reelected to the presidency. But the economic odds were stacked against his reformist government.

Colombia was another country where U.S. planners placed great hope. The president from 1958 to 1962, Alberto Lleras Camargo, was an articulate and attractive Liberal Party politician from a prominent Colombian family. His cadre of economists and technocrats prepared the kind of detailed economic and social plans called for in the Alliance for Progress. They won large loans from the U.S. government and the multilateral agencies, and Colombia was soon labeled a "showcase" for democratic reformism by enthusiastic U.S. observers. Unfortunately, the hopes proved premature, as the Colombian government achieved little reform in crucial areas such as land redistribution. Although economic growth was forthcoming, precious little of it benefited the rural sector or the mushrooming shanty towns around the cities. By the mid-1960s Colombia had become a prime example for U.S. congressional critics, such as Senator J. William Fulbright, who found far too few of the Alliance aims accomplished in this "showcase" country.

Events in the Dominican Republic brutally exposed the contradictions in U.S. policy. The assassination in 1961 of Rafael Trujillo, one of the Caribbean's most notorious dictators, opened the way for a free presidential election—won by Juan Bosch, a popular reformer whose ideas certainly fit the mold of the Alliance for Progress. Despite U.S. support, Bosch was deposed by a military coup in 1963. Another armed revolt in 1965 triggered fears in the Lyndon Johnson administration that a Castro-like regime might emerge, which would be a disaster both for Johnson's foreign policy and for his standing with U.S. domestic opinion. Johnson listened to all his advisers and then sent in

20,000 U.S. troops. They were joined by troops from Brazil, now ruled by a military government anxious to show its Cold War zeal.

A new civilian, Joaquín Balaguer, was elected president in 1966 and the U.S. and Brazilian troops left. But the United States had aroused resentment in much of Latin America for the heavy-handed manner in which the Johnson administration demanded (and just barely got) OAS blessing for the U.S. intervention. It wasn't quite the 1920s again, but it certainly wasn't the bright new era John Kennedy had envisioned in 1961.

Clearly, the Alliance for Progress had failed as of 1970. The expectations had been too high, given the political realities of the decade. Furthermore, the goal of promoting democracy soon conflicted with the goal of preventing any more Cubas. In Brazil, for example, the United States became convinced that President João Goulart was leading his country dangerously leftward and Washington therefore quickly endorsed a military-civilian conspiracy when it overthrew him. By the end of the decade the Brazilian military regime had joined the ranks of the highly repressive, with little pretense of social reform; nevertheless, it continued to be the largest beneficiary of U.S. aid.

In Argentina between 1963 and 1966, the United States found itself dealing with the ineffectual civilian government of Illia, which was succeeded by the repressive regime of General Onganía. But here too, a general had his attractions for Washington, as did his military successors, because they were cracking down on the growing guerrilla movement.

The ideology of the Alliance for Progress was set by John F. Kennedy and the "New Frontiersmen," although President Lyndon B. Johnson pledged to continue the basic policies when he took office in November 1963. The election of Richard Nixon in 1968 brought a change in U.S. policy. Although Nixon maintained a rhetorical commitment to democracy and social reform, at heart he and his advisers favored a return to the more traditional Republican stance of leaving economic development primarily to the private sector. The Nixon administration also increased military aid, acting on the advice of Governor Nelson Rockefeller, whose 1969 presidential mission to Latin America pointedly noted that "a new type of military man is coming to the fore and often becoming a major force for constructive social action in the American republics." The implications were obvious.

Nixon was the first U.S. president to have to deal with an elected Marxist head of state in Latin America. Salvador Allende's victory in 1970 was a test for the United States, publicly committed to social reform but strongly opposed to leftist movements. Although Allende's regime never reached a revolutionary stage, the Nixon administration was determined from the day of Allende's election to use every means ("make the economy scream" was one of Nixon's suggestions to CIA

Director Richard Helms) to prevent his inauguration or, failing that, to speed his overthrow.

The United States discouraged new private investment in Chile and obstructed, wherever possible, Chile's access to financing from multilateral agencies such as the International Monetary Fund, the World Bank, and the Inter-American Development Bank. At the same time, President Nixon ordered the CIA to develop and carry out a secret plan for harassing the Chilean government. At least $10 million was spent in subsidies to the Chilean opposition press (especially the militantly anti-Allende *El Mercurio*) and to opposition groups, including many strikers who helped paralyze the Chilean economy in 1972–73. Seen in the context of all Allende's problems, the U.S. effort was probably of marginal importance. It was the Chilean military and the Chilean middle and upper classes who rose up against the Popular Unity government. They needed no lessons from the United States in why and how to do it. But the United States increased the sense of siege felt by the Allende government and encouraged the opposition to believe generous U.S. aid awaited them after a coup.

Not the least important result of the Allende era was the wide publicity about U.S. covert activities. The official documentation of a U.S. Senate Select Committee in 1975 and subsequent journalistic revelations confirmed that the United States was still ready to intervene in the way Latin Americans had so long claimed. This same tendency would reappear during the 1980s, when President Ronald Reagan authorized a military invasion of Grenada and a sustained campaign of covert action against Nicaragua.

Development and Debt

By 1960 West Europe and Japan were entering Latin America as major investors and traders. Another increasingly important force was the multinational corporations (MNCs). Firms such as International Telephone and Telegraph (ITT), Philips, and Royal Dutch Shell were no longer identified with any one country. Their multicountry base of operations meant they could play off one country against another while channeling business among branch firms. Multinationals predominated in many high technology areas. In some key sectors, such as pharmaceuticals or computers, countries had little alternative but to allow the multinationals to enter.

As time passed, Latin America was unable to devise an effective strategy for this problem of technology transfer. The hard fact is that advanced technology in industrialized countries is produced by private firms which use it to make profits. They are unwilling to sell the technology, and insist on marketing it themselves. Distribution and sales require direct investment, leading to eventual profit remissions that

can, in principle, far exceed the original investment. Furthermore, the MNCs may be able to wield monopolistic power in the local market, charging prices higher than a competitive market would have allowed.

Attempts by individual countries (or even country consortia) to develop advanced technology have faced long odds. The research facilities and experience of an IBM or Bayer are hardly within the reach of hard-pressed scientists and researchers in Latin America. In consequence, the terms of transfer of technology have become one of the central issues in the relationship of the Latin American economies to the industrialized world.

By the late 1980s a far graver short-run problem weighed down the Latin American economies: their staggering foreign indebtedness. By early 1988 the combined debt to the commercial (private) banks, as well as to such multilateral lenders as the World Bank and the Inter-American Bank, was $400 billion and growing daily. The three largest debtors were Brazil ($113 billion), Mexico ($108 billion), and Argentina ($54 billion). Most of the debt had been incurred since the huge OPEC oil price hikes of 1973 and 1979, which sharply increased the import bill for most of Latin America. The loans came primarily from private banks (flush with deposits from the oil exporters), and they made possible a higher rate of growth than the borrowers could otherwise have enjoyed. In truth, however, they simply postponed payment for the increased cost of oil. The borrowers were assuming that they could increase their trade surplus enough to repay the loans. This applied to almost all of Latin America, except the net oil exporters (Bolivia, Ecuador, Venezuela, post-1975 Mexico, and post-1978 Peru) and Argentina, which was virtually self-sufficient in oil.

By the late 1970s the loans were not limited to the oil importers. Almost every Latin American country borrowed massively, partly because bankers aggressively sought out Latin American customers. One major U.S. bank, for example, anxious to emulate the giant New York banks, entered the game late and had to be content with lending to Peru, one of the poorer risks. The bankers liked these customers because they were paying high interest rates and because they were supposedly immune from bankruptcy—the United States would never allow that to happen to a sovereign country. Soon the borrowed funds were going for current consumption, enabling governments— whether military or civilian—to win short-term favor from their import-consuming urban countrymen. Thus all of Latin America, with the conspicuous exception of Colombia, became hooked on private bank loans.

Net borrowing by developing countries is perfectly normal. The United States, for example, was a net borrower until 1900. If the funds had gone for investment in Latin America to insure future growth and to promote exports, then the capacity to repay would have increased.

Instead, much of the money was squandered. In countries such as Mexico and Argentina, the rich commandeered much of the borrowed funds and promptly deposited them in bank accounts abroad.

In August 1982 Mexico declared it could no longer make payments on its foreign debt, thereby triggering a world credit crisis. Brazil, the largest borrower, joined Mexico in December. Since 1982 no major Latin debtor has been able to repay much on principal. Even to maintain the interest payments required repeated "rescue" loans, which the debtors then immediately sent back to the banks as interest payments. Such rescues offered no long-term relief for the borrowers, but they produced lucrative "fees" for the lending banks and kept U.S. banks from having to classify the loans as delinquent. This, in turn, allowed them to avoid increased bank reserves which would have reduced their profits.

Meanwhile, the major debtors faced interest payments equal to 5 or 6 percent of their GDP. Not even Weimar Germany's burden of payments on World War I reparations had reached that level. Virtually all observers agreed that the Latin American loans would never be repaid in full. Banks wishing to sell those loans in the late 1980s found the going rate to be about 50 percent of face value. Confronted with such losses, the bankers investigated other solutions, such as swapping debt for equity investments (in the same country) or for "exit bonds," which would be readily negotiable on the open market. None of these plans reduced significantly the Latin American debt service charges in the short run. Only a general write-down of the total indebtedness and a stretch-out of payments could do that. If this were not accomplished through negotiation, it seemed likely to come by default, either declared or de facto. Some U.S. banks began to accept that possibility by 1987, as they increased their reserves against potential loan losses in Latin America.

Latin America's credit crisis was accompanied by a protracted economic slump throughout the 1980s. Country after country had to adopt an orthodox IMF-style austerity economic plan to satisfy foreign creditors. Deflationary measures, combined with the capital outflow for debt payments, produced stagnation. In per capita terms, Latin America's GDP fell by 8.3 percent between 1981 and 1989. Not surprisingly, the cost was paid disproportionately by those at the bottom, as real incomes of the poor shrank most. But Latin America's economic distress also hit the United States, whose exports to the region fell from $42 billion in 1981 to $31 billion in 1986. Caterpillar factory workers in Illinois, for example, were laid off as South American road contractors had no dollars to buy the giant yellow earth movers. Dependency proved that it could be a two-way street.

The international response to Latin America's debt crisis finally took shape in the late 1980s. In 1985 U.S. treasury secretary James A. Baker

III openly acknowledged that heavily indebted countries were facing structural crises of solvency, not just "liquidity," and called for an injection of $20 billion in developing countries that were willing to undertake market reforms. The Baker Plan led to scant practical results, mainly because the $20 billion proved to be unavailable, but it marked an important shift in approaches toward the problem.

In March 1989 Nicholas F. Brady, Baker's successor as treasury secretary, proposed a broad portfolio of debt reduction and restructuring alternatives and offered U.S. government support to countries undertaking market-based economic policies. The Brady Plan had two distinct features: one was its flexibility and open-endedness; another was its explicit recognition that debt reduction would have to be an integral element in any lasting resolution of the debt crisis. By February 1990 the Brady Plan resulted in the restructuring of nearly $50 billion of Mexican debt, and reduced the face value of the country's external commercial debt by $5.5 billion. Negotiations under the Brady Plan also led to agreements with Costa Rica, Venezuela, Uruguay, Argentina—and, in mid-1992—Brazil. With the exceptions of Costa Rica and Mexico, implementation of the plan did not bring much significant debt reduction. Through export expansion and economic growth, indebted countries gradually improved their capacity to service the loans, however, and the Brady agreements helped alleviate the sense of crisis that surrounded this issue throughout the 1980s.

The End of the Cold War

Near the end of the 1980s the international order underwent a sudden, fundamental change: the Cold War came to an end. The Berlin Wall collapsed in November 1989, the U.S.S.R. retreated from East Europe, and the dynamic Soviet leader Mikhail Gorbachev sought a rapprochement with the West. There was an immediate impact in Europe, where East Germany was incorporated into a "reunified" Germany and East European countries plunged headlong into dramatic political transitions. But there would be impacts elsewhere as well. No longer would the United States and the Soviet Union, as two global superpowers, engage in a nuclear standoff—and in unceasing struggle throughout the world. No longer would developing countries of the Third World provide a battleground for this conflict. No longer would capitalism and communism compete for ideological and economic supremacy. It was the passing of an era: some even called it "the end of history."

By the mid-1990s the implications for Latin America, and for U.S.–Latin American relations, appeared only in vague outline. The United States would no longer perceive its major interest in Latin America as the prevention of a communist threat. In principle, this should have obliged the United States to revise its policy in the late 1980s toward El

Salvador, where it continued to support a right-wing government against a rebel movement, and toward Cuba, where it continued to express implacable hostility toward the Castro regime throughout the 1990s. Change in policy was slow. The question was whether the U.S. government would abandon its distrust of reformist movements and politicians, which had long antedated the arrival of the Cold War.

At the same time, the end of the Cold War implied a change in the ideological content of reformist and radical movements in Latin America. The collapse of international communism struck a near-fatal blow to socialist and communist doctrines and organizations throughout the region. Adherence to Marxism appeared to be a certain prescription for rejection and defeat. To be sure, some movements remained unflinching in the face of doctrinal adversity—such as *Sendero Luminoso* in Peru—but most began to revise their outlooks in accordance with the "social democratic" ideas spread in Europe and elsewhere. Radicalism, in the meantime, took on a new appearance, moving away from socialism and toward nationalism and/or populism. In short, global events had far-reaching impacts on the political landscape in Latin America.

Yet another consequence of the conclusion of the Cold War was the disappearance of any great-power rivalry to U.S. hegemony within the Western Hemisphere. Almost immediately, the U.S.S.R. lowered its profile in the Americas—and its support to Fidel Castro and to rebel movements. One of the most remarkable features of the U.S. invasion of Panama, in December 1989, was the absence of forceful condemnation from major powers throughout the world. West European officials were preoccupied with events in East Europe. And while Japan showed increasing interest in economic relations with Latin America, its spokesmen repeatedly clarified and emphasized their unwillingness to challenge U.S. supremacy in the region. Leaders in Asia and Europe appeared to concede that Latin America was, after all, the "backyard"—and the responsibility—of the United States.

It was initially unclear how the United States might exercise its new hegemony throughout the hemisphere. Some observers argued that Washington should "rediscover" Latin America and develop a natural harmony of economic and social interests. Others predicted that, in the absence of East-West conflict, the United States would no longer concern itself with the region. "After so many years of worrying about excessive U.S. involvement," one commentator wrote in 1990, "Latin America may soon suffer from U.S. indifference, compounded by the rest of the world's traditional, relative lack of interest. . . . The hemisphere could well face the prospect of 'Africanization'—condemnation to the margins of world financial and trade flows but also, inevitably, to neglect and irrelevance." Without the Cold War, Latin America could face the prospect of marginalization.

Regional Economic Integration

Ultimately, economic considerations led Washington to advocate the adoption of "free trade" throughout the Americas. The goal was to develop expanding markets for U.S. exports, enhance efficiency for U.S. manufacturing (mainly through access to low-cost labor), and, in a variety of ways, strengthen America's "competitiveness" in the international economy. Regional integration in the Americas would also strengthen Washington's hand in negotiations with Europe, Japan, and other major powers.

In 1990 the Bush administration opened formal negotiations with Canada and Mexico for the creation of a North American free trade area. Also in mid-1990, President Bush proposed the development of a free trade zone embracing the entire Western Hemisphere. Asserting that "prosperity in our hemisphere depends on trade—not aid," Bush envisioned the creation of a free trade zone "stretching from the port of Anchorage to Tierra del Fuego. . . ."

The first step in this process would be the Mexican accord, signed in 1992. As described in Chapter Seven, the North American Free Trade Agreement (NAFTA) went into effect in January 1994 and created one of the largest trading blocs in the world. Essentially, NAFTA promoted the free flow of goods between Canada, Mexico, and the United States by eliminating duties, tariffs, and trade barriers over a period of fifteen years. NAFTA also opened Mexico to U.S. investments in a variety of ways. Ultimately, NAFTA marked a turning point in U.S. economic policy and in relations with Mexico. For the first time in history, Washington was pursuing an explicit strategy of economic integration with its neighbor to the south.

The prospects for a free trade zone embracing all of the Americas nonetheless remained unclear. President Bush had spoken of a free trade zone stretching from Alaska to Argentina, and in December 1993 Vice President Al Gore endorsed the prospect of "a Western Hemisphere Community of Democracies" that would consolidate political and economic harmony throughout the region, with NAFTA as "a starting point." Eager to gain access to this privileged circle, Latin American leaders came to regard eventual accession to NAFTA as a key part of their development strategy. Expectations were soon running high.

Yet formation of a hemispheric free-trade area faced formidable obstacles, both economic and political. The commercial rationale was less compelling than in the case of NAFTA. The partners in NAFTA, especially Mexico, might be hesitant to admit new members and share their status as countries with unique access to the U.S. market. And while NAFTA had strong political motivations, as sketched out in Chapter Seven, there was no such pressing incentive for a hemispheric accord.

Despite these odds, the Clinton administration promoted and hosted

a grandiose "Summit of the Americas" in Miami in December 1994. Attended by thirty-four heads of state, with the conspicuous exception of Cuba's Fidel Castro, this was the first such gathering since 1967, when the United States and Latin America concurred on a stillborn plan for economic integration. The ostensible goal of the Miami summit was to develop a blueprint for hemispheric collaboration into the twenty-first century. An implicit purpose, from Washington's perspective, was to provide assurances that the United States would neither neglect nor abandon Latin American countries outside of Mexico. After intense behind-the-scenes negotiation, the Miami agenda eventually came to focus on a Free Trade Area of the Americas (FTAA).

What happened in fact was that signatories in Miami designated the year 2005 as a deadline for the conclusion of *negotiations* for a free trade area—with implementation to follow in subsequent years. This was an ambiguous result. Advocates hailed the agreement for its high-minded principles and ambitious goals. Skeptics lamented its vagueness and its drawn-out timetable, which meant that official talks could drag on for a decade or more. Ironically enough, the principal resistance to the practical realization of an FTAA was likely to come not from Latin America but from the United States. As one observer noted, "The real pitfalls are the U.S. Congress and the U.S. public. . . . The public is not ready for a free-trade agreement with Latin America. In the post–Cold War environment, they don't understand what the dominant values are."

In the meantime, governments in Latin America responded to this uncertainty by pursuing strategies of subregional integration—projects for economic cooperation among groups of Latin American countries, rather than for the continent as a whole. The Central American Common Market was resuscitated, CARICOM was reinvigorated, and the Andean Pact was reshaped and revitalized. But the most ambitious and influential of these schemes emerged in South America, where the "Common Market of the South" (MERCOSUR) linked the economic fortunes of Argentina, Brazil, Uruguay, and Paraguay. Under the Treaty of Asunción, reached in March 1991, the member countries committed themselves to construct by December 1994 a customs union, with a common external tariff, and to move onward to a full-fledged common market in subsequent years. Especially in view of longstanding rivalries among its constituent members, MERCOSUR was a truly remarkable development. Its partner countries comprised nearly one-half of Latin America's gross domestic product, more than 40 percent of its total population, and about one-third its foreign trade.

More important than its size, however, was its strategic orientation. According to one of the project's original architects, the principal innovation of MERCOSUR stemmed from its commitment to "outward-oriented integration"—that is, from its determination to make member states more competitive in the international arena, rather than to rely

on closed markets via import-substitution industrialization. MERCO-
SUR also had clear political goals: the consolidation of democracy and
the maintenance of peace throughout the southern cone. At the same
time that MERCOSUR was taking shape, agreements were reached in
the nuclear field between Argentina and Brazil, countries that shared
significant nuclear capacity as well as historic rivalry. In a sense, MER-
COSUR would provide civilian democrats throughout the subregion
with a regular opportunity for consultation and mutual support, thus
offsetting the long-established conclaves for representatives of the
armed forces.

From the start, MERCOSUR's designers saw it as a dynamic institution, one that would evolve rapidly over time and also crystallize relations with advanced industrial economies. Once President Bush announced the Enterprise for the Americas Initiative, MERCOSUR
became a potential instrument for collective bargaining: as a Brazilian
observer recalled, "There was the perception that it would be interesting to concentrate efforts to negotiate *en bloc.*" Contradictory tendencies also emerged: having undergone a harsh program of structural
adjustment, Argentina expressed eagerness to negotiate its own FTA
with the United States (and/or seek membership within NAFTA), an
act that would logically lead to the dismantlement of MERCOSUR.

Taking a different approach, Brazil launched its proposal in April
1994 for a South American Free Trade Area, or SAFTA. The goal of
SAFTA was to create a free trade zone throughout the continent over
a ten-year period from 1995 to 2005. Public intentions behind SAFTA
were manifold: to capitalize on the experience of MERCOSUR, which
led to growth of intraregional trade; to avoid the "isolation" of MER-
COSUR, especially from Chile and the Andean Group; and to accumulate negotiating power for dealing with the possibility of broader integration schemes in the Americas. Not coincidentally, SAFTA would
confirm Brazil's historic claim to be the dominant power throughout
South America.

As regional integration schemes moved forward in the 1990s, one
central question arose: What would be their worldwide implications?
Would they promote cooperation—or would they provoke hostility
among competing regional blocs? Optimists asserted that regional integration necessarily engenders multilateral consultation and collaboration. Skeptics expressed the fear that, even as they achieve internal cooperation among their member countries, integration schemes could
stimulate external conflict with other regional groups. In particular,
they warned about the eventual emergence of three massive economic
blocs:

- a European bloc, eventually including East Europe and possibly part of the
  former Soviet Union

- a North American bloc, including Canada and Mexico as well as the United States (as with NAFTA), or a Western Hemisphere bloc, including all countries of the Americas (as envisioned by FTAA)
- a Pacific/Asia bloc, including Japan and neighboring countries and eventually China.

The creation of hostile trading blocs could have dire consequences for the Americas and for the world at large. A central challenge of the 1990s would be to assure that liberalization and integration at the regional level would promote liberalization and cooperation at the global level.

The Social Agenda

Aside from economics, the inter-American agenda in the 1990s came to concentrate on social questions. One of the most contentious of these issues was migration, especially illegal immigration. By 1994 the proportion of foreign-born Americans had climbed to 8.7 percent, the highest level in more than fifty years. (It was estimated that 6.2 million of these residents, legal and illegal, came from Mexico; more than 800,000 were from Cuba; and more than 700,000 were from El Salvador.) According to another count, there may have been as many as 2.5–4.0 million "illegal aliens" in the United States from all parts of the world in the early 1990s—with over half from Mexico, and many others from Central America and the Caribbean.

Undocumented migrants came to play important roles in the U.S. economy, especially in such areas as Florida and California, where they provided skilled labor at low rates of pay. In addition to working as field hands in agriculture, illegal aliens from Latin America filled jobs in construction, restaurants, hotels, and other service sectors. While some U.S. workers saw the mgirants as a source of competition, American employers tended to regard them as a source of docile and dependable labor.

This influx of foreigners nonetheless ignited a political backlash. As noted in Chapter Seven, on Mexico, the Simpson-Rodino law of 1986 did not significantly reduce illegal ("undocumented") migration into the United States. As the U.S. economy struggled to climb out of recession in the 1990s, politicians began to accuse the migrants of taking jobs from U.S. citizens and of abusing social services. Voters in California approved a measure that would deny social services (such as health care and public schools) to illegal aliens and their children. The military action in Haiti was prompted at least in part by a desire to prevent large-scale immigration from Haiti. And in 1995, a Texas judge found a woman guilty of child abuse because she spoke only Spanish at home. Antiforeign sentiment was clearly on the rise, and the atmosphere became tense in many American communities.

Another social issue concerned illicit drugs. The traffic from Latin America to the United States of illegal drugs, especially cocaine, continued throughout the 1980s and early 1990s. According to some estimates, this commerce amounted to as much as $110 billion per year, and it was said to be the most important economic activity for organized crime in the United States. Coca leaf was cultivated in the Andes, especially Peru and Bolivia, then shipped to Colombia where it was transformed into cocaine and prepared for export to the U.S. market. Marijuana flourished in Mexico, Central America, and elsewhere (including the continental United States). Heroin initially came from other parts of the world, but responded to laws of supply and demand: as supplies from the Middle East dried up in the early 1970s (with the breakup of the infamous "French connection"), resourceful traffickers encouraged production in Latin America; and as heroin regained popularity among U.S. consumers in the 1990s, Latin American output substantially increased. In the meantime other countries of the region—Jamaica, Panama, Venezuela, Brazil, Argentina, and Chile—became centers for the trans-shipment of drugs and the laundering of funds.

One of the most corrosive effects of this trade was to build up powerful trafficking rings, which came to mount serious challenges to legitimate political authority. This was most apparent in Colombia, where organized criminal "cartels" exercised de facto power through a combination of corruption and intimidation. During the 1980s the "Medellín cartel" attempted to take the country hostage by unleashing a massive wave of violence, assassinating scores of government officials and a presidential candidate. In the early 1990s the "Cali cartel" employed more subtle and sophisticated tactics, relying on bribery and co-optation rather than coercion; this erupted in political scandal in 1995, when it became apparent that the nation's president had accepted large sums of drug money to help finance his electoral campaign. In Mexico and in other countries of the region, too, drug barons and their criminal organizations were accumulating substantial amounts of political and economic leverage.

The ultimate cause of drug trafficking in Latin America was the strength of consumer demand, especially in the United States. Throughout the early 1980s, there were estimated to be nearly 25 million Americans making habitual use of illicit drugs. By the early 1990s this figure had been reduced by half, as middle-class suburbanites turned away from drugs, but hard-core use of "crack" and heroin persisted in U.S. inner cities. Demand for illegal substances also mushroomed in Europe, Asia, and the former Soviet Union. This posed an irresistible opportunity for drug lords and a challenge to governmental authorities. As a former president of Colombia once lamented, "The only law the narco-terrorists do not break is the law of supply and demand."

In response to these developments, U.S. policy attempted to stop drug trafficking by shutting down sources of supply. President Ronald Reagan declared a "war on drugs" that concentrated largely on eradicating crops and seizing shipments: the idea was to cut off supplies, drive up prices, and dissuade consumers from purchasing drugs. George Bush continued this emphasis, launching the 1989 military invasion of Panama as part of the war on drugs. Under President Bill Clinton this policy shifted, but only slightly, with budgetary increases for treatment and prevention of drug addiction. But the primary emphasis was still on law enforcement, and these efforts had little effect. Some observers called for the creation of an international police force; others spoke out for decriminalization (or legalization) of drug possession in the United States. Only one thing was apparent: as long as the illicit traffic continued, it would pose a serious threat to social and political order in Latin America and a major obstacle to U.S.-Latin American relations.

Hispanic Culture within the United States

The United States in the 1990s found itself facing a new dimension in its relationship with Latin America. Suddenly the United States discovered it had one of the largest Spanish-speaking populations in the hemisphere. Census Bureau data placed the "Hispanic" population in 1990 at 22.3 million, but most knowledgeable observers thought it was higher if all the illegal immigrants (labeled "undocumented" in more circumspect official language) were included. That meant that only Mexico, Argentina, and Colombia had larger Spanish-speaking populations in the Western Hemisphere.

According to 1990 data, for example, New York City had 1.8 million Hispanics, primarily from Puerto Rico and the Caribbean. This was an increase over the 1.2 million in 1970, achieved despite the fact that the total New York City population declined during the 1970s. In the Los Angeles and Chicago public school systems Hispanic students came to outnumber whites. Hispanics composed 26 percent of the population in California, the nation's most populous state, and 26 percent in Texas, the third most populous state. In much of south Florida, especially in the greater Miami area, Hispanic influence (essentially Cuban) clearly predominates. Most significant of all, between 1970 and 1980 the U.S. Hispanic population increased by more than 50 percent, and another 53 percent by 1990. In many cities in the Southwest the Hispanic population is growing faster than either the black or the "Anglos" (as non-Hispanic whites have been labeled). The increase occurs not only by birth but also by the influx of new immigrants.

There may be reason to believe that these Hispanics will retain their language and culture longer than did previous generations of non-English speaking immigrants into the United States. The painful ques-

tion of how—if at all—to preserve non-U.S. culture has been faced by millions of non-English speaking immigrants to America. Waves of Europeans were filtered through the public schools of New York, Philadelphia, Chicago, and Cleveland, where everyone assumed that the only language at school was English. Indeed, immigrant parents often forbade their children to speak the Old World language because they were so anxious to make their children into Americans. The result, of course, was that usually the second and certainly the third generation immigrant lacked any knowledge of his ancestral language—German, Polish, Italian, or Greek. All wanted to be Americanized, and language was the first instrument at hand.

The Hispanics in the United States, on the other hand, have reached center stage at a different moment in U.S. history. They followed the 1960s activisim of the black movement, which belatedly generated a new legitimacy for the cultural traditions of ethnic minorities. Operating in this climate of opinion (which mixes Anglo guilt and sympathy), Hispanics have successfully campaigned to institute bilingualism in schools and public facilities. The logic is that the Spanish-speaking Americans, by their number and by their past discrimination, deserve the opportunity to continue using their language. Most important, it is argued that Hispanic children are often disoriented and demoralized when faced with an all-English-language school. Maintaining some instruction in Spanish will help build their confidence and forge links between their culture and the wider U.S. society.

Bilingualism in public schools is officially intended to enable Spanish-speaking children to have their subjects in Spanish, at the same time as learning English. The argument is that children can learn biology or mathematics just as well in Spanish as in English. The final objective is for the child to finish high school fully competent in both English and Spanish. Critics of the system charge that all too often, especially in schools with a high percentage of Spanish speakers, students never learn correct English and may not learn Spanish grammar either. If true, they enter the job market with a grave disadvantage.

Whatever the case, there are millions of Spanish-speaking Americans who cannot command enough English to conduct essential business. In response the U.S. federal government now publishes in Spanish as well as English a wide range of official announcements and forms. City, county, and state authorities in key areas—including northern cities such as Chicago and Kansas City—have had to hire Spanish-speaking staff. In effect, the United States now has large regions and urban pockets where Spanish has official sanction as a second language.

This is bound to have implications for U.S. society. New York City, Newark, Miami, and Los Angeles all have regular television programming in Spanish, and the entire U.S. Southwest from Texas to Los Angeles has a cable TV network connected directly to Mexican national

television. Television, like bilingualism, will help maintain a live Hispanic culture.

Politics is another area where Hispanic power has begun to be felt. Like blacks, Hispanics have tended to be apathetic about politics. Long suspicious of the Anglo world, they often did not see any point in voting. Sometimes, however, Hispanic voters have begun to bring their numbers to bear. In San Antonio, Texas, for example, Henry Cisneros was elected mayor in 1981, the first ever of Mexican-American descent. New Mexico elected a Mexican-American governor, Jerry Apodaca, and Arizona did the same in electing Raúl Castro to the governorship. In New York City the Hispanics are steadily increasing their influence in the Democratic Party. In Miami, the Cuban-American community has achieved prominence in sectors of commerce, banking, and real estate and has become a predominant political force as well. And after his election in 1992, President Bill Clinton appointed two Hispanics to cabinet positions.

This growing Hispanic presence in the United States will influence how the North American public thinks about key issues in its relationships with Latin America. For a broad variety of complex reasons, U.S. citizens therefore can no longer afford to ignore Latin America.

Since the 1960s Latin America has been drawn increasingly into the wider world. U.S. hegemony in Latin America, at its height from 1940 to 1960, seemed to be thereafter on the wane, then revived as the Cold War came to an end. Even so, pluralism may yet be an option for Latin America. Governments may be able to negotiate their way into gaining help from a broad spectrum of foreign sources. Past experience suggests that the only way to escape from dependency is to increase the options. Gabriel García Márquez, one of Latin America's most famous contemporary writers, stated the issue eloquently in his 1982 Nobel Prize acceptance speech: "Why is the originality so readily granted us in literature so mistrustfully denied us in our different attempts at social change? Why think that the social justice sought by progressive Europeans for their own countries cannot also be a goal for Latin America, with different methods for dissimilar conditions?" As the United States asserted its hemispheric preeminence in the post-Cold War environment, Latin Americans would continue their quest for an autonomous pathway to equity, prosperity, and independence.

WHAT FUTURE FOR LATIN AMERICA?

Predicting the future is always risky, more so for Latin America. The continent has repeatedly been described as on the verge of miraculous development, only to disappoint the optimists. In 1912 Lord Bryce, after a tour of South America, predicted its temperate area "will be the home of rich and populous nations, and possibly of great nations." In 1910 another English traveler located Brazil "upon the road that leads, surely though slowly, to a future of great prosperity." More than eighty years later the certainty had faded.

It has been frequently maintained that education would solve all of Latin America's problems. Ignorance and illiteracy held back the Latin Americans. If they could only follow the U.S. and West European example with state-supported mass education! But which have been the most educated nations in Latin America? Argentina, Chile, and Uruguay. And those countries produced Latin America's most brutally repressive military dictatorships of the 1970s. Their elegant constitutions were torn up, their congresses closed, their courts rendered a sham. How could this be so?

But if prediction is treacherous, it is also necessary. As we follow the historical development of Latin America, we cannot help but wonder about the rest of the story. And since the United States will continue to have vital interests in Latin America, we cannot ignore the possible implications of future Latin American developments for the United States and its citizens.

Prediction has other virtues too. As we strain to think ahead, we have to think again about the past. To estimate a future balance of forces, one has to assess their past and present weight. If urban labor has been subject to manipulation in Brazil, will it continue to be that way? Why did the Chilean coup of 1973 take thousands of lives, whereas the Brazilian coup in 1964 took almost none? One must ponder these historical questions before considering possible scenarios for the 1990s and beyond. And with some care, we can hope to identify sources of change and determine a probable range of results.

Preparing to Predict: Comparative Analysis

Let us begin by a retrospective survey of the country studies presented in this book. In Chapter Two we offered a schematic outline of historical transformations in Latin America since the 1880s, and in subsequent chapters we traced the history of individual countries and regions. One of our concerns throughout has been to focus on (1) each country's location in the world economy, (2) the social structure associated with each pattern of economic activity, (3) the kind of coalition among classes or groups that might result, and (4) the political consequences deriving from all these factors.

We attempt to approach this problem through a systematic comparative analysis, and this requires a comprehensive conceptual framework. At the conclusion of Chapter Two we presented a series of questions for consideration in each of the country case-studies. Central to these interrogations was a focus on the structure and alignment of social classes. To make these issues more explicit, we now offer an abstract analytical scheme.

The central idea entails a classification of social strata along two separate dimensions: urban-rural position and class status. There are, in this view, six such groups:

- the *urban upper class,* consisting primarily of industrialists, bankers, financiers, and large-scale merchants—or, in Marxist terms, the upper bourgeoisie;
- the *rural upper class,* mainly landowners;
- the *urban middle class,* a heterogeneous stratum including professionals, teachers, shopkeepers, and so on—otherwise known as the petty bourgoisie;
- the *rural middle class,* not often noticed in Latin America, one that includes small farmers as well as merchants in rural areas;
- the *urban lower class,* principally an industrial working class or proletariat, but a stratum that also includes growing segments of unemployed migrants from the countryside; and
- the *rural lower class,* either an agrarian proletariat or a traditional peasantry—some of whose members may take part in the national economy, some of whom (especially in indigenous communities) may subsist on the fringes of the marketplace.

The groupings in the "lower class," often known as the "popular classes" in Latin America, represent, by far, the largest segments in society. These are poor people, undereducated and sometimes malnourished, and they have been systematically deprived of the benefits of development. Many of them participate in the rapidly emerging "informal sector," working at odd jobs outside the formal economy. (The informal sector is an unusually amorphous group, including peddlers and beggars and small-scale entrepreneurs, and for simplicity's sake it does not receive separate consideration in this analysis.)

One additional social actor—not a class or stratum, but a critical group nonetheless—consists of the *foreign sector.* This includes private

FIGURE 12-1   Hypothetical Array of Social Actors

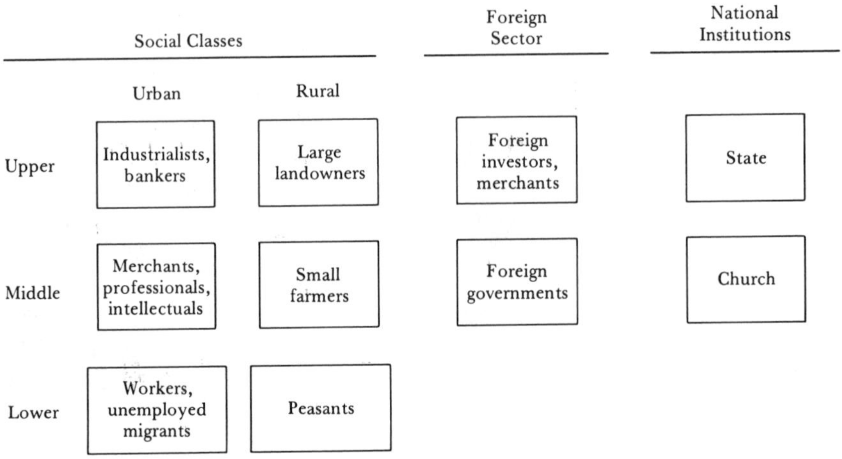

	Social Classes		Foreign Sector	National Institutions
	Urban	Rural		
Upper	Industrialists, bankers	Large landowners	Foreign investors, merchants	State
Middle	Merchants, professionals, intellectuals	Small farmers	Foreign governments	Church
Lower	Workers, unemployed migrants	Peasants		

investors and corporations as well as foreign governments and military establishments. Though sometimes divided against itself, the foreign sector has often wielded enormous power in Latin America.

To enhance their relative position, these social actors typically compete for control of major institutions. The most crucial institution, at least in recent times, has been the *state,* which commands large-scale resources and usually claims an effective monopoly on the legitimate use of force (only a government, for example, can put a citizen in jail). One key group within the state is the military; another consists of party politicians (when they exist); another is composed of technocrats and bureaucrats. Also important as an institution has been the Roman Catholic Church.

Figure 12-1 provides a general picture of these groups and institutions. It does not depict the outlines of any specific Latin American society. It is an abstract scheme, a hypothetical means of illustrating the subject of concern.

To apply the framework to any historical situation we need to ask the questions first posed in Chapter Two. In abbreviated summary they are:

- What are the principal social classes? Which ones are present and which ones are absent?
- Which social classes have the most power?
- Which groups are allied with which? On what basis?
- How autonomous is the state? Is it captive to any of the social classes, or is it independent?
- What are the predominant factors on the international scene? What, in particular, is the position of the United States?

To demonstrate these possibilities we next present schematic analyses of political and social transitions in each of the areas covered in Chapters Three through Ten. We concentrate here on relatively recent developments, though the method could just as well apply to earlier periods in time. This is an interpretive exercise, we emphasize, not a definitive statement; it requires estimates and judgments that should provoke discussion and debate. Nonetheless we think the approach provides strong confirmation of our basic arguments: that political outcomes in Latin America derive largely from the social class structure, that the class structure derives largely from each country's position in the world economy, and that a comparative perspective on these phenomena can help elucidate the variations and the regularities in Latin American society and politics.

Our first application deals with Argentina, where the economic dominance of beef and wheat produced two major social results: the absence of a peasantry, especially in the pampa region, and the importation of working-class labor from Europe. In the years before Perón the state and the foreign sector were mostly in league with landed interests, as shown in Figure 12-2. (Solid arrows represent relatively firm alliances; broken arrows represent fragile or partial coalitions.) Even the Radicals who governed with urban middle-class support in 1916–30 tended to favor the cattle-raising oligarchs.

For economic and demographic reasons Argentina's urban working class suddenly began exerting pressure on the political system in the 1930s, but there was no possibility of a class-based alliance with a peasantry; the most likely allies, instead, were newly emergent industrialists ready to challenge the landowning aristocracy and its foreign connections. The preconditions thus existed for an urban, multiclass coalition of workers, industrialists, and some segments of the middle class. It took the political instinct, the populist rhetoric, and the personal charisma of Colonel Juan Perón to make this alliance a reality, and he used a corporatist state structure to institutionalize it. One reason for its initial success was that the landowners had no peasantry with which to form a common conservative front. A reason for its ultimate failure was that limited industrial growth led to class-based worker-owner conflict within the coalition itself.

Starting in 1966 and again in 1976, the military seized the state and attempted to impose a "bureaucratic-authoritarian" regime. The dominant alliance consisted of military officers, foreign investors, local industrialists, and landowners. Workers were repressed and forcibly excluded from power. The middle sectors played a waiting game, then found their opportunity with Alfonsín's election in 1983. Their party was, in turn, displaced by a Peronist president, Carlos Menem. He soon launched an orthodox stabilization program that turned Argentine class politics on its head. The Peronists, once the implacable foes of

FIGURE 12-2   Political and Social Coalitions: Argentina

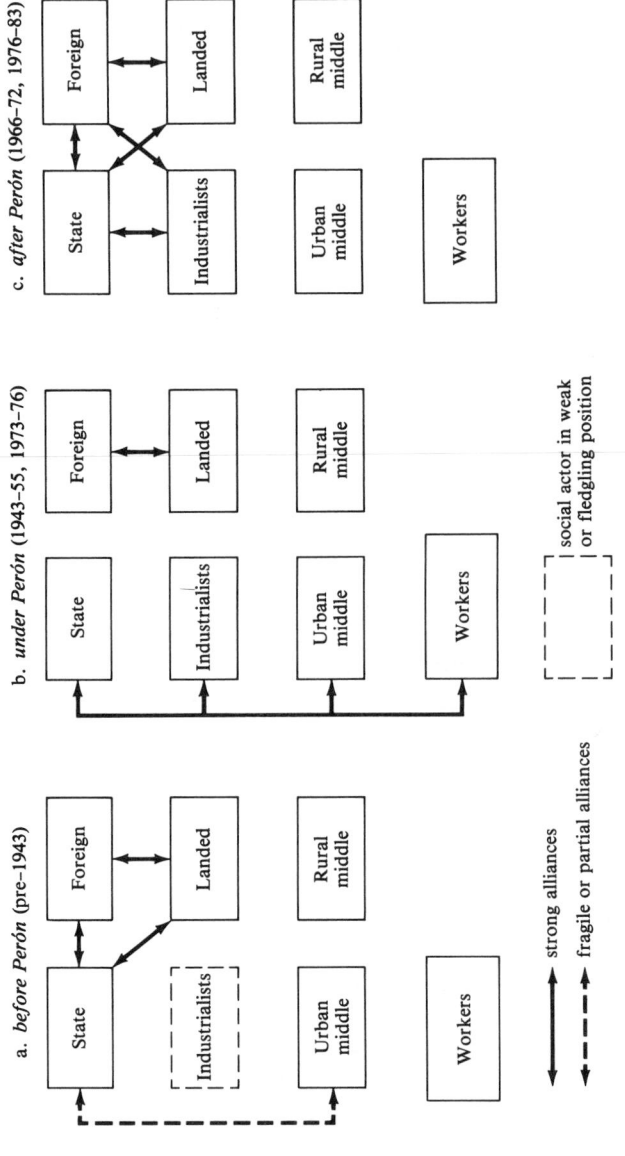

a. *before Perón* (pre-1943)   b. *under Perón* (1943–55, 1973–76)   c. *after Perón* (1966–72, 1976–83)

strong alliances

fragile or partial alliances

social actor in weak
or fledgling position

Before Perón (a) a coalition of landed and foreign interests controlled the state and obtained support from leaders of the urban middle class; under Perón (b) a populist coalition of urban workers and new industrialists—with some middle-class participation—dominated the state, to the virtual exclusion of foreigners and landed oligarchs; after Perón (c) an alliance of foreign interests, landowners, and industrialists led technocrats and the military to impose a "bureaucratic-authoritarian" regime (1966–72; 1976–83).

Note: Here and in subsequent figures, the foreign sector is depicted as a single unit, since the economic and political interests of individual foreign powers have usually tended to coincide.

FIGURE 12-3   Political and Social Coalitions: Chile

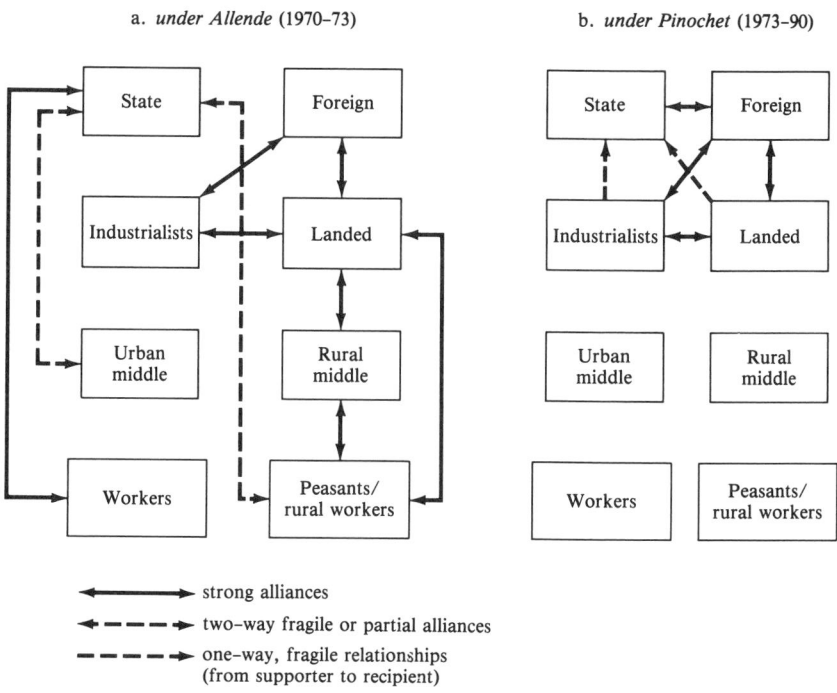

a. *under Allende* (1970–73)   b. *under Pinochet* (1973–90)

→ strong alliances

◄ − − − ► two-way fragile or partial alliances

− − − − ► one-way, fragile relationships
(from supporter to recipient)

Under Allende (a) a political movement based largely on urban workers gained control of the state—against the opposition of foreign interests, industrialists, and a multiclass array of rural groups; after Allende (b) a coalition of industrialists and landowners joined with foreign interests, most conspicuously the United States, to create a military-dominated bureaucratic-authoritarian state that showed remarkable autonomy in the domestic arena (hence the broken one-way arrows).

economic orthodoxy, now provided the congressional votes to put that doctrine, including wholesale privatization, into action.

Chile is quite a different case. It has contained every type of social actor, including a peasantry (and migratory rural proletariat) and a working class that by 1900 was well organized, at least by Latin American standards. Foreign interests, especially the copper companies, collaborated with an upper class that, in contrast to Argentina, was deeply involved in finance and industry as well as land. Though political parties represented specific social groups the state generally retained substantial independence.

So there existed elements of a powerful socialist movement. Party politics could (and did) lead to ideological polarization. The alliance of the foreign sector with the upper class added a nationalistic dimension to antiaristocratic resentment. A broad-based coalition of workers and

FIGURE 12-4   Political and Social Coalitions: Brazil

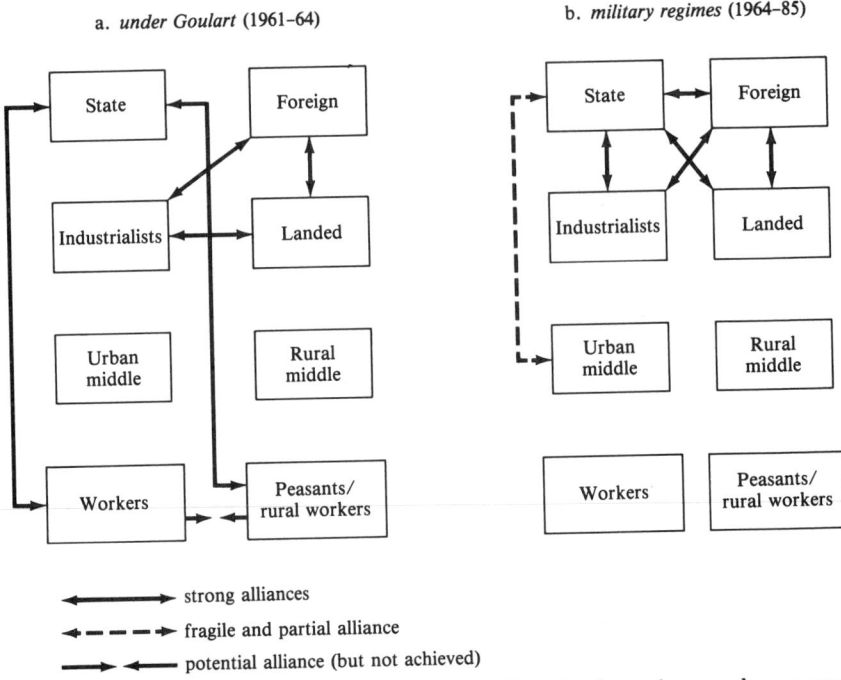

a. *under Goulart* (1961–64)   b. *military regimes* (1964–85)

→————————→ strong alliances

←— — — —→ fragile and partial alliance

——→ ←—— potential alliance (but not achieved)

Under Goulart (a) the state mobilized support from both workers and peasants, raising the specter of a potential worker-peasant alliance; after Goulart (b) a bureaucratic-authoritarian state relied on numerous constituent groups and retained considerable allegiance from the urban middle class.

peasants seemed possible: hence the triumph and euphoria of the early Salvador Allende government. Chile's socialist movement was not able, however, to expand its support much beyond its industrial working-class base. Allende supporters failed especially to convert many of the lower middle class. Urban and rural elements of the upper class, on the other hand, maintained their solidarity, partly through family connections, and landowners managed to get support from other strata in the countryside. U.S. undercover intervention further hastened the downfall of Allende's regime. (Figure 12-3.)

After 1973 the Chilean military, like its counterpart in Argentina, established a bureaucratic-authoritarian system. The ruling coalition included industrialists, landowners, foreign investors, and a state that possessed extraordinary power. Staffed by generals and technocrats, especially the "Chicago boys," the Chilean government set about its course determined to prevail over any and all opposition. In the course of financial reorganization and extensive privatization, the government

also increased the concentration of wealth, as a few rich clans and conglomerates bought the privatized state enterprises.

Brazil presented a similar picture. Under Vargas the *Estado Nôvo* organized urban workers under the auspices of state control. In the early 1960s his protégé, João Goulart, stepped up the mobilization of the workers—and also fomented (or at least permitted) the organization of peasants in the countryside. The prospect of a worker-peasant alliance antagonized both the upper class and foreign interests, depicted in Figure 12-4, and prompted the military to intervene in 1964 and to establish a prototypical bureaucratic-authoritarian regime. Despite waves of repression that hit every social sector (although to highly differing degrees), the Brazilian government succeeded in retaining more residual middle-class support than its counterparts in Argentina or Chile, and this explains in part why the process of liberalization *(abertura)* was successful at an earlier stage there.

In Peru the period from 1948 to the mid-1960s witnessed a close association between the state (especially under Odría), foreign capital, landowners, and—to the extent that they existed as an identifiable power group—local industrialists. (See Figure 12-5.) The urban middle sectors took an ambivalent stance, sometimes supporting APRA or Belaúnde's Popular Action but not challenging the overall structure of power. Left out of the ruling alliance were organized workers, migrants in the shantytowns, and, of course, the peasants who finally took up arms in the sierra.

The revolutionary military regime under Velasco Alvarado (1968–75) dismantled this coalition and built an entirely new one, based on state-sponsored mobilization and control of workers and peasants against the formerly ascendant groups: especially foreign investors and aristocratic landowners, the latter being gravely weakened by agrarian reform. The state was conspicuous for its autonomy, and no sector was immune from its intervention. Ultimately, the reformist military proved unable to institutionalize its corporate structure and thus consolidate its ties to workers and peasants. Subsequent governments—under Morales Bermúdez (1975–80), Belaúnde (1980–85)—have gradually returned significant shares of power to factions of the pre-1968 elite, although García (1985–90) attempted populist measures.

Mexico offers a different combination. Prior to the Revolution of 1910 the country had no indigenous industrial elite or rural middle sector; there was a nascent but unorganized working class. As shown in Figure 12-6 the ruling coalition, under the *Porfiriato*, included three groups: landowners, the foreign sector, and the state.

The Revolution ruptured this coalition and, through agrarian reform, weakened the rural elite. The state increased its authority and, from the 1930s onward, encouraged the formation of an industrial bourgeoisie. The postrevolutionary governments drew popular support from both workers and peasants, and under Cárdenas developed its

FIGURE 12-5   Political and Social Coalitions: Peru

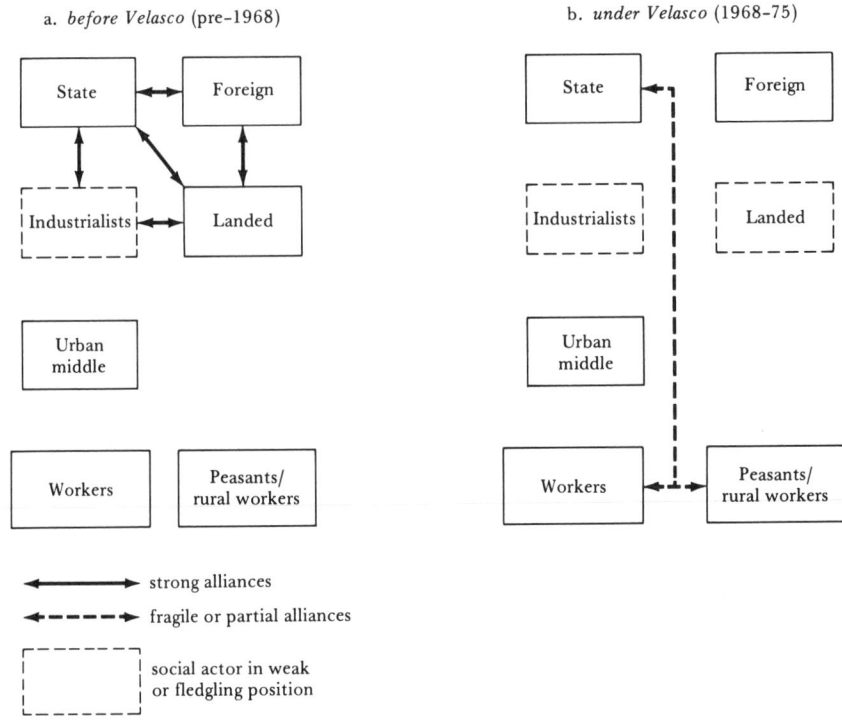

a. *before Velasco* (pre-1968)

b. *under Velasco* (1968-75)

◄─────────► strong alliances

◄─ ─ ─ ─ ─► fragile or partial alliances

social actor in weak or fledgling position

Before Velasco (a) a coalition of landowners, foreign interests, and fledgling industrialists controlled the state; under Velasco (b) the state succeeded in weakening the landed oligarchy and attempted to mobilize support from workers and peasants through a corporate structure (without forging a horizontal lower-class alliance) but could not do so on a long-term basis.

strategy for dealing with the masses: the state would organize workers and peasants in such a way as to keep them apart. The PRI developed separate "sectors" for workers and peasants, reflecting the regime's obsession with heading off any spontaneous, class-based politics. By the mid-1990s, however, the PRI was suffering major electoral defeats, especially on the state and local level. Furthermore, top-level feuds were threatening to destroy the party's supposedly multiclass hegemony.

Cuba's monocultural society reveals still another profile. Foreign (that is, U.S.) domination of the sugar industry meant that, for all practical purposes, there was hardly any local upper class. Workers in the mills and on plantations formed an active proletariat, as pictured in Figure 12-7, and migration strengthened ties between laborers in the cities and the countryside. Unions were weak, the army was corrupt, and the state, under Batista, was a pitiful plaything of U.S. interests.

FIGURE 12-6   Political and Social Coalitions: Mexico

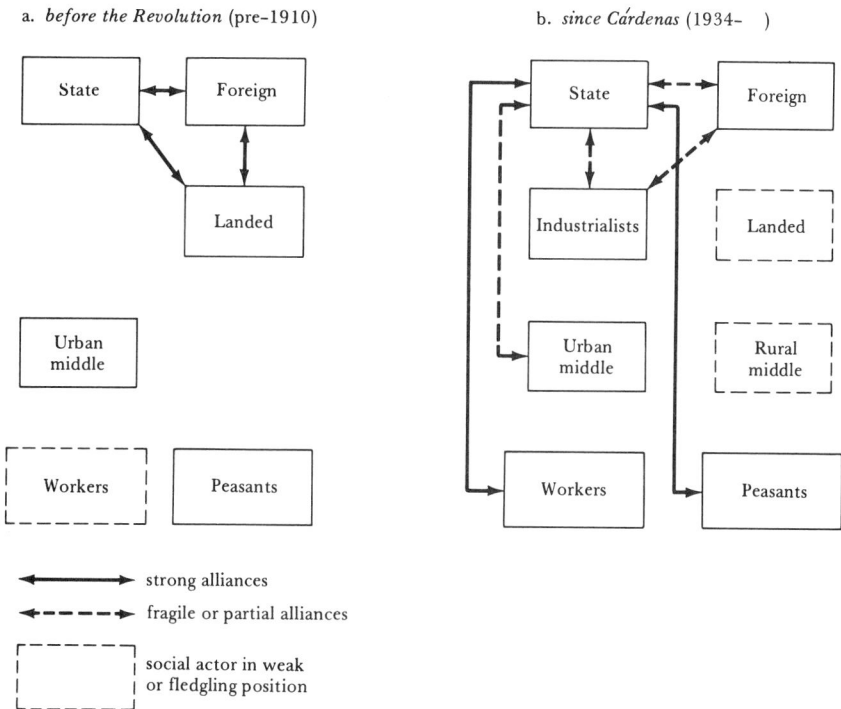

a. *before the Revolution* (pre–1910)

b. *since Cárdenas* (1934–   )

Before the Revolution (a) landowners and foreign interests combined with the military and contemporary technocrats (the *científicos*) to gain control of the state; after the 1930s (b) the ruling coalition consisted of a delicate balance among foreign interests, local industrialists and financiers, and the state—which kept a firm hand over workers and peasants while cultivating some support from the urban middle class.

Cuba possessed elements of a socialist movement, one that could capitalize on anti-imperialist sentiments. There was another secret to Fidel's eventual success: his movement would meet very little resistance, except for the foreign sector—whose proconsuls did not use all the resources at their disposal. Since 1959 Fidel and his lieutenants have revamped the island's social structure, eliminating vestiges of the old upper class, organizing middle- and lower-class groups in cities and the countryside, and implementing a "command" economy. It was achieved, however, only with massive Soviet support. This dependency became painfully apparent when the Soviet Union and its subsidy both disappeared in the early 1990s.

In partial resemblance to Cuba, most of Central America prior to the 1970s typified a traditional "plantation" society: landlords (but here resident, not absentee) and peasants in the rural sector, a fledgling ur-

FIGURE 12-7   Political and Social Coalitions: Cuba

a. *before the Revolution* (pre-1959)   b. *since the Revolution* (1959–)

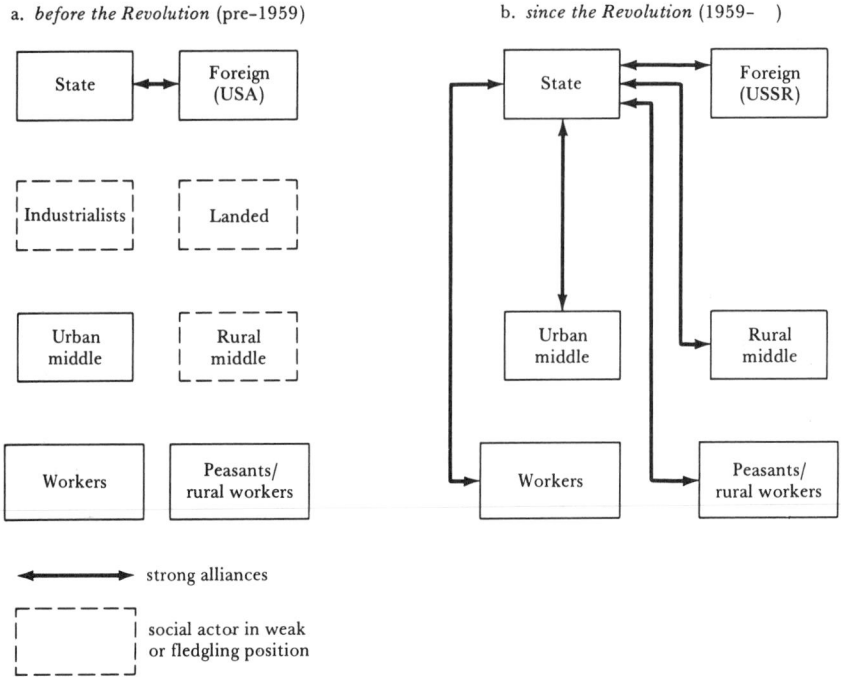

→→→ strong alliances

social actor in weak
or fledgling position

Before the Revolution (a) a weak state relied on the United States for suste-
nance and support; the post-revolutionary state (b) has dislodged the upper
class (much of which migrated to the United States) and mobilized support
throughout the rest of Cuban society, but still received assistance and protection from a foreign power until the early 1990s.

ban middle class, and a ruling alliance consisting of an aristocracy, foreign interests, and a dictatorial state fully supported by the church. In
the last decade or so, two major developments have taken place, especially in Nicaragua and El Salvador (the latter is portrayed in Figure
12-8). First, political leaders from the middle classes have sought support from peasants. Second, perhaps even more important, the Roman
Catholic Church has openly and courageously espoused the cause of
the poor. But large-scale U.S. intervention helped to defeat leftist guerrillas in El Salvador and to reverse the revolution in Nicaragua. In general, the alliance of landed elite and foreign interests has successfully
defended its hold over the state in Central America.

   In addition to placing each of the country studies within a comparative framework, this overview can offer several basic clues for predicting trends and outcomes. One is that *a large share of influence on
Latin America's future development will continue to come from outside the region.* Economic growth (or decline) at the industrial center of the capi-

FIGURE 12-8   Political and Social Coalitions: El Salvador

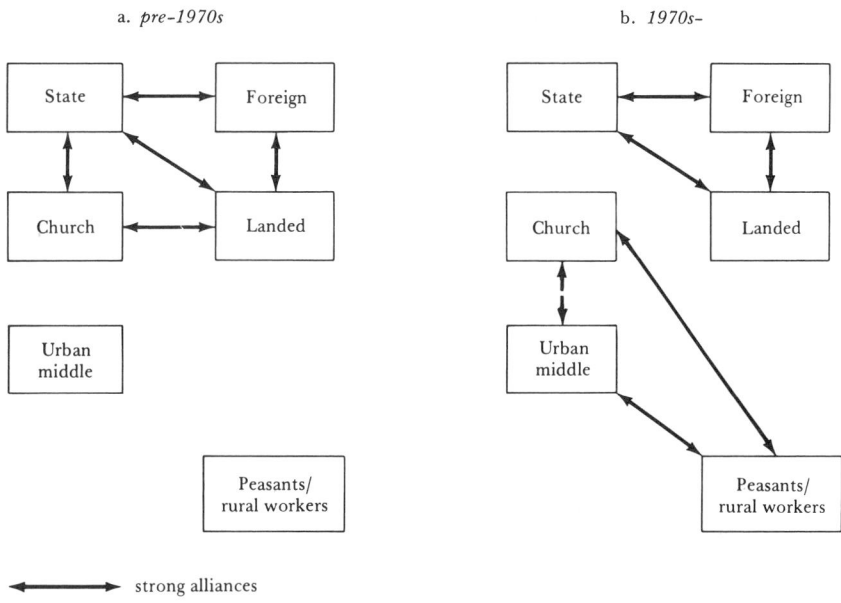

a. *pre-1970s*   b. *1970s-*

◄──────►   strong alliances

◄─ ─ ─ ►   fragile and partial alliance

Prior to the 1970s (a) El Salvador was dominated by an alliance of landowners, foreign interests, and the military, with the approval of a conservative church hierarchy; by the mid-1970s (b) a middle-class reformist movement was seeking support from peasants and leading clerics pronounced their solidarity with the dispossessed.

talist world-system will have major effects on demand for Latin American goods, and could thus affect the power relationships between groups in the producing countries. Whether Cuba or Central America embarks on industrialization, for instance, depends not only on local planning and resources; it also depends on the actions of the United States, the European Community, and possibly Japan and other leading powers. As the world becomes ever more interdependent, it is less and less likely that this situation will change.

Second, the action of any particular social group or class will depend not only on its own growth and strength. *It will depend on what is happening among other social groups.* It is reasonable to suppose, for instance, that the urban working class will expand in most countries of Latin America. But this alone will not determine political outcomes. Those will depend in large part on the other groups in each society: on what other groups are present, on the nature of alliances, and on the resulting power arrangements. A socialist movement succeeded in Cuba and failed in Chile not only because of their differing degrees of internal cohesion but also because of different bases and degrees of opposition.

This stricture applies especially to middling groups, often called middle classes or sectors. They are bound to grow. But their political activity is likely to depend, in very large part, on the power relationships among other major groups in each society. *Latin America's middle classes have tended to react to political opportunities, rather than to initiate structural transformation, and there is no reason to expect this to change.* The middle-class movements of the 1890–1910 period represented attempts to gain access to power, not to implement structural change; even Francisco Madero, the "apostle" of the Mexican Revolution, had limited goals of his own. Ever since that era, middle-class parties—in Argentina, Chile, Brazil, and elsewhere—have taken a responsive stance. To be sure, some *individuals* from these strata (and the upper-class elites) have assumed leadership of popular and revolutionary movements, but *collective* behavior has been cautious, tentative, and often incoherent. By itself, the expansion of the middle sectors will not determine patterns of political change. In conjunction with other factors, though, it could become a decisive force.

The interplay among key social groups will have a critical influence on what happens in Latin America between now and the year 2000 and presumably beyond. To assess the implications of this fact, we must first anticipate the social and economic environment that is most likely to prevail.

Dimensions of Change: Demography and Economics

Among the most important factors are the size and growth of Latin America's population. They will determine the overall demand for resources (especially food) and for employment, services, and political participation. The specter of a "population explosion" has long haunted visions of the continental future, and not without reason. According to one estimate, for instance, the population of Latin America—about 453 million in 1992—could swell to more than 515 million by the year 2000, almost twice the projected size of the United States. An expansion of this magnitude would place enormous strain on Latin American society, and many observers have predicted such dire outcomes as famine, disorder, and stagnation.

It is important to realize, however, that the prospects vary widely from country to country. Between 1970 and 1980 the population of Mexico increased at an average annual rate of 2.2 percent. The rate for Chile was 1.7 percent, and for Cuba it hovered around 1.3 percent. On the basis of recent growth rates, Table 12-1 displays a set of projections for major countries of the region. Brazil, it appears, may have 172 million inhabitants by the end of this century, when it will be almost two-thirds the size of the U.S. population of about 276 million. Argentina, Cuba, and Chile need not worry about inordinant demands on resources, at least in respect to population growth. Mexico and Cen-

TABLE 12-1   Population Growth and Projections, 1980–2025

	Average Annual Growth Rate (%)		Population Size (millions)	Population Projections (millions)	
	1980–1992	*1992–2000*	*1992*	*2000*	*2025*
Argentina	1.3	1.0	33	36	43
Brazil	2.0	1.4	154	172	224
Chile	1.7	1.3	14	15	19
Mexico	2.0	1.9	85	99	136
Peru	2.1	1.8	22	26	36

Source: World Bank, *World Development Report 1994* (New York: Oxford University Press, 1994), Table 25.

tral America offer cause for concern because of the growing demand for jobs.

Changes in population growth rates have multiple causes. Modern history has shown that urbanization and rising incomes are generally accompanied by reduced birth rates. Important also are social attitudes and availability of contraceptives. The most dramatic case of recent change in demographic growth rate in Latin America is Cuba, which between 1958 and 1980 saw its birth rate decline 46 percent. At fourteen births per thousand population, Cuba's rate is comparable to or lower than that of most developed nations. This can be explained by the transformation in social structure and economic conditions, as well as by the scarcity of housing and the cost of child care. The Cuban government has made contraceptives freely available and permitted abortion on demand, although women are strongly advised against using abortion as a means of birth control.

Elsewhere in Latin America the picture on birth control is more complicated. There has been much resistance to it—not so much because of Catholic teachings, but because poor people in traditional rural society tend to see large numbers of children as a benefit. High infant mortality rates induce parents to have lots of babies so that some, at least, will survive. And children are not only mouths to feed; at an early age they can begin to work, in fields or elsewhere, and contribute to the family income. And parents have generally looked forward to being supported in their old age by their children. Within this traditional outlook a reluctance to use contraceptives—even if they are available—is entirely reasonable.

Attitudes toward childbearing have nonetheless changed over recent decades. Urbanization and rising living standards, among other factors, have led to significantly declining population growth rates, falling from an annual average of 2.8 percent in the 1960s to 1.9 percent in the early 1990s. Mexico's crude birth rate dropped from forty-five per

thousand in 1965 to twenty-eight in 1988, while Brazil's dropped from thirty-nine to twenty-eight.

The problem, moreover, does not merely involve the number of births in coming years. A critical concern must be young people who are already here and who will be seeking jobs within the predictable future. In countries such as Mexico and Brazil nearly half the population is under the age of fifteen. Within the next two decades, therefore, the pressure for employment will be immense. Demographic trends convert quickly into social realities.

Will the regional economy be able to support this kind of population? The 1980s were hardly encouraging. Latin America's per capita gross domestic product *declined* by almost 10 percent over the decade. Brazil's fell more than 5 percent, Mexico's more than 8 percent, and Peru's a staggering 30 percent. The 1980s had truly been a "lost decade." The early 1990s brought a modest annual growth rate of 3.5 percent, not much more than the population growth rate of 1.9 percent.

And capitalist economic growth, even when it accelerates, seldom produces economic equality, especially in the early phases. On the contrary, it often tends to concentrate wealth in small sectors of the population—especially in "dependent" societies, where economic expansion so often takes place within restricted enclaves or "pockets." By the end of the century Latin America will be overwhelmingly urban. But because of natural growth, migration from the countryside, and the shortage of jobs, city dwellers without formal-sector employment may well amount to *nearly half* the total population. This group may remain politically inactive for a while, but the long-run prospects are nonetheless unsettling. The cities may well become seedbeds of discontent.

Moreover, most of Latin America was by the mid-1990s still bearing a heavy burden of foreign debt payments. During the 1980s Latin America transferred to its foreign creditors over $200 billion. The net burden declined because of debt renegotiation, export growth, and a return of capital inflow. Nonetheless, in 1993 almost one in every three dollars earned by exports was going to service past loans.

Looking Ahead: Political Responses

Economic misery alone does not create revolution—otherwise Haiti would long ago have become a cockpit of revolution. And in the 1970s, it was children of the middle and upper classes, not the poorest segments of society, who joined guerrilla movements in Uruguay and Argentina. By the early 1990s the Latin American revolutionary potential, so celebrated by the left in the wake of the Cuban Revolution, seemed minimal. Communist parties, which had rarely been in the forefront of armed action, were in full disarray, often in dissolution, as their Soviet and East European models scrambled to discard their ideological and institutional trappings. Even the more radical left, once militant in its

admiration of Mao and Ché Guevara, was almost everywhere shrinking or even disappearing.

Organized labor was fighting merely to protect or restore traditional material gains. Argentina's labor movement, for example, has shown an extraordinary ability to survive, but, because of its populist and Peronist orientation, it has never shown much interest in revolution. Its concerns are bread-and-butter issues—wages and working conditions—and it will continue to be a powerful actor on the Argentine political stage. Chile also had a thriving union movement before the 1973 coup and, despite continuous repression under the military government, is reemerging as a substantial force.

In Brazil the experience has been different. Brazil has been a labor surplus country throughout this century. That has worked against unionization, even in the dynamic center-south region. The Brazilian government has followed a shrewd mixture of repression and co-optation to keep the major unions under control. The years 1979 and 1980 brought a new worker militancy in São Paulo that threatened this hegemony. But nothing since then suggests that the Brazilian labor movement has reached the degree of class consciousness or organizational experience evident in Argentina and Chile. In fact, in the 1989 presidential election, São Paulo, the stronghold of Brazilian unionism, was one of the few state capitals to vote against Lula, the ex-metalworker presidential candidate who ran on a radical leftist platform. In the 1994 presidential election, Lula's support in São Paulo was even weaker.

Mexico, like Brazil, is a labor surplus economy. Urban workers know that if they strike there are plenty of new arrivals from the countryside ready to take their jobs. As in Brazil, the government has been shrewd in using co-optive measures to corrupt the union leadership. Where it has come to confrontation, the Mexican government has not hesitated to repress the workers and imprison their leaders for long periods. Mexico appears to have tighter control over the working class than does any other major Latin American country. It seems safe to assume that Mexican labor will prove unable to bend history to its will.

None of the major political shifts in Latin America has been directly brought about primarily by workers. They have been able, once mobilized, to throw their weight, as in Argentina. But the working class per se has not been able to seize control of events. It tried in Chile and failed. And in Cuba, the Fidelista rebellion was carried out independently of organized labor, which was dominated by the communists. The guerrillas were predominantly middle class, with no initial links to the organized working class. To say that organized labor is unlikely to take the initiative in Latin America is not to deny that it will fight for the bread-and-butter rights of its members. It will do so, at great cost to the leadership, as has occurred under the military governments in Argentina, Chile, and Brazil. But this is not the same as revolution.

And the peasantry? The revolutionary potential of rural workers is obviously difficult to measure. It fueled the Mexican Revolution at crucial stages, and it has left its mark on Chile (land invasions in the Frei and Allende presidencies), Bolivia (in the revolution of 1952), and Peru (the guerrilla movement that helped provoke the military revolution of 1968 and the far more serious *Sendero Luminoso* movement), to mention only a few cases. The Chiapas revolt of early 1994 in southern Mexico threatened political stability in a presidential election year. Yet by mid-1995, the rebels appeared to be effectively contained. Farther south, the Guatemalan military had liquidated their guerrilla opponents in a brutal campaign, and *Sendero Luminoso* had been reduced to a minimal security threat in Peru. The only country with significant surviving guerrilla forces was Colombia.

What about the middle classes? In the 1950s the middle classes were repeatedly "discovered" by U.S. scholars, who solemnly declared that the growing middle class would serve as a ballast to ensure a gradual reformist approach to the deep-seated problems of the region. A sizable middle stratum did emerge after World War II, especially in Argentina, Chile, Mexico, and Brazil. In the latter two the middle class was much smaller in proportion to the total population, but that still meant a significant number in absolute figures.

The problem with the Latin American middle classes was that they resided in Latin America, not the United States or Europe. That meant that their relation to other classes was completely different from that of the United States or Europe. Above them was a rich and powerful upper class, whose lifestyle they often envied. Below was an immense lower class—in Mexico, Brazil, and Chile perhaps 65 or 75 percent, in Argentina 50 percent. In crises the middle classes tended to identify with the upper class, as happened in Chile in 1973, in Brazil in 1964, and in Argentina in 1976. The middle classes apparently became frightened at the prospect of losing income, status, and property.

In calmer times these middle classes could be expected to vote for representative government and centrist leaders. The tendency is to favor coups in a crisis but elections when the dust clears. This has posed a constant problem for the military officers who carry out the coups. The middle classes will remain important, not least because they produce so many of the technocrats who frame policy in virtually all governments, military-dominated or civilian.

What about industrialists? In country after country the business community has proved to be timid and uncertain. Though manufacturing output will increase, Latin American entrepreneurs are so preoccupied with surviving—against the formidable odds of inflation, government regulations, and foreign competition—that they have not been a major political force. In times of crisis they have sided with the military and the middle classes. Businessmen have only rarely played the reformist role of the "progressive national bourgeoisie" envisioned in Marxist

theory. Instead, they have faced increasing pressure both from the state sector and from foreign firms. In many cases they have chosen to associate with foreign companies to get capital and technology, thereby undermining a potential role as independent national spokesmen. Furthermore, they have recently been battered by the neo-liberal economic policies that have slashed tariffs, reduced subsidies, and tightened credit in an effort to promote productivity and thus improve Latin America's competitive status in the world economy. They are vulnerable and on the defensive, unlikely to take the initiative.

What about the church? This institution bears close watching. In Brazil, Chile, and Central America the church, after the 1960s, had created, among the lay population, an extraordinary new consciousness and mobilization. The momentum lay not with the clergy, but with the congregations. The "theology of liberation" was the church's more dramatic reaction to the "social question" in Latin America. The subsequent experience of dictatorships, with repression directed especially at the liberal churchmen, exerted a profound effect on the politically active classes in Latin America.

But opposing torture has proved simpler than articulating a viable stand on the complex social and economic issues that inevitably divide more open societies. The Catholic progressives have also come under institutional siege on two very different fronts. One is in Rome, where Pope John Paul II has skillfully used his powers to silence liberation theologians and appoint conservative bishops throughout Latin America. The other front is at home, where the church's onetime monopoly on Christianity has been undermined by the rapid inroads of Protestantism, led by highly organized Evangelicals.

The military is another key group. Today it is difficult to remember the enthusiasm generated by the "progressive" Peruvian military after their coup of 1968. Given the experience of the 1970s, the Latin American military are now remembered as the repressive praetorians protecting the privileged in all too many countries. The return of civilian governments in Argentina and Brazil has left the military in the shadows. And the military-supported coup in Peru in 1992 showed how quickly the generals could turn on the civilians.

Latin America: Closing the Socialist Route

Between the late 1940s and the early 1990s Latin Americans found themselves the target of ideological competition between the United States and the Soviets. When Cuba went Marxist-Leninist in 1961, it gave the Soviets a potential "showcase" for socialism in the Americas. The United States responded with its own offensive in Latin America, promoting both reform-oriented economic growth and counterinsurgency. Now that the Cold War has faded from Latin America, has the prospect of revolution from the left also faded?

In nineteenth-century Europe the Marxist-oriented political movements argued passionately over the proper tactics for achieving a socialist society. A radical wing urged revolutionary methods, including violence when necessary. The capitalists and their apologists would never be willing to give up a society from which they profited so handsomely, argued the radical Marxists. Moderates maintained that it was possible to move steadily toward socialism by working within the legal structure, especially where there was representative government.

This argument deeply split the European left, creating a division between the revolutionaries and the democratic socialists. Out of the latter arose such important political parties as the German Social Democrats, the British Labour Party, and the French Socialist Party. The revolutionaries remained fragmented until the Russian Revolution in 1917 gave a new focus. The victorious Bolsheviks created an international command structure to coordinate (and, as it soon turned out, to dominate) newly formed communist parties around the world. Most of the revolutionaries entered the communist parties, although important dissident factions, such as the Trotskyists, survived.

This argument over the revolutionary versus the "peaceful" path to socialism carried over to Latin America. Interestingly, the communist parties of Latin America up to 1959 promoted the "peaceful" path in domestic politics, with two exceptions: El Salvador in 1932 and Brazil in 1935. The communist legacy in Latin America was therefore not revolutionary. The Allende election in Chile in 1970 was in keeping with this tradition.

By contrast, the revolutionary path toward socialism was traversed in Cuba and Nicaragua. Both of these regimes had to face unremitting hostility from the United States, including overt (or covert) military action. And while they could claim substantial improvements in health care and education, especially in adult literacy, they alienated substantial sectors of their own populations. In varying degrees, they also became excessively dependent on the largess of the former Communist bloc. The end of the Cold War and the subsequent collapse of the U.S.S.R. not only led to the disappearance of Soviet patronage; it also led to widespread disenchantment with Marxist ideology. As a result, the war-weary citizens of Nicaragua voted the Sandinistas out of power in the 1990 elections. And Fidel Castro's Cuba, isolated and abandoned, lost its once-prized status as a vanguard of continental revolution.

Instead, peoples of Latin America have turned away from utopian ideologies toward practical efforts at the grass-roots level. They are less involved in conquering the state for revolutionary purposes and more interested in applying power for the practical solution of local or specific problems. At the same time, nationalism has lost much of its appeal throughout the region, especially among the new ruling elites. By

the mid-1990s, the socialist route for Latin America appeared to have reached a dead end.

The Prospects for Development Under Capitalism

Capitalism has had a checkered history in Latin America. The colonial era brought a classic mercantilist system, with the colonies forced to produce the maximum economic surplus for the Spanish and Portuguese crowns. In the late eighteenth century, cracks began to appear in this system. Smuggling, abetted by the English, eroded the Iberian monopoly on trade, and elementary free markets began to emerge alongside, often in spite of, the officially sanctioned economy.

In the nineteenth century a fraction of the elites sought to eliminate the vestiges of colonial privilege and introduce a market-oriented economy, geared to foreign trade. The more radical of these reformers wanted to reduce all social relationships to a market definition. Since this was easiest in countries without Indians, it was far simpler in Argentina than in Mexico. The attempt to accelerate that process helped provoke the rural rebellions in the Mexican Revolution.

This transition to capitalism has dominated Latin America in our century. It was sharply modified after the 1930s, as state intervention in the economies increased. In every major country the central government used such instruments as state oil companies, government marketing institutes, and special loan programs. The state role was so great by the early 1970s that one could no longer speak of textbook capitalism in Latin America, but of a *hybrid* capitalism. It had three sources of capital: private national, state, and foreign (usually multinational). When foreign capital came in to supplement national capital, national elites saw a danger that foreigners would gain too much economic power. So the state gained an increasing share of responsibility. In countries such as Argentina and Brazil, the military greatly reinforced this trend. The result was a market economy with many more artificial constraints than the nineteenth-century liberals would have ever envisioned.

Latin American politicians also used the state in the 1930s and 1940s to institutionalize a deep division in the work force. They created a network of social benefits (minimum wage, paid vacations, health care, job security) for workers in the formal labor market, i.e., civil servants, professionals, members of labor unions. They were overwhelmingly urban-based, which made them, in most of Latin America, a minority of the labor force. But they were the most politically active, and therefore the most important for politicians seeking votes.

This hybrid capitalism aroused the fury of twentieth-century economic liberals. Throughout Latin America, but especially in the southern cone, economists and businessmen of extreme laissez-faire views

had long fought against the growing state role. Thanks to military coups in Argentina in 1976 and Chile in 1973, they won control of economic policy making. They sought to reduce the government sector drastically by selling off state enterprises and opening the national market by slashing tariffs and regulations.

By the late 1980s these policy views, often labeled "neo-liberal," had also been adopted by such multilateral financial agencies as the World Bank and the Inter-American Development Bank, as well as the U.S. government (hence the tag "Washington consensus"). Needless to say, such views had long been held dear by foreign investors in New York and London. The neo-liberal wave thus swept Latin America. Brazil was the only holdout, and was climbing aboard by mid-1995.

Thus newly redemocratized Latin America faced harsh medicine: "adjustment" policies aimed not only at correcting the unprecedented imbalance in the external accounts but also at imposing the discipline of market mechanisms on societies long steeped in cozy corporatism and the favoritism of extended families. Might capitalism finally be arriving in Latin America?

If so, with what political impact? By the mid-1990s, the neo-liberal policymakers were riding high (even in Chile the second elected centrist government shied from altering many essentials of Pinochet economic policy), with the left, much less the potential revolutionaries, in disarray. But if adjustment—aided by an expanding world economy relatively free of trade barriers—succeeds in restoring growth, might the left find a new lease on life? As De Tocqueville reminded us, discontent is most likely to rise when material conditions, and therefore expectations, are improving. And it is worth remembering that Latin America's most sustained economic growth since the Great Depression occurred under nationalistic policies that concentrated on production for domestic markets. The appeal of that model has waned but is not dead. It can be counted on to revive among intellectuals and technocrats whose ranks are now dominated by more orthodox doctrines.

No less important is the issue of social justice. By such measures as income distribution (although the data vary in quality), most of Latin America has grown more unequal in the last two decades. Domestic spending on social services—education, health, housing—has plummeted, even as international agencies such as the World Bank have tried to compensate.

Our inquiry leads us back to the connection between political regime type and economic policies. Chile's highly successful application of the neo-liberal model was carried out by a military government, not a democracy. Now we see democratically elected governments attempting to follow the same model. Unfortunately, these democratic systems are all too often manipulated by the "nonpoor" (in the World Bank's euphemism), who show scant interest in improving the social welfare of

their societies. On the contrary, they excel at using the state to promote their own interests. They seem bent on producing a primitive capitalism reminiscent of late nineteenth-century Europe and the United States. In those halcyon days, would-be capitalists were told *"Enrichez-vous"* (get rich!). They did so and left most of their societies waiting decades for state intervention to correct the gross inequalities. Could it be that the Latin American capitalists of the late twentieth century, like the Bourbon monarchs of France, have learned nothing and forgotten everything?

What Will Happen to the Non-European Cultures in Latin America?

Few can study Latin America without becoming fascinated with its kaleidoscopic mixture of races and peoples. A question soon arises: Will the unusual, the different, the exotic, be homogenized into national amalgams that lack the originality of the Indian, African, or provincial cultures? What about the Mexican Indians of Chiapas, or the Indians of the Peruvian altiplano, or the blacks of Brazil's Bahia? Will they or their ethnic identities disappear?

It is not easy to argue that their way of life will survive. Latin America is hardly immune from the process of social homogenization so familiar in the industrial world. Television and radio have eroded regional and provincial barriers and concentrated attention on "national" models, as in the highly popular television soap operas *(telenovelas)*. There is also relentless economic pressure on the ethnic holdouts to learn the national language and adopt national cultural ways. In Mexico, for example, the proportion of the population still speaking only an Indian language had dropped to less than 2 percent by the mid-1990s. Incorporating these Indians into the Spanish-speaking population has been a prime goal for Mexican leaders. In the process, however, ethnic traditions have been lost. Could it be otherwise? The Mexican government, more than any other in Latin America, has attempted to preserve indigenous traditions (partly to earn tourist dollars, since "native" customs are an attraction). There has also been a genuine desire to preserve a measure of Mexico's unique pre-Hispanic culture. The pursuit of social development, however, tends to work against these goals.

In Brazil the most important non-European tradition has been African. Indeed, African slaves so penetrated every region of Brazil that its modern culture is indelibly stamped with the African presence. Can it survive the relentless process of national cultural assimilation? It seems to persist as an underground influence, especially in religion. The context is Brazil's multilevel religious world. On top is the Roman Catholic Church, official in every respect. Underneath lie the worlds of *umbanda*, spiritism, and Afro-Brazilian cults such as *candomblé* and *ma-*

cumba. In these counterestablishment religions the African element has so permeated the national culture that its survival, if in an assimilated mode, seems assured.

The only areas where Indian cultures seem likely to survive for long are in the Andes, southern Mexico, and Guatemala, where Indian populations appear sufficiently concentrated to preserve traditional social identities. In general, however, urbanization in Latin America is engulfing or liquidating the rural and the provincial. The most likely outcome is that these nations will all emerge with primarily European cultures, with scattered remnants of the indigenous and African influence. This seems no more surprising than the annihilation of American Indian culture in the United States or the relentless obliteration of regional culture in North America. In the last analysis, few elites inside or outside Latin America set store by the preservation of African or Indian culture. And where does it rank in the ratings of the World Health Organization or the International Monetary Fund? The picturesque may interest tourists, but it becomes an impediment to reducing illiteracy or infant mortality. In the modern world, cultural heterogeneity seems to count for little.

The Potential for Regional Conflict

Latin America has not seen frequent wars in the twentieth century, although some longstanding conflicts have remained alive. Among the most important are the Falklands-Malvinas dispute between Argentina and Britain, the Argentina-Chile clash over the Beagle Straits, the Peru-Chile tension in 1979–82 over the lands taken by Chile in the War of the Pacific, and the Venezuela-Guyana boundary conflict. In the Andes the thriving drug traffic continues to offer potential for interstate clashes, as has happened recently on the Amazonian border of Peru and Brazil. A boundary clash erupted in 1995 on the Peruvian-Ecuadorian frontier. In the earlier border war of 1942, Peru had emerged victorious with large chunks of Ecuadorian territory, said to harbor quantities of oil and silver. This new conflict, which lasted only weeks, was quickly contained by multilateral intervention of other Latin American nations.

What about possible conflict on the U.S.-Mexican border? Can the huge flow of undocumented migrants continue without a further U.S. reaction? This volatile question must be seen in the context of U.S. policies toward the entire Caribbean basin. Here the question is the degree to which the United States is willing to continue serving as the escape valve for those who flee from poverty and/or political repression. The key variable will be the attitude of the U.S. public.

Signs from the U.S. Congress have been contradictory. In 1986 the United States passed a law toughening penalties against employers who knowingly hire aliens without legal documentation. Despite the predic-

tions of both supporters and opponents, the law proved to have little lasting effect on the flow of undocumented migration, as the U.S. southern borders continued to be notoriously porous and many "illegals" easily obtained forged identity papers once they were in the United States. Meanwhile, the U.S. economy went on readily absorbing "illegals" into the low-paying service sector. In California, however, a powerful backlash occurred in the form of a voter-approved 1994 referendum that would deny many public benefits to "illegal" and even to some legal aliens. A lengthy court fight about the measure was soon under way.

Another source of continuing tension along the U.S.-Mexican border has been drug traffic. Despite the intermittent cooperation of U.S. and Mexican authorities, drug flow across the border has increased because of more effective interdiction in the Caribbean. Repeated denunciations of high-level corruption in Mexico by sanctimonious U.S. politicians only added fuel to the fires of aggravation.

There is one final source of regional conflict: intervention by external powers. The historic culprit has been the United States. Buoyed by low-cost (to U.S. military forces, not the natives) invasions of Grenada in 1983 and Panama in 1989, the undermining of Sandinista popularity through support for the Contras, plus the dramatic victory in the 1991 Gulf War against Iraq, the Bush White House boasted that it had "kicked the Vietnam syndrome." Most Latin Americans find that hardly reassuring. The Cuban-U.S. face-off retains the greatest explosive potential. The U.S.-U.S.S.R. agreement ending the 1962 missile crisis included a U.S. pledge not to invade Cuba. The Reagan administration seriously considered rescinding that agreement. With Russian economic support for Cuba gone, living standards there plunging, and the demise of communism in Europe, the chances for internal conflict in Cuba increase. Should it become violent, U.S. public opinion could easily be mobilized to support intervention.

Latin America's Contribution to the World

Given its political and economic constraints, what is Latin America likely to contribute to the human experience? It has already made its mark in literature. Gabriel García Márquez, Carlos Fuentes, Jorge Luis Borges, Jorge Amado—the "boom" in Latin American literature has led to its translation into major European languages. Mass paperback editions have facilitated wide distribution. Music is another artistic realm of Latin American excellence. Afro-Brazilian and Afro-Cuban music have penetrated North American popular music, giving an unmistakable style and rhythm.

Latin Americans have been conspicuous for their impressive contributions in athletics. In soccer, the most universally played sport, Brazil is the first country to win the World Cup four times. Argentina won

the cup in 1978 and 1986. Even small Latin American countries such as Uruguay have taken home the World Cup. European soccer clubs bid into the millions of dollars to lure Latin American stars.

Theology and church organization have seen impressive innovation in Latin America. The highly controversial "theology of liberation" is largely a Latin American phenomenon, an attempt by Latin American theologians to reconcile their religious tradition with the political and economic pressures around them. No less important are the rapidly burgeoning laymen's groups ("ecclesiastical base communities"), which represent a reawakening in Latin American nations that had long seemed sunk in religious apathy. It is worth remembering that the church, through both its clergy and its laity, played a key role in Brazil's redemocratization in the 1970s. In Chile the church also served as a rallying point for resistance to a repressive military regime. This oppositionist role brought the church back onto center stage and reminded democrats everywhere that one of Latin American society's most traditional institutions can still be highly relevant in the modern world.

Latin America has also made a great contribution in the field of race relations. Notwithstanding persistent cruelty to those of non-European descent, Latin America has produced societies in which mixed-bloods have enjoyed great mobility. The *mestizos* of Mexico, Central America, and the Andean region represent a new social category born out of the mixture of European and Indian. Although racism still exists in many guises, mobility has been remarkable. The same can be said for the mulatto in Brazil, Cuba, Colombia, and the Caribbean nations. To find a contrast one need only look to North America. Of course, prejudice and discrimination still occur in Latin America, especially against the "pure blood" Indians and very dark-skinned people in general. Yet the relative social harmony is notable, especially in light of the miserable record left by the Europeans in so many parts of today's developing world.

In the future, as in past centuries, the fate of Latin America will depend largely on its relationship to the centers of international power. In the meantime it must mobilize its own resources for sustained economic growth and seek a more equitable distribution of the results. The region will also have to continue to cope with the implications of subordination and dependency. Meanwhile, outsiders will continue to be startled and fascinated by what García Márquez called "the unearthly tidings of Latin America, that boundless realm of haunted men and historic women, whose unending obstinacy blurs into legend."

STATISTICAL APPENDIX

TABLE 13-1   Selected Social Indicators

	Population Size (millions)		Annual Population Growth Rate (%)		Urbanization[a] (%)		Life Expectancy (Years)
	1900	1992	1900–10	1980–92	1900	1980	1992
Argentina	4.6	33.1	4.3	1.3	24.9	70.2	71
Brazil	18.0	153.9	2.2	2.0	8.7	45.7	66
Chile	3.0	13.6	1.2	1.7	19.9	67.9	72
Mexico	13.6	85.0	1.0	2.0	9.2	42.5	70
Peru	3.0	22.4	3.0	2.1	6.0	47.2	65
Central America							
Costa Rica	0.3	3.2	1.5	2.8	8.5	30.1	76
El Salvador	0.8	5.4	2.1	1.4	6.0	24.9	66
Guatemala	0.9	9.7	2.1	2.9	—	18.9	65
Honduras	0.4	5.4	2.8	3.3	—	23.8	66
Nicaragua	0.4	3.9	2.6	2.7	—	36.9	67
Panama	0.3	2.5	2.4	2.1	6.8	40.9	73
Caribbean							
Cuba	1.6	10.8	1.0	1.0	25.0	47.5	76
Dominican Republic	0.6	7.3	2.1	2.1	3.6	40.8	68
Grenada	n.a.	(91,000)[b]	n.a.	n.a.	n.a.	n.a.	71
Haiti	1.25	6.7	3.2	2.0	8.2	16.5	55
Jamaica	n.a.	2.4	n.a.	1.0	—	n.a.	74
United States	76.1	255.4	—	1.0	28.9	64.0[c]	77

n.a. = not available.

[a] Population in communities of 20,000 or more, as percentage of national total.

[b] Total figure (not in millions).

[c] Figure for 1970.

Sources: *Statistical Abstract of Latin America*, 19 (Los Angeles: UCLA Latin American Center, 1978) Table 623; 21 (1981), Tables 104, 623, 634; and 25 (1987), Tables 600, 661; World Bank, *World Development Report 1994: Infrastructure for Development* (New York: Oxford University Press, 1994), Tables 1, 1a, 25, 31.

TABLE 13-2   Selected Economic Indicators

	GNP per capita (dollars)	Average Annual Economic Growth (GDP)		External Debt, 1992	
		(%) 1965–80	(%) 1980–92	Dollar Amount (billions)	Interest Payments as % of Export Earnings
	1992				
Argentina	6,050	3.3	0.4	67.6	18.7
Brazil	2,770	9.0	2.2	121.1	9.2
Chile	2,730	1.9	4.8	19.4	10.4
Mexico	3,470	6.5	1.5	113.4	16.4
Peru	950	3.9	−0.6	20.3	10.7
Central America					
Costa Rica	1,960	6.3	3.3	4.0	9.1
El Salvador	1,170	4.4	1.3	2.1	5.0
Guatemala	980	5.9	1.4	2.7	7.9
Honduras	580	4.1	2.8	3.6	15.3
Nicaragua	340	2.6	−1.7	11.1	12.6
Panama	2,420	5.5	0.9	6.5	4.3
Caribbean					
Cuba	1,370[a]	n.a.	n.a.	n.a.	n.a.
Dominican Republic	1,050	7.3	1.7	4.6	5.4
Grenada	2,310	n.a.	n.a.	n.a.	n.a.
Haiti	340[b]	2.9	n.a.	0.8	n.a.
Jamaica	1,340	1.5	1.8	4.3	8.5.

n.a. = not available.

[a] Figure for 1989.

[b] Figure for 1991.

Sources: World Bank, *World Development Report 1994: Infrastructure for Development* (New York: Oxford University Press, 1994), Tables 1, 1a, 2, 20, 23, 24; and Central Intelligence Agency, *World Factbook 1994–95* (Washington: Brassey's, 1994).

TABLE 13-3   Structure of the Labor Force, Early 1990s

	Percentage of Labor Force in		
	Agriculture	*Industry*	*Services*
Argentina[a]	12	31	57
Brazil	29	25	44
Chile	18	27	56
Mexico[b]	28	21	51
Peru	33	17	50
Central America			
Costa Rica	24	25	51
El Salvador[a]	40	15	45
Guatemala	28	21	51
Honduras[a]	62	12	26
Nicaragua	30	16	40
Panama	27	13	59
Caribbean			
Cuba[b]	20	32	48
Dominican Republic[b]	31	20	49
Grenada[a]	24	13	63
Haiti[a]	66	9	26
Jamaica	27	19	54
United States	3	30	67

Note: The *agricultural sector* comprises agriculture, forestry, hunting, and fishing; in some countries the practice of subsistence farming (outside the market economy) may lead to underestimation of the sector's size. *Industry* includes mining, manufacturing, construction, electricity, water, and gas. All other branches of economic activity are categorized as *services*.

[a]Data from late 1980s.

[b]Data from 1990.

Sources: Economist Intelligence Unit, *Country Profiles, 1993–94* and *1994–95* (London: Economist Intelligence Unit, 1993 and 1994); and Central Intelligence Agency, *World Factbook 1994–95* (Washington: Brassey's, 1994).

HEADS OF STATE*

Argentina

1862	Bartolomé Mitre
1868	Domingo F. Sarmiento
1874	Nicolás Avellaneda
1880	Julio Argentino Roca
1886	Miguel Juárez Celman
1890	Carlos Pellegrini
1892	Luis Sáenz Peña
1895	José E. Uriburu
1898	Julio Argentino Roca
1904	Manuel Quintana
1906	José Figueroa Alcorta
1910	Roque Sáenz Peña
1914	Victorino de la Plaza
1916	Hipólito Yrigoyen
1922	Marcelo Torcuato de Alvear
1928	Hipólito Yrigoyen
1930	José Félix Uriburu
1932	Augustín P. Justo
1938	Roberto M. Ortiz
1940	Ramón S. Castillo
1943	Arturo Rawson, June 5–7
	Pedro P. Ramírez, June 7–March 9, 1944
1944	Edelmiro J. Farrell
1946	Juan Domingo Perón

1955	Eduardo Lonardi, September 23–November 13
	Pedro Eugenio Aramburu, November 13–May 1, 1958
1958	Arturo Frondizi
1962	José María Guido
1963	Arturo Illia
1966	Juan Carlos Onganía
1970	Roberto Marcelo Levingston
1971	Alejandro A. Lanusse
1973	Héctor Cámpora, May 27–July 13
	Juan Domingo Perón, October 12–July 1, 1974
1974	María Estela (Isabel) Martínez Perón
1976	Jorge Rafael Videla
1981	Roberto Viola, March 29–December 22
	Leopoldo Fortunato Galtieri, December 22–June 17, 1982
1982	Reynaldo Benito Antonio Bignone
1983	Raúl Alfonsín Foulkes
1989	Carlos Saúl Menem
1995	Carlos Saúl Menem

Barbados

1966	Errol Walton Barrow
1976	J. M. G. M. ("Tom") Adams
1985	H. Bernard St. John

1986	Errol Walton Barrow
1987	Erskine Sandiford
1994	Owen Arthur

Brazil

1831	Dom Pedro II
1889	Deodoro da Fonseca
1891	Floriano Peixoto
1894	Prudente de Moraise e Barros

1898	Manuel Ferraz de Campos Sales
1902	Francisco de Paula Rodrigues Alves
1906	Afonso Augusto Moreira Pena

*Countries listed here include only those which are covered in this book.

429

1909 Nilo Peçanha
1910 Hermes da Fonseca
1914 Wenceslau Brás Pereira Gomes
1918 Delfim Moreira da Costa Ribeiro
1919 Epitácio da Silva Pessòa
1922 Arthur da Silva Bernardes
1926 Washington Luís Pereira de Sousa
1930 Júlio Prestes, October 24–November 4
       Getúlio Dornelles Vargas, November 4–October 31, 1945
1945 José Linhares
1946 Eurico Gaspar Dutra
1951 Getúlio Vargas
1954 João Café Filho
1955 Carlos Luz, November 8–11
       Nereu Ramos, November 11–January 31, 1956
1956 Juscelino Kubitschek
1961 Jânio Quadros, January 31–August 25

       Ranieri Mazzilli, August 25–September 8
       João Goulart, September 8–April 2, 1964
1964 Ranieri Mazzilli, April 2–15
       Humberto de Alencar Castello Branco, April 15–March 15, 1967
1967 Artur da Costa e Silva
1969 Augusto Hamann Rademaker Gruenewald, August 31–October 25
       Emílio Garrastazu Medici, October 30–March 15, 1974
1974 Ernesto Geisel
1979 João Baptista Figueiredo
1985 José Sarney
1990 Fernando Collor de Mello
1992 Itamar Franco
1995 Fernando Henrique Cardoso

Chile

1876 Aníbal Pinto
1881 Domingo Santa María
1886 José Manuel Balmaceda
1891 Jorge Montt
1896 Federico Errázuriz
1901 Germán Riesco Errázuriz
1905 Rafael Rayas
1906 Pedro Montt
1910 Elías Fernández Albano, August 16–September 6
       Emiliano Figueroa, September 6–September 18, 1911
1911 Ramón Barros Luco
1915 Juan Luis Sanfuentes
1920 Luis Barros Borgoño, June 25–December 23
       Arturo Alessandri Palma, December 23–September 8, 1924
1924 Luis Altamirano
1925 Carlos Ibañez del Campo, January 23–March 21
       Arturo Alessandri Palma, March 21–October 1, 1925
       Luis Barros Borgoño, October 1–December 1925
       Emiliano Figueroa Larraín, December–May 4, 1927
1927 Carlos Ibáñez del Campo
1931 Pedro Opazo Letelier, July 26–27
       Juan Esteban Montero Rodríguez, July 27–August 18, 1931

       Manuel Trucco Franzani, August 18–November 15
       Juan Esteban Montero Rodríguez, November 15–June 4, 1932
1932 Arturo Puga, June 4–12
       Marmaduke Grove, June 12–16
       Carlos Dávila Espinoza, June 17–September 13
       Bartolomé Blanche Espejo, September 13–October 2
       Abraham Oyanedel, October 2–December 24
       Arturo Alessandri Palma, December 24–December 24, 1938
1938 Pedro Aguirre Cerda
1941 Geronimo Méndez Arancibia
1942 Juan Antonio Ríos Morales
1946 Alfredo Duhalde Vázquez, June 27–October 17
       Juan A. Irabarren, October 17–November 3
       Gabriel González Videla, November 3–November 3, 1952
1952 Carlos Ibáñez del Campo
1958 Jorge Alessandri Rodríguez
1964 Eduardo Frei Montalva
1970 Salvador Allende Gossens
1973 Augusto Pinochet Ugarte
1990 Patricio Aylwin
1994 Eduardo Frei Ruiz-Tagle

Costa Rica

1876	Tomás Guardia	1940	Rafael Angel Calderón Guardia
1882	Próspero Fernández	1944	Teodoro Picardo Michalski
1885	Bernardo Soto y Alfaro	1948	Santos León Herrera, April 19–
1890	José Joaquín Rodríguez		May 8
1894	Rafael Iglesias Castro		José Figueres Ferrer, May 8–November 8, 1949
1902	Ascensión Esquivel Ibarra		
1906	Cleto González Víquez	1949	Otilio Ulate Blanco
1910	Ricardo Jiménez Oreamuno	1952	Alberto Oreamuno Flores
1912	Cleto González Víquez	1953	José Figueres Ferrer
1914	Alfredo González Flores	1958	Mario Echandi Jiménez
1917	Federico Tinoco Granados	1962	Francisco José Orlich
1919	Julio Acosta García, May 7–August 13	1966	José Joaquín Trejos Fernández
		1970	José Figueres Ferrer
	Juan Bautista Quirós, August 13–May 8, 1920	1974	Daniel Oduber
		1978	Rodrigo Carazo
1920	Julio Acosta García	1982	Luis Alberto Monge Alvarez
1924	Ricardo Jiménez Oreamuno	1986	Oscar Arias Sánchez
1928	Cleto González Víquez	1990	Rafael Angel Calderón
1932	Ricardo Jiménez Oreamuno	1994	José María Figueres
1936	León Cortés Castro		

Cuba

1899	John R Brooke, January 1–December 23 Leonard Wood, December 23–May 20, 1902 } U.S. Occupation		
1902	Tomás Estrada Palma		
1906	William Howard Taft, September 29–October 13 Charles Edward Magoon, October 13–January 28, 1909 } U.S. Occupation		
1909	José Miguel Gómez	1935	José A. Barnet y Vinageras
1913	Mario García Menocal	1936	Miguel Mariano Gómez Arias, May 20–December 24
1921	Alfredo Zayas		
1925	Gerardo Machado Morales		Federico Laredo Bru, December 24–October 10, 1940
1933	Carlos Manuel de Céspedes, August 12–September 5		
		1940	Fulgencio Batista y Zaldívar
	Council of Five, September 5–September 10	1944	Ramón Grau San Martín
		1948	Carlos Prío Socarrás
	Ramón Grau San Martín, September 10–January 15, 1934	1952	Fulgencio Batista y Zaldívar
		1959	Manuel Urrutia Lleo, January 2–July 17 [Fidel Castro Ruz de facto head from here on]
1934	Carlos Hevía, January 15–18		
	Manuel Márquez Sterling, January 18		
			Osvaldo Dorticós Torrado, July 17–December 3, 1976
	Carlos Mendieta Montefur, January 18–December 11, 1935		
		1976	Fidel Castro Ruz

Dominican Republic

1868	Buenaventura Báez	1887	Ulíses Heureaux
1874	Ignacio (María) González	1889	Juan Wenceslao Figuereo, August 1–31
1876	Ulíses Francisco Espaillat, June 29–November		
			Horacio Vázquez, September 1–November 14
	Ignacio González, November–December		
			Juan Isidro Jiménez, November 19–May 2, 1902
	Buenaventura Báez, December–February 24, 1878		
		1902	Horacio Vázquez
1878	Cesário Guillermo	1903	Alejandro Wos y Gil, April 27–November 20
1879	Gregorio Luperón		
1880	Fernando Arturo de Meriño		Juan Isidro Jimenez, December 28–April 2, 1904
1884	Ulíses Heureaux		
1885	Francisco Gregorio Billini	1904	Carlos Morales

1906 Ramón Cáceres
1911 Eladio Victoria
1912 Adolfo Nouel y Bobadilla
1913 José Bordas y Valdés
1914 Ramón Baéz, August 27–December 5
Juan Isidro Jiménez, December 5–May 8, 1916
1916 Francisco Henríquez y Carvajal } U.S. Occupation
1922 Juan Batista Vicini Burgos } U.S. Occupation
1924 Horacio Vázquez
1930 Rafael Estrella Ureña, March 2–August 16
Rafael Leonidas Trujillo y Molina, August 18–June 18, 1938 [De facto head until 1961]
1938 Jacinto Bienvenido Peynado
1940 Manuel de Jesús Troncoso de la Concha
1942 Rafael Leonidas Trujillo y Molina
1952 Héctor Bienvenido Trujillo
1960 Joaquín Balaguer
1962 Rafael (Filiberto) Bonnelly, January 1–17
Huberto Bogaert, January 17–19
Rafael Bonnelly, January 19–February 27, 1963
1963 Juan Bosch Gavino, February 27–September 26
Emilio de los Santos, September 26–December 22
Donald Reid Cabral, December 22–April 25, 1965
1965 Disturbances and civil war, April 25–September 3, 1965 } U.S. Occupation
Elías Wessin y Wessin, April 28–May 7 } U.S. Occupation
Antonio Imbert Barreras, May 7–August 30 } U.S. Occupation
Francisco Caamaño Deñó, April 25–September 3 } U.S. Occupation
Héctor García Godoy Cáceres, September 3–July 1, 1966 } U.S. Occupation
1966 Joaquín Balaguer
1978 Silvestre Antonio Guzmán Fernández
1982 Jacobo Majluta, July 4–August 14
1982 Salvador Jorge Blanco
1986 Joaquín Balaguer
1990 Joaquín Balaguer
1994 Joaquín Balaguer

El Salvador

1876 Rafael Zaldívar y Lazo
1885 Francisco Menéndez
1890 Carlos Ezeta
1894 Rafael Gutiérrez
1898 Tomás Regalado
1903 Pedro José Escalón
1907 Fernando Figueroa
1911 Manuel Enrique Araújo
1913 Carlos Meléndez
1914 Alfonzo Quiñones Molina
1915 Carlos Meléndez
1919 Jorge Meléndez
1923 Alfonzo Quiñones Molina
1927 Pío Romero Bosque
1931 Arturo Araujo, March 1–December 4
Maximiliano Hernández Martínez, December 4–August 29, 1934
1934 Andrés Ignacio Menéndez
1935 Maximiliano Hernández Martínez
1944 Andrés Ignacio Menéndez, May 9–October 21
Osmín Aguirre y Salinas, October 21–March 1, 1945
1945 Salvador Castañeda Castro
1948 Manuel de J. Córdova
1949 Oscar Osorio, January 4–October 22
Oscar Bolaños, October 22–September 14, 1950
1950 Oscar Osorio
1956 José María Lemus
1960 Miguel Angel Castillo
1961 Aníbal Portillo
1962 Eusebio Rodolfo Cordón, January 25–July 1
Julio Adalberto Rivera, July 1–July 1, 1967
1967 Fidel Sánchez Hernández
1972 Arturo Armando Molina

1977 Carlos Humberto Romero	1982 Alvaro Alfredo Magaña
1979 Jaime Abdul Gutiérrez, Adolfo	1984 José Napoleón Duarte
Arnoldo Majano & junta	1989 Alfredo Cristiani
1980 José Napoleón Duarte	1994 Armando Calderón Sol

Guatemala

1873 Justo Rufino Barrios
1885 Alejandro Sinibaldi, April 2–15
Manuel Lisandro Barillas, April 15–March 15, 1892
1892 José María Reina Barrios
1898 Manuel Estrada Cabrera
1920 Carlos Herrera
1922 José María Orellana
1926 Lázaro Chacón
1930 Baudillo Palma, December 13–16
Manuel Orellana C., December 16–31
José María Reyna Andrade, December 31–February 14, 1931
1931 Jorge Ubico
1944 Federico Ponce Vaidez, July 1–October 21
Jacobo Arbenz Guzmán, December 12–March 1, 1945
1945 Juan José Arévalo Bermejo

1951 Jacobo Arbenz Guzmán
1954 Carlos Díaz, June 27–29
Elfego J. Monzón, June 29–July 8
Carlos Castillo Armas, July 8–July 26, 1957
1957 Luis Arturo González López, July 27–October 25
Guillermo Flores Avendaño, October 28–March 2, 1958
1958 Miguel Ydígoras Fuentes
1963 Enrique Peralta Azurdia
1966 Julio César Méndez Montenegro
1970 Carlos Arana Osorio
1974 Kjell E. Langerud García
1978 Fernando Romeo Lucas García
1982 Efraín Ríos Montt
1983 Oscar Humberto Mejía Victores
1986 Marco Vinicio Cerezo Arévalo
1991 Jorge Serrano Elías
1993 Ramiro de León Carpio

Haiti

1867 Sylvain Salnave
1870 Nissage Saget [fnu]
1874 Michel Dominique
1876 Boisrond Canal [fnu]
1879 Étienne Félicité Salomon
1888 Télémaque, August 24–September 19
François Denis Légitime, October 23–August 22, 1889
1889 Louis Mondastin Floréal Hippolyte
1896 P. A. Tirésias Simon Sam
1902 Boisrond Canal [fnu], May 9–December 21
Alexis Nord, December 21–December 2, 1908
1908 Antoine Simon
1911 Michel Cincinnatus Leconte
1912 Tancrède Auguste
1913 Michel Oreste
1914 Oreste Zamor, February 8–October 29
Joseph Davilmare Théodore, November 7–February 23, 1915
1915 Jean Velbrun-Guillaume, March 4–July 26

1941 Élie Lescot
1946 Frank Lavaud, January 12–August 15
Dumarsais Estimé, August 15–May 10, 1950
1950 Frank Lavaud, May 10–December 6
Paul Eugène Magloire, December 6–December 12, 1956
1956 Joseph Nemours Pierre-Louis
1957 François Sylvain, February 7–April 2
Executive Committee of 13 Ministers under the protection of the Army, April 6–May 20
Léon Cantave, May 20–26
Daniel Fignolé, May 26–June 14
Antoine Kebreau, June 14–October 15
François Duvalier, October 22–April 21, 1971
1971 Jean-Claude Duvalier
1986 Henri Namphy
1988 Leslie Manigat, February 7–June 20

Philippe Sudre Dartiguenave, August 12–May 15, 1922
1922 Joseph Louis Borno
1930 Étienne Roy, May 15–November 18
Sténio Vincent, November 18–April 1941

} U.S. Occupation 1915–1934

Henri Namphy, June 20–September 18
Prosper Avril, September 18–March 13, 1990
1990   Ertha Pascal-Trouillot

1991   Jean-Bertrand Aristide
1991   Raoul Cédras
1994   Émile Jonaissaint
1995   Jean-Bertrand Aristide

Honduras

1876   Marco Áurelio Soto
1883   Luis Bográn
1891   Pariano Leiva
1893   Domingo Vázquez
1894   Policarpo Bonilla
1899   Terencio Sierra
1903   Manuel Bonilla
1907   Miguel R. Dávila
1911   Francisco Beltrán
1912   Manuel Bonilla
1913   Francisco Bertrand
1915   Alberto Membreño
1916   Francisco Bertrand
1920   Rafael López Gutiérrez
1924   Fausto Dávila, March 27–31
        Vincente Tosta, April 1–February 2, 1925

1925   Miguel Paz Barahona
1929   Vicente Mejía Colindres
1933   Tiburcio Carías Andino
1949   Juan Manuel Gálvez
1956   Roque I. Rodríguez
1957   Ramón Villeda Morales
1963   Oswaldo López Arellano
1971   Ramón Ernesto Cruz
1972   Oswaldo López Arellano
1975   Juan Alberto Melgar Castro
1978   Policarpo Paz García
1982   Roberto Suazo Córdova
1986   José Simeón Azcona Hoyo
1990   Rafael Leonardo Callejas
1994   Carlos Roberto Reina

Jamaica

1962   William Alexander Bustamante
1967   Donald Burns Sangster [from 1965 acting Prime Minister], February 22–April 11
        Hugh Lawson Shearer, April 11–March 1, 1972

1972   Michael Norman Manley
1980   Edward Phillip George Seaga
1989   Michael Norman Manley
1992   P. J. Patterson

Mexico

1876   Porfirio Díaz
1880   Manuel González
1884   Porfirio Díaz
1911   Francisco León de la Barra, May 25–November 6
        Francisco Madero, November 6–February 18, 1913
1913   Victoriano Huerta
1914   Venustiano Carranza
1920   Adolfo de la Huerta, May 21–December 1
        Álvaro Obregón, December 1–December 1, 1924
1924   Plutarco Elías Calles

1928   Emilio Portes Gil
1930   Pascual Ortiz Rubio
1932   Abelardo Rodríguez
1934   Lázaro Cárdenas
1940   Manuel Ávila Camacho
1946   Miguel Alemán
1952   Adolfo Ruiz Cortines
1958   Adolfo López Mateos
1964   Gustavo Díaz Ordaz
1970   Luis Echeverría Alvarez
1976   José López Portillo
1982   Miguel de la Madrid Hurtado
1988   Carlos Salinas de Gortari
1994   Ernesto Zedillo Ponce de León

Nicaragua

1879   Joaquín Zavala
1883   Adán Cárdenas
1887   Evaristo Carazo

1889   Roberto Sacaza
1893   Innere Wirren
1893   José Santos Zelaya

1909 José Madriz
1910 José Dolores Estrada
1911 Juan José Estrada, January 1–May
      9
      Adolfo Díaz, May 9–December 31, 1916 ⎫
1917 Emiliano Chamorro Vargas   │
1919 Diego Manuel Chamorro   ⎬ U.S. Occupation
1923 Bartolomeo Martínez   │
1925 Carlos Solórzano   │
1926 Emiliano Chamorro Vargas, Janu- ⎭
      ary 15–November 11
      Adolfo Díaz, November 11–January 1, 1929 ⎫
1929 José María Moncada   ⎬ U.S. Occupation
1933 Juan Bautista Sacasa   ⎭
1936 Carlos Brenes Jarquín
1937 Anastasio Somoza García   1966 Lorenzo Guerrero Gutiérrez
      ("Tacho")   1967 Anastasio Somoza Debayle
1947 Leonardo Argüello, May 1–26   ("Tachito")
      Benjamín Lacayo Sacasa, May 26– 1972 Governing junta for Somoza family
      August 15   1974 Anastasio Somoza Debayle
      Victor Román y Reyes, August 15–   ("Tachito")
      May 6, 1950   1979 Francisco Urcuyo junta: Daniel
1950 Anastasio Somoza García ("Tacho")   Ortega and others
1956 Luis Somoza Debayle   1985 José Daniel Ortega Saavedra
1963 René Schick Gutiérrez   1990 Violeta Barrios de Chamorro

Panama

1904 Manuel Amador Guerrero   1939 Agusto S. Boyd
1908 José Domingo de Obaldia   1940 Arnulfo Arias
1910 Carlos Antonio Mendoza, March   1941 Ernesto Jaén Guardia, October 9
      1–October 1   Ricardo Adolfo de la Guardia,
      Federico Boyd, October 1–October   October 9–June 15,
      4   1945
      Pablo Arosemena, October 4–   1945 Enrique Adolfo Jimenez
      February 2, 1912   1948 Domingo Díaz Arosemena
1912 Rodolfo Chiari, February 2–March 1949 Daniel Chanis, July 28–November
      7   20
      Pablo Arosemena, March 7–   Roberto Chiari, November 20–25
      October 1   Arnulfo Arias, November 25–May
      Belisario Porras, October 1–   10, 1951
      October 1, 1916   1951 Alcibíades Arosemena
1916 Ramón Maximiliano Valdés   1952 José Antonio Remón Cantera
1918 Ciro Louis Urriola, June 3–   1955 José Ramón Guizado, January 3–
      October 1   15
      Pedro Antonio Díaz, October 1–12   Ricardo M. Arias Espinosa, January
      Belisario Porras, October 12–   15–October 1, 1956
      January 30, 1920   1956 Ernesto de la Guardia, Jr.
1920 Ernesto Tisdel Lefevere, January 1960 Roberto F. Chiari
      30–October 1   1964 Marco Aurelio Robles
      Belisario Porras, October 1–   1968 Arnulfo Arias Madrid, October 1–
      October 1, 1924   October 11
1924 Rodolfo Chiari   Omar Torrijos Herrera, October
1928 Florencio Harmodio Arosemena   11–September, 1978
1931 Harmodio Arias, January 2–16   1978 Aristedes Royo
      Ricardo J. Alfaro, January 16–Octo- 1982 Ricardo de la Espriella
      ber 1, 1932   1984 Jorge Illueca
1932 Harmodio Arias   Nicolás Ardito Barletta Vallarino
1936 Juan Demóstenes Arosemena   1985 Eric Arturo Devalle Henriquez

| 1988 | Manuel Solís Palma (Manuel Antonio Noriega de facto head) | 1989 | Guillermo Endara |
| 1989 | Francisco Rodriguéz, September December | 1994 | Ernesto Pérez Balladares |

Peru

1886 Andrés Avelino Cáceres
1890 Remigio Morales Bermúdez
1894 J. Borgoño, May 1–August 10
Andrés Avelino Cáceres, August 10–March 19, 1895
1895 Manuel Candamo, March 21–September 8
Nicolás de Piérola, September 8–September 8, 1899
1899 Eduardo López de Romaña
1903 Manuel Candamo
1904 Serapio Calderón, May 7–September 24
José Pardo y Barreda, September 24–September 24, 1908
1908 Augusto B. Leguía
1912 Guillermo E. Billinghurst
1914 Oscar R. Benavides
1915 José Pardo
1919 Augusto B. Leguía
1930 Manuel Ponce, August 25–27, 1930
Luis M. Sánchez Cerro, August 27–March 1, 1931
1931 Ricardo Leoncio Elías, March 1–5
Gustavo A. Jiménez, March 5–11

David Sámanez Ocampo, March 11–December 8
Luis M. Sánchez Cerro, December 8–April 30, 1933
1933 Oscar R. Benavides
1939 Manuel Prado y Ugarteche
1945 José Luis Bustamante y Rivero
1948 Zenón Noriega, October 29–30
Manuel A. Odría, October 31–June 1, 1950
1950 Zenón Noriega, June 1–July 28
Manuel A. Odría, July 28–July 28, 1956
1956 Manuel Prado y Ugarteche
1962 Ricardo Pérez Godoy
1963 Nicolás Lindey López, March 3–July 28
Fernando Belaúnde Terry, July 28–October 3, 1968
1968 Juan Velasco Alvarado
1975 Francisco Morales Bermúdez
1980 Fernando Belaúnde Terry
1985 Alan García Pérez
1990 Alberto Fujimori
1995 Alberto Fujimori

Puerto Rico

1948 Luiz Muñoz Marín
1964 Roberto Sánchez Vilella
1968 Luis A. Ferré
1972 Rafael Hernández Colón

1976 Carlos Romero Barceló
1984 Rafael Hernández Colón
1988 Rafael Hernándes Colón
1992 Pedro J. Rosselló

Trinidad and Tobago

1962 Eric Eustace Williams
1981 George Chambers

1987 A. N. Raymond Robinson
1991 Patrick Manning

SUGGESTIONS FOR
FURTHER READING

These reading suggestions are arranged by the chapter sequence in the text, which necessarily leads to overlap in coverage. Some suggestions listed for the country and regional chapters are also relevant for the Prologue; Chapters One, Two, and Eleven; and the Epilogue. Readers interested in a specific country within the context of these chapters should look also at the suggestions by country chapter.

The following titles were chosen to provide interesting and profitable reading for the novice student of modern Latin America. For that reason only books in English are included. We have avoided unwieldy "definitive" works. Our objective is to suggest books that will actually be read, and the emphasis is on recently published titles. The listing correlates closely with the themes developed in each chapter of the text. Readers who wish greater detail will find a wealth of leads in most of the suggested books.

Prologue. Why Latin America?

U.S. stereotypes about Latin America as seen in our newspaper and magazine cartoons since the late nineteenth century are laid out in John J. Johnson, *Latin America in Caricature* (Austin: University of Texas Press, 1980). One of the best attempts to explain contemporary Latin America to the nonspecialist is Peter Winn, *Americas: The Changing Face of Latin America and the Caribbean* (New York: Pantheon, 1992). For a rather leaner approach to the same task, see Oxford Analytica, *Latin America in Perspective* (Boston: Houghton Mifflin, 1991). A multidisciplinary reference work on Latin America is Simon Collier, Thomas E. Skidmore, and Harold Blakemore, eds., *The Cambridge Encyclopedia of Latin America and the Caribbean*, rev. ed. (Cambridge: Cambridge University Press, 1992). Broad treatments of selected topics appear in Alfred Stepan, ed., *The Americas: New Interpretive Essays* (New York; Oxford University Press, 1992), and discussion of current research trends can be found in Peter H. Smith, ed., *Latin America in Comparative Perspective: New Approaches to Methods and Analysis* (Boulder: Westview, 1995).

Dependency analysis has spawned much literature and considerable controversy. The classic formulation remains Fernando Henrique Cardoso and Enzo Faletto, *Dependency and Development in Latin America*, trans. Marjory Mattingly

Urquidi (Berkeley: University of California Press, 1979). The most systematic attempt to discredit the dependency approach is Robert A. Packenham, *The Dependency Movement: Scholarship and Politics in Development Studies* (Cambridge: Harvard University Press, 1992).

An excellent starting point for understanding the Chicano experience in the United States is Rodolfo O. de la Garza et al., eds., *The Mexican American Experience: An Interdisciplinary Anthology* (Austin: University of Texas Press, 1985). For a penetrating social history of northernmost Mesoamerica, see Ramón A. Gutiérrez, *When Jesus Came, the Corn Mothers Went Away: Marriage, Sexuality, and Power in New Mexico, 1500–1846* (Stanford: Stanford University Press, 1991). The complicated question of Latino identity is skillfully treated in Suzanne Oboler, *Ethnic Labels, Latino Lives: Identity and the Politics of (Re)Presentation in the United States* (Minneapolis: University of Minnesota Press, 1995).

Chapter One. The Colonial Foundations, 1492–1880s

Too often the Conquest is seen only from the European viewpoint. An excellent corrective is Miguel León-Portilla, ed., *The Broken Spears: The Aztec Account of the Conquest of Mexico,* trans. Lysander Kemp (Boston: Beacon Press, 1961). For a superb synthesis of this drama, see Hugh Thomas, *Conquest: Montezuma, Cortes, and the Fall of Old Mexico* (New York: Simon & Schuster, 1993). A pioneering study of how the native peoples adjusted to European rule while also preserving a sense of community is found in Steve J. Stern, *Peru's Indian Peoples and the Challenge of Spanish Conquest: Huamanga to 1640,* 2nd ed. (Madison: University of Wisconsin Press, 1993). An excellent parallel study for an earlier era in Mexico is Inga Clendinnen, *Ambivalent Conquests: Maya and Spaniard in Yucatan, 1517–1570* (Cambridge: Cambridge University Press, 1987).

The finest synthesis of current scholarship on the colonial Iberian world is Leslie Bethell, ed., *The Cambridge History of Latin America,* vols. I and II, *Colonial Latin America* (Cambridge: Cambridge University Press, 1984). For a cogent brief treatment, see Mark A. Burkholder and Lyman L. Johnson, *Colonial Latin America,* 2nd ed. (New York: Oxford University Press, 1994). For colonial Brazil there is no substitute for Gilberto Freyre's classic *The Masters and the Slaves* (New York: Alfred A. Knopf, 1946). The finest monographic study of Brazilian colonial slavery is Stuart B. Schwartz, *Sugar Plantations in the Formation of Brazilian Society: Bahia, 1550–1835* (Cambridge: Cambridge University Press, 1985). New research on family structure is given in Alida C. Metcalf, *Family and Frontier in Colonial Brazil: Santana de Parnaíba 1580–1822* (Berkeley: University of California Press, 1992).

The origins and travail of independence in Spanish America are described in John Lynch's broad-ranging *The Spanish-American Revolutions, 1808–1826,* 2nd ed. (New York: W. W. Norton, 1986). The postindependence era is well treated in Leslie Bethell, ed., *The Cambridge History of Latin America,* vol. III, *From Independence to c. 1870* (Cambridge: Cambridge University Press, 1985) and in David Bushnell and Neill Macaulay, *The Emergence of Latin America in the Nineteenth Century,* 2nd ed. (New York: Oxford University Press, 1994). The story for Mexico is told in copious detail in D. A. Brading, *The First America: The Spanish Monarchy, Creole Patriots, and the Liberal State 1492–1867* (Cambridge: Cambridge University Press, 1991). The creation of national identities

through literature is traced in Doris Sommer, *Foundational Fictions: The National Romances of Latin America* (Berkeley: University of California Press, 1991).

Chapter Two. The Transformation of Modern Latin America, 1880s–1990s

The authoritative history of much of this period is in Leslie Bethell, ed. *The Cambridge History of Latin America*, vols. IV and V, *c. 1870 to 1930* (Cambridge: Cambridge University Press, 1986). Victory Bulmer-Thomas, *The Economic History of Latin America Since Independence* (Cambridge: Cambridge University Press, 1994), the first comprehensive economic history in English, tells the story of the shift from export orientation to import substitution and thence to neoliberalism. A critique of the liberal ideology is given in Joseph L. Love and Nils Jacobsen, eds., *Guiding the Invisible Hand: Economic Liberalism and the State in Latin American History* (New York: Praeger, 1988). A pioneering study of the creation of the modern Latin American state is Florencia E. Mallon, *Peasant and Nation: The Making of Postcolonial Mexico and Peru* (Berkeley: University of California Press, 1995). Industrialization, one of the most important topics, is given an enlightening comparative analysis in Gary Gereffi and Donald L. Wyman, eds., *Manufacturing Miracles: Paths of Industrialization in Latin America and East Asia* (Princeton: Princeton University Press, 1990). A crucial decade in this process is analyzed in Leslie Bethell and Ian Roxborough, eds., *Latin America Between the Second World War and the Cold War 1944–1948* (Cambridge: Cambridge University Press, 1992) and in David Rock, ed., *Latin America in the 1940s: War and Postwar Transitions* (Berkeley: University of California Press, 1994).

The foreign debt burden, which proved overwhelming in the 1980s and left a heavy legacy for the 1990s, produced a spate of analysis, as in Robert Devlin, *Debt and Crisis in Latin America: The Supply Side of the Story* (Princeton: Princeton University Press, 1989). The human costs of debt adjustment are laid out in Howard Handelman and Werner Baer, eds., *Paying the Costs of Austerity in Latin America* (Boulder: Westview, 1989). For a fine comparative study, see Stephan Haggard and Robert R. Kaufman, eds., *The Politics of Economic Adjustment: International Constraints, Distributive Conflicts, and the State* (Princeton: Princeton University Press, 1992).

The indispensable starting point for an understanding of the political left is Jorge G. Castañeda, *Utopia Unarmed: The Latin American Left After the Cold War* (New York: Alfred A. Knopf, 1993). A comprehensive analysis of the historical interaction between state and labor is given in Ruth Berins Collier and David Collier, *Shaping the Political Arena: Critical Junctures, the Labor Movement, and Regime Dynamics in Latin America* (Princeton: Princeton University Press, 1991). There has been a rapid increase in research and publishing on women in Latin American history. For examples see Rae Lesser Blumberg, Cathy A. Rakowski, Irene Tinker, and Michael Monteón, eds., *EnGendering Wealth and Well-Being: Empowerment for Global Change* (Boulder: Westview, 1995), Christine E. Bose and Edna Acosta-Belén, eds., *Women in the Latin American Development Process* (Philadelphia: Temple University Press, 1995), Carmen Diana Deere and Magdalena León de Leal, eds., *Rural Women and State Policy: Feminist Perspectives on Latin American Agricultural Development* (Boulder: Westview, 1987), Jane S. Jaquette, ed., *The Women's Movement in Latin America: Participation and Democ-*

racy, 2nd ed. (Boulder: Westview, 1994), Emilie Bergmann et al., *Women, Culture, and Politics in Latin America* (Berkeley: University of California Press, 1990), and Gertrude M. Yeager, ed., *Confronting Change, Challenging Tradition: Women in Latin American History* (Wilmington: Scholarly Resources, 1994).

The argument in our chapter on the transformation (1880s–1990s) has been much influenced by Guillermo O'Donnell, *Modernization and Bureaucratic-Authoritarianism: Studies in South American Politics* (Berkeley: Institute of International Studies, University of California, 1979, first published in 1973), the thesis of which is critiqued in David Collier, ed., *The New Authoritarianism in Latin America* (Princeton: Princeton University Press, 1979). The frightful human cost of Latin America's dictatorships is described in Juan E. Corradi, Patricia Weiss Fagen, and Manuel Antonio Garretón, eds., *Fear at the Edge: State Terror and Resistance in Latin America* (Berkeley: University of California Press, 1992). The transition to democracy in much of Latin America in the 1980s is well analyzed in Guillermo O'Donnell, Philippe C. Schmitter, and Laurence Whitehead, eds., *Transitions from Authoritarian Rule*, 4 vols. (Baltimore: Johns Hopkins University Press, 1986), and in Scott Mainwaring, Guillermo O'Donnell, and J. Samuel Valenzuela, eds., *Issues in Democratic Consolidation: The New South American Democracies in Comparative Perspective* (Notre Dame: University of Notre Dame Press, 1992). The growing role of television is analyzed in Thomas E. Skidmore, ed., *Television, Politics, and the Transition to Democracy in Latin America* (Washington: The Woodrow Wilson Center Press, 1993).

Chapter Three. Argentina: Prosperity, Deadlock, and Change

The best synthesis is David Rock, *Argentina, 1516–1987: From Spanish Colonization to the Falklands War and Alfonsín*, rev. ed. (Berkeley: University of California Press, 1987). To understand nineteenth-century Argentina and the origins of modern Argentine nationalism one must start with the famous strongman Rosas, who has received his best scholarly biography in John Lynch, *Argentine Dictator: Juan Manuel de Rosas, 1829–1852* (Oxford: Oxford University Press, 1981). The crucial role of labor in Argentina is traced in Ronaldo Munck, Ricardo Falcón, and Bernardo Galitelli, *Argentina: From Anarchism to Peronism* (London: Zed, 1987). The interaction of interest groups and government policy making for one sector between 1900 and 1946 is explored in Peter H. Smith, *Politics and Beef in Argentina: Patterns of Conflict and Change* (New York: Columbia University Press, 1969). The manner in which the United States systematically misread Argentine foreign policy intentions is revealed in Ronald C. Newton, *The "Nazi Menace" in Argentina, 1931–1947* (Stanford: Stanford University Press, 1992).

Carlos F. Díaz Alejandro, *Essays on the Economic History of the Argentine Republic* (New Haven: Yale University Press, 1970) is the premier economic history on modern Argentina. Later scholarship is incorporated in Guido di Tella and D.C.M. Platt, eds., *The Political Economy of Argentina, 1880–1946* (New York: St. Martin's Press, 1986), and Guido di Tella and Rudiger Dornbusch, eds., *The Political Economy of Argentina, 1946–83* (Pittsburgh: University of Pittsburgh Press, 1989). For the more recent period, see Felipe A.M. de la Balze, *Remaking the Argentine Economy* (New York: Council on Foreign Relations Press, 1995).

Juan and Eva Perón aroused such emotions that objective history writing about them has been slow to appear. Joseph A. Page's ambitious *Perón: A Biography* (New York: Random House, 1983) is both well written and well documented. There is a similar study of Eva Perón, if on a smaller scale, by Nicolas Fraser and Marysa Navarro, *Eva Perón* (New York: W. W. Norton, 1980). For a negative view of Perón's economic policies, see Paul H. Lewis, *The Crisis of Argentine Capitalism* (Chapel Hill: The University of North Carolina Press, 1990). The important role of Frondizi's government is covered in Celia Szusterman, *Frondizi and the Politics of Developmentalism in Argentina, 1955–62* (London: Macmillan, 1993).

One of the most persistent questions about Argentina is why such an economically developed country should have been plagued by authoritarian regimes since 1945. The origins of that apparent breakdown are laid out in Peter H. Smith, *Argentina and the Failure of Democracy: Conflict among Political Elites, 1904–1955* (Madison: University of Wisconsin Press, 1974). For the analysis of an Argentine sociologist, see Carlos Waisman, *Reversal of Development in Argentina: Postwar Counterrevolutionary Policies and Their Structural Consequences* (Princeton: Princeton University Press, 1987). Guillermo O'Donnell has deepened his earlier analysis (see reference in Chapter Two) in *Bureaucratic Authoritarianism: Argentina, 1966–1973, in Comparative Perspective* (Berkeley: University of California Press, 1988). The sordid end of the Galtieri dictatorship is told entertainingly in Jimmy Burns, *The Land that Lost Its Heroes: Argentina, the Falklands and Alfonsín* (London: Bloomsbury, 1987). More recent events are recounted in E. Epstein, ed., *The New Argentine Democracy: The Search for a Successful Formula* (Westport: Praeger, 1992).

Chapter Four. Chile: Socialism, Repression, and Democracy

Brian Loveman has produced a reliable general history in *Chile: The Legacy of Hispanic Capitalism*, 2nd ed. (New York: Oxford University Press, 1988). The history of the agrarian sector between 1850 and 1940 is traced in Arnold J. Bauer, *Chilean Rural Society from the Spanish Conquest to 1930* (Cambridge: Cambridge University Press, 1975). One aspect of that story is brought up to date in D.E. Hojman, *Change in the Chilean Countryside: From Pinochet to Aylwin and Beyond* (London: Macmillan, 1993).

The interpretation of nineteenth-century Chile revolves around perceptions of President Balmaceda and the civil war of 1891. The best documented and best argued—although not universally accepted—interpretation is to be found in Harold Blakemore, *British Nitrates and Chilean Politics, 1886–1896: Balmaceda and North* (London: Athlone Press, 1974). Labor's struggle for bargaining rights, as well as bread-and-butter gains, is well portrayed in Peter DeShazo, *Urban Workers and Labor Unions in Chile, 1902–1927* (Madison: University of Wisconsin Press, 1983), which emphasizes the anarchist-led unions.

One of the most penetrating studies of any Latin American political party is given in Paul W. Drake, *Socialism and Populism in Chile, 1932–52* (Urbana: University of Illinois Press, 1978). The party is the Socialists, who attracted many former anarchists and later moved to the left of the Chilean Communist party. A detailed history of the Allende era can be found in Edy Kaufman, *Crisis in Allende's Chile: New Perspectives* (New York: Praeger, 1988).

One of the most balanced pictures of the process leading to the coup of 1973 is found in Arturo Valenzuela, *The Breakdown of Democratic Regimes: Chile* (Baltimore: Johns Hopkins University Press, 1978). There is a beautifully written account of a factory seizure by its workers in 1971 in Peter Winn, *Weavers of Revolution: The Yarur Workers and Chile's Road to Socialism* (New York: Oxford University Press, 1986). The story of how the technocratic opposition planned for a post-Pinochet era is well told in J.M. Puryear, *Thinking Politics: Intellectuals and Democracy in Chile, 1973–1988* (Baltimore: Johns Hopkins University Press, 1994). Chile's record under neo-liberal policies is analyzed in Barry P. Bosworth, Rudiger Dornbusch, and Raúl Labán, eds., *The Chilean Economy: Policy Lessons and Challenges* (Washington: The Brookings Institution, 1994), Joseph Collins and John Lear, *Chile's Free-Market Miracle: A Second Look* (Oakland: Food First, 1995), and D.E. Hojman, *Chile: The Political Economy of Development and Democracy in the 1990s* (London: Macmillan, 1993).

Few would deny that the Catholic church has played a pivotal role in twentieth-century Chile. The history of that role is lucidly analyzed in Brian H. Smith, *The Church and Politics in Chile: Challenges to Modern Catholicism* (Princeton: Princeton University Press, 1982).

For a wealth of information and analysis on Chile under the military, see Pamela Constable and Arturo Valenzuela, *A Nation of Enemies: Chile Under Pinochet* (New York: W.W. Norton, 1991), and Genaro Arriagada, *Pinochet: The Politics of Power* (London: Unwin Hyman, 1988). The autobiography of Chile's Nobel laureate in poetry is a fascinating journey through the artistic worlds of Latin America and Europe: Pablo Neruda, *Memoirs* (New York: Farrar, Straus and Giroux, 1977). It is also an eye-opener about political persecution, since Neruda was a long-time Communist Party militant.

Chapter Five. Brazil: Development for Whom?

E. Bradford Burns, *A History of Brazil*, 3rd ed. (New York: Columbia University Press, 1993), is a sure-handed survey, with extensive suggestions for further reading. The most important general interpretation of Brazil's economic history is Celso Furtado, *The Economic Growth of Brazil* (Berkeley: University of California Press, 1963), which stops in the early 1950s. The more recent years are covered in Werner Baer, *The Brazilian Economy: Growth and Development,* 4th ed. (Westport: Praeger, 1995), which includes a wealth of data, as does Edmar L. Bacha and Herbert S. Klein, eds., *Social Change in Brazil, 1945–1985: The Incomplete Transition* (Albuquerque: University of New Mexico Press, 1989). Brazil's modern culture is given an intriguing analysis in Roberto A. Da Matta, *Carnivals, Rogues, and Heroes: An Interpretation of the Brazilian Dilemma* (Notre Dame: University of Notre Dame Press, 1991), and in David J. Hess and Roberto A. Da Matta, eds., *The Brazilian Puzzle: Culture on the Borderlands of the Western World* (New York: Columbia University Press, 1995). Joseph A. Page, *The Brazilians* (Reading: Addison-Wesley Publishing Company, 1995) is an overview with much information.

There is a brilliant analysis of the 1822–1889 era in Emilia Viotti da Costa, *The Brazilian Empire: Myths and Histories* (Chicago: University of Chicago Press, 1985). One of the finest microstudies for any area of Latin America is Stanley

J. Stein, *Vassouras: A Brazilian Coffee County, 1850–1900* (Cambridge: Harvard University Press, 1957). A later study, focusing on São Paulo industrial unions, highlights the crucial role of women workers: Joel Wolfe, *Working Women, Working Men: São Paulo and the Rise of Brazil's Industrial Working Class, 1900–1955* (Durham: Duke University Press, 1993).

The intricate functioning of the monarchy's parliamentary system is analyzed in Richard Graham, *Patronage and Politics in Nineteenth-Century Brazil, 1850–1914* (Stanford: Stanford University Press, 1990). For a fascinating case study of foreign investment, see Marshall C. Eakin, *British Enterprise in Brazil: The St. John d'el Rey Mining Company and the Morro Velho Gold Mine, 1830–1960* (Durham: Duke University Press, 1989). The tragedy (immortalized by Euclides da Cunha) of a government-directed slaughter in the Northeast is recounted in Robert M. Levine, *Vale of Tears: Revisiting the Canudos Massacre in Northeastern Brazil, 1893–1897* (Berkeley: University of California Press, 1992). A similar social conflict is depicted in Todd A. Diacon, *Millenarian Vision, Capitalist Reality: Brazil's Contestado Rebellion, 1912–1916* (Durham: Duke University Press, 1991). A model study of the key role of coffee in pre-1930 economics and politics is Mauricio A. Font, *Coffee, Contention, and Change in the Making of Modern Brazil* (Cambridge: Basil Blackwell, 1990).

For an analysis of the political process from Vargas' overthrow in 1945 to the military coup of 1964, see Thomas E. Skidmore, *Politics in Brazil, 1930–1964: An Experiment in Democracy* (New York: Oxford University Press, 1967). The subsequent military role in the post-1964 authoritarian system is analyzed in Skidmore, *The Politics of Military Rule in Brazil, 1964–85* (New York: Oxford University Press, 1988). The human cost of the repression is related in Joan Dassin, ed., *Torture in Brazil: A Report by the Archdiocese of São Paulo* (New York: Vintage Books, 1986). The historical role of Brazilian women is finally being studied, as in June E. Hahner, *Emancipating the Female Sex: The Struggle for Women's Rights in Brazil, 1850–1940* (Durham: Duke University Press, 1990), and Sonia E. Alvarez, *Engendering Democracy in Brazil: Women's Movements in Transition Politics* (Princeton: Princeton University Press, 1990).

An excellent analysis of the labor relations system in the pre-1964 era can be found in Kenneth Paul Erickson, *The Brazilian Corporative State and Working-Class Politics* (Berkeley: University of California Press, 1977). Updating that topic for the later 1970s is John Humphrey, *Capitalist Control and Workers' Struggle in the Brazilian Auto Industry* (Princeton: Princeton University Press, 1982). For an anthropologist's view of the varieties of religion, see Rowan Ireland, *Kingdoms Come: Religion and Politics in Brazil* (Pittsburgh: University of Pittsburgh Press, 1991). For a broader view of post-1964 trends, see John D. Wirth et al., eds., *State and Society in Brazil: Continuity and Change* (Boulder: Westview, 1987). Environmental issues have drawn increasing attention, as in Warren Dean, *With Broadax and Firebrand: The Destruction of the Brazilian Atlantic Forest* (Berkeley: University of California Press, 1995), and Susanna Hecht and Alexander Cockburn, *The Fate of the Forest: Developers, Destroyers and Defenders of the Amazon* (London: Verso, 1989).

President Kubitschek's construction of Brasília in 1957–61 captured the world's imagination, but the reality proved a mixed blessing, as is made clear in James Holston, *The Modernist City: An Anthropological Critique of Brasília* (Chicago: University of Chicago Press, 1989).

The finest introduction to Brazilian race relations is Carl N. Degler, *Neither*

Black nor White: Slavery and Race Relations in Brazil and the United States (New York: Macmillan, 1971), although his hypothesis of the "mulatto escape hatch" has drawn fire. Of the many works on abolition the most penetrating is Robert Conrad, *The Destruction of Brazilian Slavery, 1850–1888* (Berkeley: University of California Press, 1972). The intriguing life of a late nineteenth-century Afro-Brazilian is recaptured in Eduardo Silva, *Prince of the People: The Life and Times of a Brazilian Free Man of Colour*, trans. Moyra Ashford (London: Verso, 1993). For a Gramscian analysis of contemporary race relations, see Michael George Hanchard, *Orpheus and Power: The Movimento Negro of Rio de Janeiro and São Paulo, Brazil 1945–1988* (Princeton: Princeton University Press, 1994).

The Brazilian elite's attempt to reconcile racist science and the reality of their multiracial society is described in Thomas E. Skidmore, *Black into White: Race and Nationality in Brazilian Thought*, rev. ed. (Durham: Duke University Press, 1993). For the reflection of the "whitening" ideal in Brazilian literature, see David T. Haberly, *Three Sad Races: Racial Identity and National Consciousness in Brazilian Literature* (Cambridge: Cambridge University Press, 1983).

Chapter Six. Peru: Soldiers, Oligarchs, and Indians

An able survey of Peruvian history is given in Fredrick B. Pike, *The United States and the Andean Republics: Peru, Bolivia, and Ecuador* (Cambridge: Harvard University Press, 1977). Economic history is covered superbly in Paul Gootenberg, *Between Silver and Guano: Commercial Policy and the State in Postindependence Peru* (Princeton: Princeton University Press, 1989), and in his *Imagining Development: Economic Ideas in Peru's "Fictitious Prosperity" of Guano, 1840–1880* (Berkeley: University of California Press, 1993), as well as in Rosemary Thorp and Geoffrey Bertram, *Peru 1890–1977: Growth and Policy in an Open Economy* (New York: Columbia University Press, 1978).

The effects of rapid economic change on the highland population are sympathetically described in Florencia E. Mallon, *The Defense of Community in Peru's Central Highlands: Peasant Struggle and Capitalist Transition, 1860–1940* (Princeton: Princeton University Press, 1983). For a study emphasizing a later period, see Carmen Diana Deere, *Household and Class Relations: Peasants and Landlords in Northern Peru* (Berkeley: University of California Press, 1990). Peru's most famous twentieth-century politician is given an unorthodox analysis in Fredrick B. Pike, *The Politics of the Miraculous in Peru: Haya de la Torre and the Spiritualist Tradition* (Lincoln: University of Nebraska Press, 1986).

Much writing about Peru has centered on the military regime which seized power in 1968. One of the most careful portraits of the post-1968 regime is the collection edited by Cynthia McClintock and Abraham F. Lowenthal, *The Peruvian Experiment Reconsidered* (Princeton: Princeton University Press, 1983). Peru's huge "informal" economic sector received a world-famous analysis in Hernando de Soto, *The Other Path: The Invisible Revolution in the Third World* (New York: Harper & Row, 1989). The enormous impact of *Sendero Luminoso* is given contrasting interpretations in Deborah Poole and Gerardo Renique, *Peru: Time of Fear* (London: Latin American Bureau, 1992), and David Scott Palmer, ed., *The Shining Path of Peru*, 2nd ed. (New York: St. Martin's Press, 1994).

Chapter Seven. Mexico: The Taming of a Revolution

Mexico is fortunate in having a detailed, well-balanced, and up-to-date one-volume history in Michael C. Meyer and William L. Sherman, *The Course of Mexican History*, 5th ed. (New York: Oxford University Press, 1995). It includes chapter-by-chapter bibliographies in both English and Spanish. The best over-all introduction to Mexico is Alan Riding, *Distant Neighbors: A Portrait of the Mexicans* (New York: Alfred A. Knopf, 1985), despite the author's frequent obiter dicta about Mexican character. An excellent lead into the mentality and style of the Mexican elite may be found in Larissa Adler Lomnitz and Marisol Pérez-Lizaur, *A Mexican Elite Family, 1820–1980: Kinship, Class, and Culture* (Princeton: Princeton University Press, 1987).

Mexico's most famous liberal reformer (and full-blooded Indian) of the nine-teenth century is portrayed in Brian Hamnett, *Juárez* (New York: Longman Publishing, 1994). The subject of industrialization in the Porfiriato and after-ward is treated in Stephen H. Haber, *Industry and Underdevelopment: The Indus-trialization of Mexico, 1890–1940* (Stanford: Stanford University Press, 1989). There is a marvelous social history of this era in William H. Beezley, *Judas at the Jockey Club and Other Episodes of Porfirian Mexico* (Lincoln: University of Ne-braska Press, 1987).

Explanations of the Revolution have come to dominate twentieth-century Mexican historiography. A rich and highly readable synthesis on the revolu-tion's first decade is to be found in Alan Knight, *The Mexican Revolution* (Cam-bridge: Cambridge University Press, 1986). The atmosphere of the era can be sensed from the magnificent illustrations in Anita Brenner, *The Wind That Swept Mexico: The History of the Mexican Revolution of 1910–1942* (Austin: Uni-versity of Texas Press, 1971). One of the most rewarding approaches to the Revolution is through regional history, as can be seen in Gilbert M. Joseph and Daniel Nugent, eds., *Everyday Forms of State Formation: Revolution and the Negotia-tion of Rule in Modern Mexico* (Durham: Duke University Press, 1994).

The finest study of the agrarian revolution is John Womack, Jr., *Zapata and the Mexican Revolution* (New York: Alfred A. Knopf, 1968). Whereas Womack concentrated on Morelos, a countrywide analysis and a longer historical context are offered in John Tutino, *From Insurrection to Revolution in Mexico: Social Bases of Agrarian Violence, 1750–1940* (Princeton: Princeton University Press, 1986), and Friedrich Katz, ed., *Riot, Rebellion, and Revolution: Rural Social Conflict in Mexico* (Princeton: Princeton University Press, 1988). Lázaro Cárdenas has long been seen as a major hero of the Revolution, but his presidency has been sub-jected to a revisionist critique in Nora Hamilton, *The Limits of State Autonomy: Post-Revolutionary Mexico* (Princeton: Princeton University Press, 1982).

An excellent overview of Mexico's neo-liberal economic adjustment is given in Nora Lustig, *Mexico: The Remaking of an Economy* (Washington: The Brook-ings Institution, 1992). An earlier and more pessimistic assessment is David Barkin, *Distorted Development: Mexico in the World Economy* (Boulder: Westview, 1990). Any analysis of the PRI must begin with its methods of recruitment, which are given their first systematic analysis in Peter H. Smith, *Labyrinths of Power: Political Recruitment in Twentieth-Century Mexico* (Princeton: Princeton University Press, 1979). The predominant role of the technocrats is analyzed in Miguel Ángel Centeno, *Democracy Within Reason: Technocratic Revolution in*

Mexico (University Park: Pennsylvania State University Press, 1994). For a less studied group, see Roderic Ai Camp, *Generals in the Palacio: The Military in Modern Mexico* (New York: Oxford University Press, 1992).

Central to Mexican history has been its relationship with the United States, which is traced in Josefina Zoraída Vásquez and Lorenzo Meyer, *The United States and Mexico* (Chicago: University of Chicago Press, 1985). For a stimulating dialogue between a Mexican and a U.S. political scientist, see Robert A. Pastor and Jorge G. Castañeda, *Limits to Friendship: The United States and Mexico* (New York: Alfred A. Knopf, 1988). Of continuing relevance is the policy report of the blue-ribbon Bilateral Commission on the Future of U.S.-Mexican Relations, *The Challenge of Interdependence: Mexico and the United States* (Lanham, Md.: University Press of America, 1988), whose findings are supported by scholarly background papers in *Dimensions of United States-Mexican Relations*, 5 vols. (La Jolla: Center for U.S.-Mexican Studies, 1989–90), edited by Rosario Green and Peter H. Smith.

Mexican intellectuals have written moving accounts of their collisions with their national reality, as in José Vasconcelos, *A Mexican Ulysses: An Autobiography*, trans. William Rex Crawford (Bloomington: Indiana University Press, 1963). The most famous interpretation of Mexican national character is Octavio Paz, *The Labyrinth of Solitude: Life and Thought in Mexico*, trans. Lysander Kemp (New York: Grove Press, 1961). Paz produced a sequel, *The Other Mexico: Critique of the Pyramid*, trans. Lysander Kemp (New York: Grove Press, 1972).

Chapter Eight. *Cuba: Late Colony, First Socialist State*

Anyone studying the history of Cuba becomes rapidly indebted to Hugh Thomas for his superbly researched and highly readable *Cuba: The Pursuit of Freedom* (New York: Harper & Row, 1971). An excellent general history is Louis A. Pérez, Jr., *Cuba: Between Reform and Revolution* (New York: Oxford University Press, 1988). The troubled history of the Afro-Cubans, which climaxed in the massacre of 1912, is told in Aline Helg, *Our Rightful Share: The Afro-Cuban Struggle for Equality, 1886–1912* (Chapel Hill: University of North Carolina Press, 1995). The dominant figure in Cuban history is still José Martí, the subject of a biography by Christopher Abel and Nissa Torrents, *José Martí: Revolutionary Democrat* (Durham: Duke University Press, 1986). For an important part of the background to the independence struggle, see Rosalie Schwartz, *Lawless Liberators: Political Banditry and Cuban Independence* (Durham: Duke University Press, 1989).

Cuba's relationship to the United States, which was crucial to the growth of Cuban nationalism, is lucidly outlined in Louis A. Pérez, Jr., *Cuba under the Platt Amendment, 1902–1934* (Pittsburgh: University of Pittsburgh Press, 1986), and Jules R. Benjamin, *The United States and the Origins of the Cuban Revolution: An Empire of Liberty in an Age of National Liberation* (Princeton: Princeton University Press, 1990). The relative neglect of women in Cuban historiography receives a corrective in K. Lynn Stoner, *From the House to the Streets: The Cuban Woman's Movement for Legal Reform, 1898–1940* (Durham: Duke University Press, 1991).

Much mythology has been generated over Batista's fall. The most painstak-

ing reconstruction of the "politico-military factors" leading to a rebel victory is Ramón L. Bonachea and Marta San Martín, *The Cuban Insurrection, 1952–1959* (New Brunswick: Transaction Books, 1974), which does not support the more extreme Fidelista claims for a totally rural guerrilla victory. The rapidly changing U.S. reaction is described in Richard E. Welch, Jr., *Response to Revolution: The United States and the Cuban Revolution, 1959–1961* (Chapel Hill: University of North Carolina Press, 1985). A pioneering work on a crucial topic is Maurice Zeitlin, *Revolutionary Politics and the Cuban Working Class* (Princeton: Princeton University Press, 1967).

One of the most interesting close-up portraits of Fidel is offered in the profusely illustrated *Castro's Cuba, Cuba's Fidel* (New York: Random House, 1969) by Lee Lockwood. For a highly critical view from a revolutionary who broke with Fidel, see Carlos Franquí, *Family Portrait with Fidel: A Memoir* (New York: Random House, 1984). A largely favorable biography is Tad Szulc, *Fidel: A Critical Portrait* (New York: William Morrow, 1986), while Georgie Anne Geyer, *Guerrilla Prince: The Untold Story of Fidel Castro* (Boston: Little, Brown, 1991) is hyperbolically anti-Castro. The best brief study is Sebastian Balfour, *Castro*, 2nd ed. (London: Longman, 1995). Despite the sensational title there is much valuable information in Andrés Oppenheimer, *Castro's Final Hour: The Secret Story Behind the Coming Downfall of Communist Cuba* (New York: Simon & Schuster, 1992).

The indispensable starting point on the economic record of the Fidelista government is Carmelo Mesa-Lago, *The Economy of Socialist Cuba: A Two-Decade Appraisal* (Albuquerque: University of New Mexico Press, 1981). For studies by economists who largely disagree with Mesa-Lago, see Andrew Zimbalist, ed., *Cuba's Socialist Economy: Toward the 1990s* (Boulder: Lynne Rienner, 1987). Post-1986 policy is analyzed in Richard Gillespie, ed., *Cuba after Thirty Years: Rectification and the Revolution* (London: Frank Cass, 1990). Cuba's response to the current economic crisis is the subject in Susan Eva Eckstein, *Back from the Future: Cuba Under Castro* (Princeton: Princeton University Press, 1994), and in Carmelo Mesa-Lago, *Are Economic Reforms Propelling Cuba to the Market?* (Miami: North-South Center, 1994).

On U.S.-Cuban relations, the views of a former U.S. diplomat highly critical of U.S. policy may be found in Wayne S. Smith, *The Closest of Enemies: A Personal and Diplomatic Account of U.S.-Cuban Relations since 1957* (New York: W.W. Norton, 1987). For a pathbreaking collaborative study by both U.S. and Cuban scholars, see Jorge I. Domínguez and Rafael Hernández, eds., *U.S.-Cuban Relations in the 1990s* (Boulder: Westview, 1989).

One area of notable success for the revolutionary government is health care, as documented in Julie M. Feinsilver, *Healing the Masses: Cuban Health Politics at Home and Abroad* (Berkeley: University of California Press, 1993). A fascinating series of individual portraits, based on in-depth interviews, are painted in Oscar Lewis, Ruth M. Lewis, and Susan M. Rigdon, *Four Women: Living the Revolution: An Oral History of Contemporary Cuba* (Urbana: University of Illinois Press, 1977). This project produced two other volumes, *Four Men* (1975) and *Neighbors* (1978). For the darker side of the Revolution, see Jorge Valls, *Twenty Years and Forty Days: Life in a Cuban Prison* (New York: Americas Watch, 1986), and Jacobo Timerman, *Cuba: A Journey* (New York: Alfred A. Knopf, 1990).

Chapter Nine. The Caribbean: Colonies and Mini-States

Franklin W. Knight, *The Caribbean: The Genesis of a Fragmented Nationalism,* 2nd ed. (New York: Oxford University Press, 1990) offers an overview of the region. An authoritative collaborative history can be found in Leslie Bethell, ed., *The Cambridge History of Latin America,* vol. VII, *Latin America since 1930: Mexico, Central America, and the Caribbean* (Cambridge: Cambridge University Press, 1990). The Caribbean's historically most important export is given a witty and erudite history in Sidney W. Mintz, *Sweetness and Power: The Place of Sugar in Modern History* (New York: Viking Penguin, 1985).

Among the classic works on Caribbean history are C.L.R. James, *The Black Jacobins: Toussaint-L'Ouverture and the San Domingo Revolution,* 2nd ed., rev. (New York: Vintage, 1989), and Sidney W. Mintz, *Worker in the Cane: A Puerto Rican Life History* (New Haven: Yale University Press, 1960). There is an excellent collection of articles on Caribbean themes in Sidney W. Mintz and Sally Price, eds., *Caribbean Contours* (Baltimore: Johns Hopkins University Press, 1985).

The former British colonies have been the subject of numerous studies, such as David Watts, *The West Indies: Patterns of Development, Culture, and Environmental Change since 1492* (Cambridge: Cambridge University Press, 1987). No West Indian was more eloquent on the colonial experience than C.L.R. James, as in his classic book on cricket, *Beyond a Boundary* (London: Stanley Paul, 1963). Among other country studies are Frank Moya Pons, *The Dominican Republic: A National History* (New Rochelle: Hispaniola Books, 1995), and David Nicholls, *From Dessalines to Duvalier: Race, Colour, and National Independence in Haiti* (Cambridge: Cambridge University Press, 1979).

Puerto Rico has suffered from an ambiguous political identity, but much of its experience parallels that of the rest of the Caribbean. Economic aspects are covered in James L. Dietz, *Economic History of Puerto Rico: Institutional Change and Capitalist Development* (Princeton: Princeton University Press, 1986). For the study of an English academic analyzing the Puerto Rican-U.S. nexus, see Raymond Carr, *Puerto Rico: A Colonial Experiment* (New York: Vintage Books, 1984).

Chapter Ten. Central America: Colonialism, Dictatorship, and Revolution

Central America has been relatively neglected by North American historians until recently. With the explosion of guerrilla movements in the late 1970s, along with the deepening involvement of the United States, we have been deluged with "instant" books often lacking in historical perspective and documentation. The best starting point for this region is Ralph Lee Woodward, Jr., *Central America; A Nation Divided,* 2nd ed. (New York: Oxford University Press, 1985), which includes a forty-three-page guide to the literature on Central America. For the modern period see James Dunkerley, *Power in the Isthmus: A Political History of Modern Central America* (London: Verso, 1988) and also his *The Pacification of Central America* (London: Verso, 1994), as well as Leslie Bethell, ed., *The Cambridge History of Latin America,* vol. VII, *Latin America since 1930:*

Mexico, Central America, and the Caribbean (Cambridge: Cambridge University Press, 1990). The origins of U.S. economic penetration are discussed in Thomas D. Schoonover, *The United States in Central America, 1860–1911: Episodes of Social Imperialism and Imperial Rivalry in the World System* (Durham: Duke University Press, 1991). Recent Central American economic history has been given a sophisticated analysis in Victor Bulmer-Thomas, *The Political Economy of Central America since 1920* (Cambridge: Cambridge University Press, 1987). Important detail on the interwar period may be found in Rodolfo Cerdas-Cruz, *The Communist International in Central America, 1920–1936* (Basingstoke: Macmillan, 1993).

Guatemala, the largest and potentially richest country of Central America, has recently attracted attention from first-rate historians, such as David McCreery, *Rural Guatemala 1760–1940* (Stanford: Stanford University Press, 1994), and Jim Handy, *Revolution in the Countryside: Rural Conflict and Agrarian Reform in Guatemala, 1944–1954* (Chapel Hill: University of North Carolina Press, 1994). The U.S. role in the overthrow of President Arbenz in 1954 has been superbly documented and described in Piero Gleijeses, *Shattered Hope: The Guatemalan Revolution and the United States, 1944–1954* (Princeton: Princeton University Press, 1991).

Nicaraguan history of the past century is still dominated by the Somoza dynasty, whose origins are depicted in Knut Walter, *The Regime of Anastasio Somoza, 1936–1956* (Chapel Hill: University of North Carolina Press, 1993). Essential for understanding the durability of the Somoza dynasty is Richard Millett, *Guardians of the Dynasty: A History of the U.S.-Created Guardia Nacional de Nicaragua and the Somoza Family* (Maryknoll: Orbis Books, 1977). For a key institution, see John M. Kirk, *Politics and the Catholic Church in Nicaragua* (Gainesville: University Press of Florida, 1992). Among the many books describing Nicaragua since 1979 is Dennis Gilbert, *Sandinistas: The Party and the Revolution* (New York: Basil Blackwell, 1988), David Close, *Nicaragua: Politics, Economics and Society* (London: Pinter, 1988), and Rose J. Spalding, *Capitalists and Revolution in Nicaragua: Opposition and Accommodation 1979–1993* (Chapel Hill: University of North Carolina Press, 1994).

The unhappy republic of El Salvador descended into a bloody civil war, with much meddling by outside powers. El Salvador has a longstanding tradition of repression of popular opposition, as can be seen in James Dunkerley, *The Long War: Dictatorship and Revolution in El Salvador* (London: Verso, 1982). Recent trends are discussed in Joseph S. Tulchin with Gary Bland, eds., *Is There a Transition to Democracy in El Salvador?* (Boulder: Lynne Reinner, 1992).

The U.S. presence hovers over Central America, making every domestic political decision a possible conflict with Uncle Sam. For the attempt by a distinguished authority in U.S. foreign policy analysis to explain the context, see Walter LaFeber, *Inevitable Revolutions: The United States and Central America* (New York: W.W. Norton, 1983). An up-to-date and insightful analysis of the subject appears in John H. Coatsworth, *Central America and the United States: The Colossus and the Clients* (New York: Twayne, 1994). LaFeber has also written the best overview of an issue that long bedeviled U.S.-Central American relations: *The Panama Canal: The Crisis in Historical Perspective,* updated ed. (New York: Oxford University Press, 1989).

Chapter Eleven. Latin America, the United States, and the World

Twentieth-century world wars have powerfully affected the region, as can be seen in Bill Albert, *South America and the First World War: The Impact of the War on Brazil, Argentina, Peru, and Chile* (Cambridge: Cambridge University Press, 1988). The Soviet role in the region is studied in Nicola Miller, *Soviet Relations with Latin America: 1959–1987* (Cambridge: Cambridge University Press, 1989), and Eusebio Mujal-León, ed., *The USSR and Latin America: A Developing Relationship* (Boston: Unwin Hyman, 1989).

U.S.-Latin American relations have generated an enormous bibliography, mostly from a conventional diplomatic history point of view. For the recent era, see John D. Martz, ed., *United States Policy in Latin America: A Quarter Century of Crisis and Challenge, 1961–1986* (Lincoln: University of Nebraska Press, 1988), and Gaddis Smith, *The Last Years of the Monroe Doctrine, 1945–1993* (New York: Hill and Wang, 1994). The triumph of Cold War doctrines is clearly traced in Stephen G. Rabe, *Eisenhower and Latin America: The Foreign Policy of Anticommunism* (Chapel Hill: University of North Carolina Press, 1988). Recent studies of U.S. relations with individual countries or regions include Joseph S. Tulchin, *Argentina and the United States: A Conflicted Relationship* (Athens: University of Georgia Press, 1990), Louis A. Pérez, Jr., *Cuba and the United States: Ties of Singular Intimacy* (Athens: University of Georgia Press, 1990), W. Michael Weis, *Cold Warriors & Coups D'etat: Brazilian-American Relations, 1945–1964* (Albuquerque: University of New Mexico Press, 1993), and Anthony P. Maingot, *The United States and the Caribbean* (Boulder: Westview, 1994).

For a critical survey of U.S. policy toward Latin America see Abraham F. Lowenthal, *Partners in Conflict: The United States and Latin America in the 1990s*, rev. ed. (Baltimore: Johns Hopkins University Press, 1990). Also useful are the reflections and analyses in Robert A. Pastor, *Whirlpool: U.S. Foreign Policy Toward Latin America and the Caribbean* (Princeton: Princeton University Press, 1992). An up-to-date and original overview of the subject is Peter H. Smith, *Talons of the Eagle: Dynamics of U.S.-Latin American Relations* (New York: Oxford University Press, 1996).

World War II virtually finished off any major role for Britain in Latin America, as can be seen in R.A. Humphrey, *Latin America and the Second World War*, 2 vols. (London: Athlone, 1981–1982). The first volume covers 1939–1942 and the second covers 1942–1945. The story is brought up to date in Victor Bulmer-Thomas, ed., *Britain and Latin America: A Changing Relationship* (Cambridge: Cambridge University Press, 1989). The key role of the Rockefeller family, which was prominent in the U.S. economic penetration of Latin America, is documented in Elizabeth A. Cobbs, *The Rich Neighbor Policy: Rockefeller and Kaiser in Brazil* (New Haven: Yale University Press, 1992), and Gerard Colby with Charlotte Dennett, *Thy Will Be Done: The Conquest of the Amazon: Nelson Rockefeller and Evangelism in the Age of Oil* (New York: HarperCollins Publishers, 1995).

In the 1970s human rights became a highly controversial issue in U.S.-Latin American relations. The most comprehensive study of this question is Lars Schoultz, *Human Rights and United States Policy Toward Latin America* (Princeton: Princeton University Press, 1981). Schoultz favored greater emphasis on human rights, a position roundly condemned by Jeane J. Kirkpatrick in *Dictatorships and Double Standards: Rationalism and Reason in Politics* (New York: Simon

& Schuster, 1982). For an eye-opening analysis of the views of U.S. officials who have made Latin American policy, see Lars Schoultz, *National Security and United States Policy Toward Latin America* (Princeton: Princeton University Press, 1987). For highly critical analyses of U.S. Latin American policy making in the 1980s, see Eldon Kenworthy, *America/Americas: Myth in the Making of U.S. Policy Toward Latin America* (University Park: Pennsylvania State University Press, 1995), and Thomas Carothers, *In the Name of Democracy: U.S. Policy Toward Latin America in the Reagan Years* (Berkeley: University of California Press, 1991).

The prospects for economic integration have led to much analysis, notwithstanding the Mexican financial debacle of 1994–95. The question is discussed in a wider context in Peter H. Smith, ed., *The Challenge of Integration: Europe and the Americas* (New Brunswick: Transaction Publishers, 1993).

The immigration issue is case in perspective by Alejandro Portes and Robert L. Bach, *Latin Journey: Cuban and Mexican Immigrants in the United States* (Berkeley and Los Angeles: University of California Press, 1985), and by George J. Borjas, *Friends or Strangers: The Impact of Immigrants on the U.S. Economy* (New York: Basic Books, 1990). Illicit drug trafficking, another major question on the post–Cold War agenda, receives careful study in Peter H. Smith, ed., *Drug Policy in the Americas* (Boulder: Westview, 1992).

Epilogue: What Future for Latin America?

A very useful survey of labor throughout Latin America is given in Edward C. Epstein, ed., *Labor Autonomy and the State in Latin America* (Boston: Unwin Hyman, 1989). With democratization, political parties have returned to center stage, as pointed out in Scott Mainwaring and Timothy Scully, eds., *Building Democratic Institutions: Party Systems in Latin America* (Stanford: Stanford University Press, 1995).

The "new" social movements which emerged in the 1970s are well analyzed in Susan Eckstein, ed., *Power and Popular Protest: Latin American Social Movements* (Berkeley: University of California Press, 1989). An excellent starting point on the church is the collective volume edited by Scott Mainwaring and Alexander Wilde, *The Progressive Church in Latin America* (Notre Dame: University of Notre Dame Press, 1989).

The background to the turn toward neo-liberalism in economics is given in Rudiger Dornbusch and Sebastian Edwards, eds., *The Macroeconomics of Populism in Latin America* (Chicago: University of Chicago Press, 1991), while the application of the new policies is examined in William C. Smith, Carlos H. Acuña, and Eduardo A. Gamarra, eds., *Democracy, Markets, and Structural Reform in Contemporary Latin America: Argentina, Bolivia, Brazil, Chile, and Mexico* (Miami: North-South Center, 1994).

Argentina's most famous tango singer is empathetically portrayed in Simon Collier, *The Life, Music, and Times of Carlos Gardel* (Pittsburgh: University of Pittsburgh Press, 1986), a book that reminds us that Latin America is about much more than just generals and inflation. Another neglected topic has an excellent collective treatment in Joseph L. Arbena, *Sport and Society in Latin America: Diffusion, Dependency, and the Rise of Mass Culture* (Westport: Greenwood Press, 1988). The theme of popular culture is explored in William Rowe

and Vivian Schelling, *Memory and Modernity: Popular Culture in Latin America* (London: Verso, 1991).

For an admirably clear analysis of the economic scene at the outset of the 1990s, see Eliana Cardoso and Ann Helwege, *Latin America's Economy: Diversity, Trends, and Conflicts* (Cambridge: The MIT Press, 1992).

The debate over Latin America's future invariably begins with debate about its past. Nowhere was this more true than in the polemics stimulated by the 1992 Columbus Quincentenary, of which the opening salvo was Kirkpatrick Sale, *The Conquest of Paradise: Christopher Columbus and the Columbian Legacy* (New York: Alfred A. Knopf, 1990). For the view of a Spanish philosopher, see Xavier Robert deVentos, *The Hispanic Labyrinth: Tradition and Modernity in the Colonization of the Americas* (New Brunswick: Transaction Books, 1991). One of the most perceptive North American observers has been Richard Morse, whose *New World Soundings: Culture and Ideology in the Americas* (Baltimore: Johns Hopkins University Press, 1989) argues that his compatriots have systematically misunderstood and misinterpreted Latin America. Perhaps first-rate journalism is the best guide to the future, as in Alma Guillermoprieto's moving and witty *The Heart that Bleeds: Latin America Now* (New York: Alfred A. Knopf, 1994).

INDEX